GREAT INVENTIONS
THAT CHANGED
THE WORLD

GREAT INVENTIONS THAT CHANGED THE WORLD

JAMES WEI
Princeton University

A JOHN WILEY & SONS, INC., PUBLICATION

Published by John Wiley & Sons, Inc., Hoboken, New Jersey
Published simultaneously in Canada

For general information on our other products and services or for technical support, please contact our Customer Care Department within the United States at (800) 762-2974, outside the United States at (317) 572-3993 or fax (317) 572-4002.

Wiley also publishes its books in a variety of electronic formats. Some content that appears in print may not be available in electronic formats. For more information about Wiley products, visit our web site at www.wiley.com.

Library of Congress Cataloging-in-Publication Data

Wei, James, 1930-
  Great inventions that changed the world / James Wei, Princeton University.
    pages cm
  Includes index.
    ISBN 978-0-470-76817-4 (hardback)
1. Inventions. 2. Technological innovations. I. Title.
  T15.W45 2012
  600–dc23

                                                        2011053470

Printed in the United States of America
10 9 8 7 6 5 4 3 2 1

CONTENTS

FOREWORD

The Earth is now 4.5 billion years old. Yet, virtually all the knowledge and inventions available today appeared in the last century or so. Extrapolate that pace of change forward and accelerate it and one has an idea what life will be like a century, or millennium, in the future. Or perhaps more accurately one has *no* idea.

That is not to suggest that all previous times were Dark Ages of Innovation: On the contrary, there was the lever, wheel, wedge, stirrup, long bow, telescope, and more, but nothing like the veritable flood of innovation that engulfs this fast-forward world in which we live today.

Various studies have shown that 50–85% of the growth in US GDP over the past half century and two-thirds of the productivity increase (read standard of living gain) are due to advancements in science and engineering. The US National Academies and many other organizations have concluded that the future quality of life in developed nations that have huge competitive disadvantages in the cost of labor will depend upon their ability to innovate; that is, to create knowledge through extraordinary scientific research, to translate that knowledge into products and services through engineering leadership, and through world-class entrepreneurship shepherd those products and services across "the Valley of Death," where so many new innovations fail for economic reasons, and into the marketplace.

This is not easy. Only about 1 patent application in 100 leads to a successful product. Thomas Edison, seeking a filament for an electric light bulb, once explained, "I have not failed. I have found 10,000 ways that won't work." And 60% of new companies go out of business in less than 3 years.

Today, the populace of Earth must produce $1.5 million of goods and services each *second*, 24/7, merely to preserve the existing standard of living on the planet—a standard under which half the population still survives on less than $2 per day. The Red Queen, speaking to Alice in Lewis Carroll's *Through the Looking Glass*, offers sound advice: "*Here*, you see, it takes all the running *you* can do, to keep in the same place. If you want to get somewhere else, you must run at least twice as fast as that!"

Consider a snapshot of the lifetime of the author of this Foreword. He began his engineering training performing calculations using three sticks of wood and two pieces of glass; his youth included occasional summers of confinement to his yard due to the fear of polio; his early professional years included helping put a dozen friends on the moon; and his business years ended with him working with 82,000 engineering colleagues, along with experts in many other fields, to create $1000 of new business every second, merely to keep the firm that employed them all afloat. Craig Barrett, the retired CEO of Intel, points out that 90% of the revenues that company receives on the last day of the firm's fiscal year are derived from products that did not even exist on January 1 of that same year. It took 55 years for one-fourth of the US population to have an automobile, 35 years for the telephone, 21 years for the radio, 13 years for the cell phone, and only 7 years for the World Wide Web.

In this book, Professor Jim Wei, a superbly qualified guide, conducts us—poets and physicists alike—through a fascinating and informative tour of what it means to invent. This is a tour punctuated with risk-taking, failure, determination, insightfulness, luck, and, yes, even resounding success. We watch as scientists, engineers, and entrepreneurs collect a great deal of scar tissue—all without pain for us. It seems that success in innovation is not always where one looks for it: penicillin was discovered when Sir Alexander Fleming noticed that the bacteria he had been studying were not growing near the mold that had accumulated on a Petri dish that had been contaminated. A researcher of the Raytheon Corporation conceived the microwave oven when he noticed that a candy bar he was carrying in his pocket at the company's radar lab was melting. But, importantly, Louis Pasteur reminds us that "Chance favors only the prepared mind."

Margaret Thatcher notes that " . . . although basic science can have colossal economic rewards, they are totally unpredictable. Nevertheless, the value of Faraday's work today must be higher than the capitalization of all shares on the stock exchange . . . ." Indeed, it is doubtful that the early researchers in quantum mechanics had iPods or GPS in mind as they labored in their laboratories.

Today, the character of innovation itself is changing. While there will always be room for the Edisons, Fultons, and Whitneys, innovation—in both the science and the engineering—is increasingly becoming the province of teams—often of very large teams possessing very diverse backgrounds. This is the era of Big Science. Astronaut Buzz Aldrin observes that "It's amazing what one person can do, along with 10,000 friends." Inventions are now being found with increasing regularity at the intersection of disciplines. Plastics are made through the efforts of tiny bugs; and if one is dissatisfied with the output of these bugs, one need to only reengineer the bugs.

Science and engineering—which have brought us the Global Village by making distance increasingly less relevant—are themselves leading the parade toward globalization. It is noteworthy in this regard that America's innovation enterprise would barely function were it not for the foreign-born individuals who came to America's shores for their education, stayed, started companies, and created millions of jobs. One such individual, Jerry Yang, has said that "Yahoo! would not be an American company today if the United States had not welcomed my family and me almost 30 years ago."

Unfortunately, Americans, in particular, have been inclined to take leadership in innovation for granted. Dan Goldin tells of an incident that occurred when he was Administrator of NASA wherein the Agency was being criticized for investing so heavily in earth satellites. "Why do we need meteorological satellites?" the critic asked. "We have the Weather Channel." If we expect to get information from the Weather Channel, we need to support meteorological satellites. Of course, some innovations do seem rather humble—but that would be a dangerous generalization. A National Academy of Engineering panel led by astronaut Neal Armstrong concluded that the most important engineering accomplishment of the twentieth century was the development of household appliances—which freed the time of half the population to contribute through more rewarding pursuits. It is also noteworthy that if one were born in America at the beginning of the twentieth century, one's life expectancy was 47 years. Much of the gain since that time has been realized through advances in fields as diverse as food production and healthcare.

Unfortunately, as in so many pursuits, progress in innovation is not without its unintended consequences. It has, for example, been asserted that in spite of automotive advances, the average speed of surface travel across London today is about what it was 200 years ago. Large-scale terrorism has been made possible by developments in science and engineering that for the first time in history enable individuals or small groups, acting alone, to profoundly impact the lives of large groups. An unscientific survey by the writer of this Foreword reveals that most people believe that an invention that will inadvertently kill a quarter of a million people a year should be banned, until they hear that the invention is the automobile.

What of the future? The historical tendency has been to overestimate the near-term impact of science, engineering, and entrepreneurship, that is, "innovation," and overestimate it in the long term. A prime example of the former comes from Alexander Lewyt, founder and president of the appliance company by the same name, who in 1955 predicted, "Nuclear powered vacuum cleaners will be a reality within ten years." On the other hand, nowhere in the writings of Wilbur and Orville Wright is the suggestion that passengers equivalent in number to the entire population of Houston would each day hop aboard commercial aircraft somewhere in the United States.

Success at innovation will have a major impact on the quality of life in the years ahead as will failure. Perhaps a vaccine can be found to prevent cancer. Perhaps an effective means will be found of providing clean, inexhaustible, affordable energy for the entire planet. Perhaps there is a large asteroid hiding somewhere and intent upon destroying the Earth—a catastrophe that, through innovation, can perhaps be prevented. The quality of life in developed countries today heavily depends upon advancements in science and technology—and this is increasingly becoming the case for all the world's nations. But the benefits of scientific advancements often accrue, not simply to the individual investor but to society at large, thus making it essential that the general public support both education and research in science and technology. Only in this manner can our children and grandchildren hope to enjoy a standard of living higher than that of the generations that have preceded them.

So just turn the page for a fascinating and illuminating adventure into the world of innovation—written by an author who knows.

Norman R. Augustine

Retired Chairman and CEO
Lockheed Martin Corporation and
Chairman, National Academies "Gathering Storm"
Committee on Competitiveness

PREFACE

We have made tremendous progress in the last two million years in comparison with the natural condition of mankind that is said to be "nasty, brutish, and short." We live much longer and healthier, and we no longer need six children to ensure that two will survive to adulthood. We are no longer confined to live in tropical forests and savannas of East Africa; in fact, we can rather live almost anywhere on earth, from temperate farms to frigid cities. Inventions and technology are the most dynamic forces of change and progress in the world today, altering our lives and work at increasing speed, as well as our societies and environment. Our grandparents toiled much longer and harder compared to our 40 hours a week, and yet we produce four times as much food per acre of farmland. We can travel thousands of miles to visit friends and family, hear their voices on the telephone, and connect to the world through computers and the Internet. Every citizen of the world would benefit from knowing how inventions were made, how they have changed the world intentionally or unintentionally, and how to use and manage inventions wisely.

This book is the outcome of a freshman course at Princeton University, intended for future citizens and leaders. This book highlights some of the most important inventions in history, from the first stone axe 2 million years ago in East Africa to the current global connection through the computer and Internet. The inventions are from many geographical regions and civilizations of the earth, from East Africa to the Middle East, Europe, America, and Asia. My criteria for a great invention include satisfying a major need for many people over a long period of time, making a major advance in technology, and having long-term consequences in changing our standard of living. This book is organized around how we live and work rather than by time in history, by geographical regions and civilizations, or by types of technology. It is organized around tools and methods of production, food and shelter, health and security, transportation and information, and pleasure and culture. Major categories of inventions are listed and described, such as methods to grow food and make clothes. A few featured cases are treated in greater depth, such as the invention of penicillin by Alexander Fleming and other contributors, and the 10 year development effort to bring sufficient supply of penicillin for the D-Day invasion at Normandy and subsequently to the marketplace.

The inventors are described in their roles as creators and innovators, covering their backgrounds and preparations before their inventions, their motivations and methods, and their rewards, if any. After the invention, a few inventors remained actively involved in the entrepreneurial work of finance and development all the way to market success—like William Perkin and his invention

of synthetic dyes, but most of them turned the tasks of development over to organizations with financial resources and staff with various needed talents, like
Fleming with penicillin. When appropriate, the descriptions of the inventions
include the underlying scientific principles, the advancements in new technology,
the creation of new markets, and the major changes in the world of our work and
our lives.

Great inventions lead to new eras in human history. The stone axe released us
from our dependency on a few nature-given tools to solve a few problems into
inventing many man-made tools to solve many problems. Fire enabled us to cook
food and create ceramics and metals, as well as providing portable warmth to leave
the tropics of Africa and move to the frozen north. Agriculture gave us abundant
food, allowed us to abandon nomadic lives to settle in villages, improved nutrition
and health, and increased life expectancy and birth rate. Writing and printing
enabled us to record our history, stories, wisdom, observations and technology, and
to communicate with people far off in space and with future generations. The steam
engine provided tireless energy and power to industrialize mills and transportation.
The computer and the Internet have connected the world and made globalization
possible. The speed of the adoption of new inventions is uneven among world
populations, which gives the early adopters distinct advantages over the late adopters or nonadopters, adding to the inequalities among people. The large-scale use of
inventions can also lead to large-scale alterations in the natural ecology, favoring a
few economically useful animal and plant species and suppressing others. The
refuse and by-products of technology can also accumulate with time and cause
damage to the environment.

Our expectations of a future with increasing prosperity and better quality of
life depend on a continued stream of new inventions. We are besieged by shortages of resources like water and energy, by natural calamities of hurricanes and
earthquakes, by epidemic diseases such as malaria and AIDS, and by the threats
of terrorists and nuclear wars. We clamor for new inventions to solve these problems. However, a steady stream of great inventions is not an entitlement, but
must be paid for by society with educational programs to train future scientists
and engineers, by funding to support promising research, and by a reward system
for successful inventors and innovative entrepreneurs. There are disturbing trends
in recent years of decreasing support for inventions, so there are predictions that
future rates of invention will not be adequate for our needs. This was eloquently
pointed out by the National Academies report, "Rising Above the Gathering
Storm," and its subsequent sequel. We have a higher standard of living than our
grandparents had, and we have much to do to ensure that our grandchildren can
continue that trend.

I wish to acknowledge the many contributors to the writing of this book,
beginning with the classes of students who took my freshman course and helped me
to clarify which topics to include and how to explain them more clearly. Norman
Augustine joined me to teach at Princeton and taught me authoritative perspectives
on inventions from the views of governments and high-technology industries. Peter
Bogucki gave me many ideas about the archaeology of early humans and brought
me ancient stone axes that continue to inspire me. Tsering W. Shawa showed me

how to make geographical maps to illustrate the stories of human diaspora. My greatest thanks are to my family. My children Alexander, Christina, and Natasha edited and improved chapters of the book and also assisted with photo acquisitions. My wife Virginia gave me continued support and encouragement, and this book is dedicated to her.

James Wei

Princeton University

CHAPTER 1

INTRODUCTION

An invention is usually considered a delightful new device or method that would make life better. Let us also look at some official definitions. The Oxford English Dictionary mentions that it could be a discovery, a fabrication, introduction of a new instrument, a design or plan, a figment of imagination, or a piece of music written by Bach.

The US Patent Office requires that for an invention to receive a patent, it should be new, inventive, and useful or industrially applicable. It is possible to be new without being inventive, such as a scientific discovery that may have no immediate practical application. The Patent Office defined four general categories of inventions: (a) a process or a method, (b) a machine, (c) a manufacture, and (d) a composition of matter. A significant improvement can be patentable, but an idea or suggestion must be accompanied with a complete description of the actual machine, and reduction to practice is often required. The patent gives the owner the exclusive right to use the invention for a number of years, such as 20 years, and can license the right to another party for considerations. The owner can also sue anyone infringing on the patent without a contract and payments. In practice, it is very easy to detect infringement on a patented product when it is sold in the market to many customers, and somewhat more difficult to detect the presence of a patented matter in a manufactured product. The most difficult to enforce is a patented machine or method that is installed in a factory not open to the public without a search warrant.

We usually think of inventions as providing the means to satisfy our material needs, such as for food, clothing, and shelter. For these purposes, we have invented the tools of agriculture, of spinning and weaving, and of beams and roofs. Our spiritual needs such as knowledge, beauty, truth, and justice are also supported by inventions and technology—including the ability to record and print words and pictures, and to communicate to people far away and to future generations. There are very few revolutionary breakthrough inventions on brand new technologies, such as penicillin and transistors. Most inventions are based on making improvements on an existing technology to make it more effective or efficient such as vulcanized rubber, or finding a new use for an old material such as using ether for painless surgery and childbirth.

What is a great invention? An invention adds to the store and power of technology, which bestows benefits (and sometimes harm) on our work, lives, society, and the environment. We value an invention according to a number of criteria including: (a) the audacity of the technology over the existing technologies, (b) the expansion of our capabilities to perform tasks that were considered impossible, and

Great Inventions that Changed the World, First Edition. James Wei.
© 2012 John Wiley & Sons, Inc. Published 2012 by John Wiley & Sons, Inc.

to open doors to exciting new possibilities, and (c) the valuable and long lasting benefits that it brings to many people.

The greatest inventions make dramatic breakthroughs, and open new eras in human history. Consider the lives of early men in East Africa about 4 million years ago without the sharp teeth of lions to tear meat. The stone axe was the first great invention that allowed our ancestors to eat the food of lions, and set us on the path of independence from our meager tools of teeth and claws bestowed by nature, as we could invent a whole arsenal of new and powerful tools. Fire led to the invention of cooking and softened tough meat and cereal as food, to ceramics and metallurgy, and to the colonization of the frozen north. Agriculture led to much greater and more secure food production, and allowed people to settle in villages and cities. The steam engine led to tireless power for manufacturing and transportation and to a burst of productivity increase and the Industrial Revolution. Modern sanitation and the germ theory lowered the rate of infant mortality, so that it is no longer necessary to have six children to ensure that two would survive to adulthood. Each generation of human society inherited a much bigger toolbox of technology from the previous generations, and can enrich it by the constant addition of ever more new inventions to benefit the next generation.

A drug that cures lung cancer would benefit millions of people, and would be considered more important than a drug for a rare disease that affects only thousands, according to the Jeremy Bentham principle of greatest happiness for the greatest numbers. Bentham also specified that happiness should be ranked by intensity, duration, and certainty. An invention that keeps us alive is more valued than an invention that improves our vanity; an invention that remains in use for many years is valued more than inventions that are quickly replaced; and a drug that always works is better than one that works only some of the time.

The direct benefits of an invention can be obvious, such as fire providing warmth and light. The unintended indirect benefits (or harm) are often slower in coming but can be far more important, such as fire leading to cooking which softened tough grains, and made possible pottery, bronze, and iron. The synthetic dye mauve was invented by William Perkin in 1856 and was used for only a short period of time before it was replaced by newer and better dyes, but its success inspired many chemists and entrepreneurs, and subsequently gave rise to many new synthetic dyes and synthetic drugs, such as sulfonamide. These new drugs became the foundation of the modern pharmaceutical industry, and saved millions of lives. Freon was a refrigerant introduced in 1920 that made possible safe home refrigerators without the hazards of fire and toxic leaks, but it accumulated in the atmosphere for many decades and led to the ozone hole and global warming, which made it no longer suitable.

1.1 INVENTORS AND INVENTIONS

Consider where and when great inventions were made, who made these inventions, what motivated them, what were their methods, and how revolutionary were their inventions?

1.1.1 Cradles of Inventions

In the past several hundred years, the most inventive places on earth were in Western Europe and later in North America as well. Is there an association between climate and inventions? Let us use the Köppen Climate Classification scheme, which is based on the distribution of temperatures and rainfall of each month in the year. Mellinger, Sachs, and Gallup observed that the temperate zones within 100 km of the ocean or a sea-navigable waterway accounts for only 8% of the world land area, but has 23% of the population, and 53% of the GDP of the world. It is also the most inventive area in the modern world, with the highest standard of living.

The most inventive places in the ancient world were in the dry climate (Mesopotamia, Egypt) and located at large rivers. The more recent inventive and economically successful places tend to be in the humid temperate zone (Athens, Rome, London, Xi'an, Philadelphia), and less often in the humid cool zone (Beijing, Boston, Berlin). The tropical humid equatorial climate of the Amazon and Congo supports a large population, but is not very inventive; the cold polar and the highland climates support little population, and are not inventive.

However, the ancient cradles of the most important inventions came from the tropics, and gradually migrated to warm subtropical climates, and later to cool temperate zones. Let us look at this migration for six great inventions: the stone axe was from Olduvai Gorge in Tanzania (latitude 5°S) 2 million years ago; fire was first mastered at Zhoukoudian in China (40°N) 500,000 years ago; agriculture began in the Fertile Crescent (33°N) 10,000 years ago; writing began in Mesopotamia (33°N) 3,500 years ago; the steam engine was established in Scotland (56°N) in 1750; and the electronic digital computer began in the United Kingdom (51°N) and the United States (41°N) around 1950–2000. They exhibit a steady northward movement with time.

The climates of Eurasia and North America were not always the same throughout human history. The world climate turned distinctly colder from Pliocene to Pleistocene 2 million years ago, perhaps by as much as 2–6°C in comparison with the year 1950, and much of Northern Europe and America were under sheets of ice. The ice age ended at the beginning of Holocene 10,000 years ago, and the world began to warm up significantly.

Why did most great inventions arise in these temperate places, and why at that moment in time? The most frequently cited requirements to support inventions include the following:

(1) *Environment.* The hunter–gatherers need a healthy and agreeable climate with warmth, rainfall, and soil, suitable for the growth of plants and animals for food. The farmers need to find local plants and animals that can be domesticated, stone and clay for construction, trees for fuel, and ores for metallurgy. The temperate climate provides the stimulus of a change of seasons and cyclonic storms, and gives the residents challenges to keep them alert with problems to solve. Arnold Toynbee proposed the Golden Means theory on the genesis of a civilization. A group of people can live in comfortable torpor for a long time, and would need a stimulus or a challenge in order to respond and move into a dynamic creative state. There are various stimuli, such as living in

a hard country and environment, moving to a new habitat, external blows from enemies, internal pressures, and penalty in comparison with other groups. If the stimulus is necessary to wake one from contented torpor then would more stimuli always lead to better responses? He proposed that challenges should be large enough to be stimulating, but not too large to be overwhelming. He gave the following examples as illustration:

- The Vikings living in Scandinavia had a mild climate during the Viking Age, and were not sufficiently challenged to produce much literature; but the Vikings in Iceland were exposed to a bleak and barren environment and made greater achievements in literature; and the Vikings living in Greenland had an even more bleak and barren settlement, and were barely able to survive and had no time for literature.

- The Europeans that settled in North America at Virginia and the Carolinas had a comfortable life, and made few contributions to literature; the settlers in Massachusetts had a more harsh climate and stony soil, and were sufficiently challenged to achieve leadership in intellect and commerce; the settlers in Maine and Nova Scotia had scanty livelihoods and no time for literature.

If we rank the climate zones according to the degree of challenge, we may obtain the following from the least challenging to the most challenging, and the optimum is presumably somewhere in the middle: humid equator $<$ humid temperate $<$ humid cold $<$ dry $<$ highland $<$ polar. Environment alone is insufficient to explain why North America was dormant before Columbus, and became a world leader after 1950.

(2) *Contacts and Heritage.* Inventors build on the stimulus of previous technologies, which they inherited from their ancestors; they also learn from neighbors and visitors that they meet. They need access to transportation in order to trade and to communicate with other people, and to learn new ideas and technology. Jared Diamond argued that the various people of temperate Eurasia can travel 13,000 km from Western Europe to East Asia without a change in climate, and can learn and adopt inventions and ideas from other people. On the other hand, the peoples of North and South America became a series of isolated communities as a similar trip of 13,000 km from the Bering Strait to Tierra del Fuego would involve crossing numerous climate zones, with the need to adapt to new food sources and enemies, and to stay warm or cool. The people of Oceania were likewise isolated from the Eurasian inventive communities. This requirement of heritage and contact does not explain why Africa did not continue to be inventive after a glorious start, or the long sleep of Rome between Justinian and the Renaissance, the sudden birth of the Islamic civilization with Mohammed, and the long silence of Mongolia after Genghis and Kublai Khan.

(3) *Soul and Leadership.* The creativity and dynamism of a civilization, as well as subsequent stagnation and decline, have many causes that are difficult to analyze and explain. One can put together a long list of influences: internal

tradition, philosophy, religion, external challenges, optimism, stability and security, openness to new ideas, and willingness to adopt progressive ideas and to reward innovators. Charismatic leaders are possibly the most important requirement. Oswald Spengler believed that each great civilization has a soul, which runs through the course of a thousand years from the awakening of barbarism to growth of a new civilization, from expansion to the zenith of empire, to decline and eventual decadence. His Apollonian soul is the Greco-Roman civilization, and his Faustian soul is the Western civilization from Merovingian to now. This explanation can be viewed as an inspired oracle instead of a scientific method, as it gives no principles to predict the arrival of the future souls.

1.1.2 Creativity

We have very little knowledge of the inventors of the first stone axe, who lived in East Africa some 2–3 million years ago. After the invention of writing, we begin to have written documents and some information on the inventors. Who were the inventors, and what were their backgrounds and education; why did they become interested in inventions, and what were their methods; and how revolutionary were their inventions?

In ancient times most people toil for immediate needs, and only a few had the leisure to pursue interests that did not produce short-term benefits. Perhaps the earliest inventors known by name in history were Imhotep and Yu the Great. Imhotep dated from 2600 BCE in Egypt, and was chancellor to the pharaoh Djoser and high priest to the sun god Ra. He was also the first engineer, architect, and physician of Egypt; was credited with the invention of the papyrus scroll and the architecture columns; and was declared to be a god after his death. Yu the Great lived around 2060 BCE in China, and was the founder and first emperor of the Xia Dynasty. China was suffering from great floods, and the king Yao assigned the task of taming the flood to his minister Gun. Gun erected numerous dikes that failed, and he was executed by Yao, who turned around and assigned the same job to Gun's son Yu. Instead of erecting dikes, Yu dredged new river channels to serve as outlets to the flood, and for irrigation. Yu labored for 13 years as an inventive hydraulics engineer, tamed the flood, and was reward with the kingship by Yao. For these two inventors, the process of invention took little time in a schedule crowded with numerous important duties. No inventor since Imhotep and Yu has received as much recognition and honor as these two.

The occupations of the inventors at the time of invention can generally be divided into part-time amateurs who dabbled in inventions as a hobby or side interest, and dedicated professionals who received support to invent for wealth and fame.

Inventions require time and patience, as well as optimism from the inventors, who often must have other means of livelihood. The part-time inventors had other occupations or inherited wealth, and occasionally took time out to invent something. Archimedes (287–212 BCE) was a wealthy aristocrat, and a relative and advisor to the king of Syracuse. He was freed from the concerns of making a livelihood, and had the time to make scientific discoveries and inventions such as a method to determine the density of metals by immersion in a bath, and the Archimedes screw to raise water from a river. Other examples of amateurs as inventors include Alexander Fleming and Wilhelm Roentgen, who were professors engaged in teaching students and making scientific investigations. They also made accidental discoveries that led to great commercial success.

Independent entrepreneurs are people without the sponsorship of a government or a company, but who gave up regular occupations to dedicate themselves to inventions, hoping to gain fame and fortune. James Watt invented several much improved steam engines, and went into partnership with Matthew Bolton. Charles Goodyear gave up all other work to concentrate on finding a way to improve rubber. Alexander Graham Bell invented the telephone. Thomas Edison was among the first to take up inventions as his main occupation and source of income, when he set up his independent industrial research laboratory in Menlo Park, New Jersey in 1876 with funds from his previous inventions. The Silicon Valley is full of such independent entrepreneurs, and some of them became very wealthy at an early age.

The corporate employees are engaged by firms or governments to do research and inventions, and may be required to sign over future patent rights and profits to the employers. Leonardo da Vinci (1452–1519) was paid by the Duke of Milan and by Francis I as a painter and military engineer, when he invented methods of fortifications and siege engines, as well as flying machines. After the dye discovery of Perkin, the German companies such as Hoechst, Bayer, and BASF poured resources into research, hired highly educated graduates from universities to discover new dyes at lower cost or higher quality, and later branched out into pharmaceuticals with the discovery of the sulfa drugs. The DuPont Company set up a research laboratory in 1902 to diversify from their traditional business of making gunpowder, and hired Wallace Carothers who went on to invent nylon in that laboratory. Thomas Midgley invented the tetraethyl lead and the refrigerant chlorofluorocarbons (CFC) in the General Motors laboratory. The Bell Laboratory was set up in 1925 by the parent company AT&T, and one of their most famous inventions was the transistor by the team of Bardeen–Brattain–Shockley. Corporate funding has become the dominant support of inventors, where teams of specially trained scientists and engineers are housed in special buildings and laboratories, equipped with modern equipment and instruments, and are paid to do full-time discoveries and related activities.

The successes of inventors owe a great deal to the encouragement of society, for support before the inventions, and for rewards after the inventions. Public support includes education in science and technology, research grants, and patent protection, which gives the inventors a monopoly on exploiting their inventions for 17–20 years. Another form of encouragement is public honors in the form of prizes and recognitions, such as the Nobel Prize and the Inventors Hall of Fame.

Why was the inventor motivated to invent something? It is often said that necessity is the mother of inventions. This would imply that many inventions began with needy and dissatisfied customers, followed by deliberate searches to find solutions to important problems. There are many examples to these *Market-Pull* inventions, when the investigators were motivated by market applications. Raw rubber was used to make raincoats and balls, but they were brittle in cold weather and sticky in hot weather. Charles Goodyear searched for a method to improve the qualities, and to turn unsatisfactory raw rubber into a useful product. He spent 5 years doing trial-and-error experiments before he discovered that sulfur and heat could be used to vulcanize rubber. The Newcomen steam engine was used to remove water from flooded mines, but it was very inefficient and wasted a great deal of coal. James Watt made many improvements so that the steam engine became much more efficient, and this tireless and inexpensive engine started the Industrial Revolution. Home refrigerators originally operated with a number of refrigerants that have toxic and flammable properties, such as sulfur dioxide and ammonia, and posed real threats. Thomas Midgley was asked to find a nontoxic and nonflammable refrigerant, and he invented the chlorofluorocarbons. In a modern company or government agency, the quest for an invention may begin at the marketing department, reporting on customers with needs who are not satisfied, and demanding a better product.

Another mode of invention is to start from a technology, and then search for customers. The investigator may have created an improved or new technology, either by accident or while looking for something else. The investigator may also start from a "platform" technology that has proven successful in serving one market, and search for other markets that can be served by the same or slightly modified technology. These are sometimes called the *Technology-Push* inventions, since the investigators had the technical capability first. The Watt steam engine was so successful in pumping water from mines that Robert Fulton modified it to operate steamships, then George Stephenson modified it to drive trains, and even the textile industry adopted it to power textile mills. CFC is effective in refrigerators and air conditioners, due to its nonflammable and nontoxic properties, which also makes it suitable for other applications such as propelling aerosols and blowing dust off computers.

The most innovative event is created when an invention creates a new demand that did not exist before, and thus a brand new market. The vegetable and mineral dyes available in 1850 were few and drab, but since the public was not aware of the possibility for a greater and more vibrant variety of dyes, they did not clamor for them. Perkin was an 18 year old schoolboy on vacation at home, who started out looking for a way to synthesize quinine to treat malaria, but his oxidation of coal tar resulted in a brilliant dye. It led to a series of more brilliant dyes, as well as the modern chemical industry. From the beginning of history, humanity suffered, and died of infectious diseases, but the suffering public did not know that it was possible to invent miracle drugs. Alexander Fleming was working with staphylococcus bacteria in a London hospital when he found colonies in the Petri dishes in his musty and dusty laboratory. He observed that bacterial colonies do not grow in rings around areas that have been accidentally contaminated by a green mold, and he found the substance that he named penicillin. This led to its widespread use in medicine, and created the new market of antibiotics. In the same way, there were no market demands before the launching of the personal computer, the cellular telephone, or the Internet. Steven Jobs was famous for inventions that anticipated public demand: the public could not imagine such miraculous machinery as the iPad and iPhone before they were introduced, and could not live without them after they were introduced.

For the inventors of the stone axe 2 million years ago, a stone with a naturally broken sharp cleavage was found to be effective in cutting meat, and became used regularly. Probably the long process of improvements over the next 2 million years was guided by random tinkering, trial-and-error experiments, and remembering which methods produce better products. This process of empirical inventions had no scientific theory and systematic data for guidance. This method survives today in areas with insufficient scientific understanding.

Paul Ehrlich was a medical doctor searching for a drug to cure syphilis without severe side effects, and was convinced that arsenic held the answer. He reacted arsenic with other chemicals to form many new compounds, and hoped that one of them would not be overly toxic to humans but still effective on syphilis. He synthesized 606 arsenic compounds, and found the compound Salvarsan to have the desired properties. The random search of thousands of objects is a very slow and expensive undertaking, but is worthwhile when the goal is very important and no knowledge or theory exists to help. This method is sometimes called "Edisonian" as it was used by Thomas Edison in his search for a carbon filament in the incandescent lamp. It was also the method used to discover taxol, the recent drug for uterine and mammary cancer.

The modern scientific method arose in the Renaissance, and became the new source of the most productive inventions. Francis Bacon described the scientific method in 1620 as an endless cycle of the following steps:

(i) Make observations of a phenomenon, note the regularity and reproducibility of the observations, and confirm by more measurements.

(ii) Make hypothesis of an explanation about the cause of this phenomenon.

(iii) Based on this hypothesis, make predictions of other phenomena that can be observed and measured.

(iv) Perform experiments designed to test the predictions, and compare to the results to confirm or deny the validity of the hypothesis.

(v) If the experiments confirm the predictions, the hypothesis receives one more vote of confidence; if the experiments deny the predictions then the hypothesis needs to be revised and we return to (iii) to repeat the cycle.

The scientific method led to a set of fundamental theories that govern the physical world, such as the Newton's law of motion, the Maxwell law of electromagnetism, and the second law of thermodynamics. Systematic experiments for many years led to a body of knowledge and databases of the properties of matter. In modern times, there is a very large body of scientific knowledge and understanding, which became the foundation of many inventions. The public support of education in science and engineering creates a larger pool of workers who have the necessary background for inventions based on science.

The modern inventors who have studied the physical and biological laws can make reliable predictions on many consequences when an action is taken. For instance when you climb a peak of 20,000 ft, what would be the boiling point of water, and how long would it take to make a hard-boiled egg? Another challenge is how you would operate a pressure cooker to control the pressure so that you can hard boil an egg in 2 min instead of 5 min. Most technology today is based on reliable science and supplemented by the less reliable intuition and hunches, based on the skill and the inspiration of the artist. The rate of inventions based solely on empiricism was painfully slow for millions of years, and became dormant in the west during the middle ages; but since the scientific method, the rate of science-based inventions has been phenomenal.

Most inventions are embedded in a long evolutionary sequence of many closely related inventions, and can be studied as a continuum over time. Each invention in such a sequence can be considered to be *incremental*, as they made small and more or less obvious improvements in the technology, or adapted a product to a slightly different market application. After many years of improvements, such technologies often become mature and do not offer further opportunities for development. Revolutionary improvements come with inventions that merit the designation of *breakthrough*, as they involve unexpected and novel ideas that spawn many applications and improvements in the future.

The following table is a matrix of inventions with rows from current technology to improved technology, and finally to revolutionary technology. The columns range from serving the current market to serving additional markets, and finally to creating brand new markets. In an incremental invention, an investigator starts from a current market served by a current technology, and looks for incremental changes that would lower costs or improve quality, or serves to find a new market. For the inventor, this is a relatively safe path that poses small challenges in getting the technology ready and a receptive market. However, the rewards in fame and fortune are likely to be modest. There are three forms of this incremental invention, which are listed as:

(i) Keeping the same market and searching for an improved technology. Salicylic acid from willow bark will cure headaches, but it is very harsh on the stomach. Felix Hoffman chemically added an acetyl group to salicylic acid, and the result was aspirin, which has the same effectiveness but is less harsh on the stomach.

(ii) The term "platform" technology is sometimes used when one takes a proven technology successfully used in one application, and looks for other applications. Botox is the deadly poison from the botulism bacteria that causes paralysis of the muscle. Many years later, it was found to be effective in removing facial wrinkles.

(iii) The steam engine was successful in pumping water from flooded mines. Robert Fulton modified the Watt engine to propel the steamboats he was operating on the Hudson River, and George Stephenson modified the engine to propel railroads.

	Serve current market	Serve additional markets	Create brand new markets
Current technology	Business as usual	(ii) Botox to remove skin wrinkles	(vi) Morton, ether for anesthesia
Improved technology	(i) Hoffman modify aspirin	(iii) Fulton and Stephenson used steam engines for trains and ships	(vii) Cellular telephone
Revolutionary technology	(iv) Midgley, CFC for refrigerators	(v) Midgley, CFC for air conditioning and aerosol	(viii) Perkin, synthetic dye for textile Fleming, penicillin as antibiotics

Occasionally, we witness the excitement of an invention that takes a revolutionary leap forward in creating a *breakthrough* technology. This is more risky, as it may be difficult to make this technology effective, safe, and economical.

(iv) Thomas Midgley was asked to come up with a refrigerant that is not flammable and not toxic, thus suitable for home refrigerators. Instead of studying the currently available refrigerants, such as sulfur dioxide and ammonia, and finding ways to improve their properties by additives or substitutions, he used the periodic table of Mendeleev to discover a new class of compounds, the CFC.

(v) Since CFC became successful in refrigerators, the same nontoxic and nonflammable properties make them useful in air conditioning and aerosols.

Sometimes we create a brand new market that did not exist before. This is also risky as customers may not embrace this unfamiliar new technology, and may refuse to use it.

(vi) Ether is a chemical used for solvent and paint removal. Morton introduced diethyl ether into surgery as an anesthetic, which reduced pain and suffering. Before ether was introduced for anesthetics, there was not a market for an effective and safe anesthetic.

(vii) Many changes were made to the traditional telephone to make the cellular phone, which is portable and not tied to the wall by a cord. It created the new market of cellular telephone.

Undoubtedly the most exciting inventions involve the simultaneous creation of a revolutionary technology and a brand new market. These inventors take a doubly risky path, as the technology may not work, and the public may not embrace this new product.

(viii) Perkin's synthesis of mauve from coal tar was a brand new technology, and created colors so brilliant and enchanting beyond what occurred in nature, which started the new field of synthetic dyes. Fleming's discovery of penicillin enabled us to save millions of lives from bacterial infections, and created the new field of antibiotics. Before Perkin and Fleming, dyes and drugs were found in plants and the earth, but these two inventions awakened the world to the realization that there is unlimited potential in synthetic chemistry. The doors that they opened are even more valuable than those two original inventions of mauve and penicillin.

1.2 INNOVATION, DEVELOPMENT, DIFFUSION

Joseph Schumpeter once said that an invention that is not widely used is irrelevant to human affairs. There are thousands of ingenious and admirable inventions that were neither carried out in large scale nor were they used by millions of people to change the world. Hero of Alexandria in the first century produced a steam engine by jet action, but it was treated as a curiosity and did not lead to benefits for society. Leonardo da Vinci invented a number of flying machines, but there is no record that they were ever built to change transportation or warfare. Crawford Long of Georgia actually used ether for anesthesia in surgery a few years before William Morton in Massachusetts, but Long did not publish his results and had no influence in subsequent medical history. Out of the many thousands of inventions in history, only a few were able to travel the long and difficult path from discovery to development, to be manufactured on a large scale, and to be sold widely in the marketplace, and effect significant change in the world.

A discovery does not become a widely used technology until it has been shaped to suit the factory where it will be manufactured, to suit the customers in the marketplace, and to find sufficient financial backing to pay the bills till revenue begins to roll in. Some argue that there are two separate acts to a successful innovation: (a) the invention, which is an original act of discovery with or without an economic motive and (b) the innovation, which is driven by entrepreneurs for economic development. In the case of the atomic bomb and radar, the innovations were driven by the government for political and military reasons.

1.2.1 Development

After a discovery, the concept needs to be shaped into one or more products to suit specific needs of the marketplace, so that it can be sold at sufficient volume and price to compete with other products. It must be possible to make the product in a factory, with a suitable manufacturing process, raw material supply, equipments and plants, and at an affordable cost. The entire innovation effort must be organized under some leadership, with access to sources of finance to pay the bills till the products can be sold in volume. This sequence of events involves many people with different talents, and must be coordinated successfully.

Wallace Carothers of DuPont discovered that he could make polymer fibers by reacting a 16-carbon diacid with a 3-carbon dialcohol, which has a melting point of below 100°C. What products could DuPont make with this technology that would earn a profit for the stockholders? DuPont decided to make nylon as a luxury stocking for women, as they already had experience making the semisynthetic rayon fiber for the textile industry, and a pair of nylon stocking requires only a few grams of polymers and can be sold for a high price. This decision set up a number of development problems of production and marketing. Nylon stockings were offered to displace silk stockings from the market, so there had to be advantages to women to wear nylon instead of silk, and the price of nylon could not be too high in comparison to silk prices. A pair of silk stockings had to be ironed, so it required a sufficiently high melting point; DuPont found a solution by replacing the dialcohol with the diamine, which resulted in a higher melting product. How could DuPont acquire enough raw materials of diacids and diamines from plentiful and cheap coal tar or petroleum? After a great deal of investigation, they decided to switch to the 6-carbon diacid and the 6-carbon diamine, which could be made from the abundant supply of 6-carbon benzene, and was named nylon 66. The polymer also had to be pulled in the molten form through diamond dies and twisted into fibers of the appropriate thickness and elasticity. The DuPont Company had enough confidence in this decision to finance the development from past earnings. Ten years passed before the first satisfactory product emerged dating from the original Carothers discovery.

When Alexander Fleming of St. Mary's Hospital discovered penicillin in a Petri dish, more than a decade elapsed before it became a lifesaver for millions. Fleming was a bacteriologist, and had no idea how to manufacture and market a novel drug. Penicillin was produced in his Petri dish with a concentration of 30 ppm, and he did not have the knowledge and skill to extract and purify the drug for clinical tests with animals and humans. He could make a few milligrams in one Petri dish, but he could not manage a million Petri dishes to make 1 kg of penicillin. Some of the purification and testing were solved a decade later by the chemists Howard Florey and Ernst Chain of Oxford University. They lacked the industrial capability to produce penicillin in wartime England, and Florey sailed to the United States to seek help. Penicillin is a very unstable liquid, and decomposes in about 3 h, and he did not know how to store it and have it ready for clinical use. After these problems were solved, a big source of financing was needed to pay for the costly steps of building plants, buying machinery and raw material, recruiting and training labor, and developing a storage and distribution organization. The US Scientific Research Board became the entrepreneur to manage these developments and assigned different tasks to different organizations and investigators, and persuaded the US President and Defense Department to take the risk and finance this unproven venture.

Perhaps more than 99% of all the discoveries do not make it to the marketplace, either because they fail to find an entrepreneur with sufficient resources to undertake the expensive development–manufacturing–marketing processes, or because they run into obstacles in the process. There is an illness called "kala-azar" or black fever, which is spread by sand flies, and kills half a million people per year, making it the second largest parasitic killer in the world after malaria. It occurs mainly among the poor people in India, Bangladesh, Nepal, Sudan, and Brazil. The drug paromomycin was discovered in the 1960s and seemed very promising, but was abandoned due to the high cost of Phase III clinical trials and the low probability of making sufficient profit to recover the development costs. A number of private foundations, such as the Bill and Melinda Gates Foundation, are beginning to finance the development of such neglected drugs. Even after the drug has been tested and found to be effective with negligible side effects, distribution looms as the next hurdle as getting the drug to remote villages at the end of pothole-pocked roads will be difficult.

In the last two centuries, there have been several very successful models of managing development from discovery to the marketplace:

- *The Discoverer–Entrepreneur Model.* Synthetic mauve was accidentally discovered in 1856 by William Perkin when he was a schoolboy of 18, who realized that its rich purple color could be used to dye textiles. He dropped out of school, formed a partnership with his father and brothers, developed the process for manufacturing, procured the necessary raw material, visited textile manufacturers to persuade them to use his dye, built the factory, and supervised the manufacturing and shipment of dyes. Besides doing the discovery and development, he was also the entrepreneur who found financial support and managed the entire enterprise from the beginning to the end. Thomas Edison also followed his discoveries with entrepreneurial work, participated in forming companies and kept much of the profit. Alfred Nobel, the inventor of dynamite, was also a discoverer–entrepreneur. The modern equivalents of this super solo model are some of the Silicon Valley information technology start-up entrepreneurs, especially those in software who do not require large capital investments in plants and equipment.

- *The Company Acquisition Model.* Fritz Haber informed the BASF Company that he had found a way to synthesize ammonia from air and water, but Haber did not have the means or the knowledge to turn it into a process. The company purchased the rights to Haber's patents, and retained him as a consultant. Then the company proceeded to develop the process under the direction of Carl Bosch, built the plants, sold the product to the German government, and financed the effort. Many start-up companies in the Silicon Valley sold their companies to large corporations, instead of remaining independent and taking more risks for more control and a chance to make bigger profits.

- *The Central Command Model.* The DuPont Company hired Wallace Carothers to do research. After the discovery of nylon in 1935, the company decided that it should be developed for stockings. DuPont assigned many talented researchers to solve problems of manufacturing and marketing as well as financing the entire project. This model can be appropriate for a company with a strong staff and deep pockets.

- *The Consortium Model.* The development of Fleming's penicillin was first carried out by Florey and Chain, and later by numerous American industrial organizations with the chemical engineering skills to make large-scale production, purification, concentration, and stabilization of the product for storage and shipment. This process was financed by the US government as part of the efforts of fighting the World War.

1.2.2 Factory Production and Market Penetration

No invention can be manufactured in large enough quantity at reasonable cost unless a number of production factors are available. For instance, the Sahara and Greenland cannot adopt agriculture as a way of production because their climate and soil are not suitable. Diamonds are mined in South Africa, which lacks skilled diamond workers or the right marketing outlets. Consequently, the raw diamonds are shipped to the Netherlands and to Israel, to be polished into gems and sold at luxury showrooms. The principles of the atomic chain reaction was demonstrated by Enrico Fermi in Chicago, but the atomic bomb could not be manufactured till sufficient uranium-235 had been separated and enriched, and plutonium had been created by neutron bombardment.

The economists often list the "factors of production" as land, labor, capital, and technology.

- The term "land" includes the natural resources that come with the land—the bounties or paucity of climate, temperature, rainfall, animals, plants, soil, and minerals.

- The term "labor" includes considerations of the total number of workers, their age distribution, the states of health and vigor, the education and skill, and the intangible aspects of diligence and creativity. Another indispensable factor is "entrepreneurship," which is the leadership with an overall vision of the enterprise, the ability to find financial resources, and the courage to take charge and make risk decisions.

- The term "capital" includes capital investments into the technology of production machinery, tools, buildings, inventions, patents, and methods of production. It also includes society's investment in the public infrastructure for transport, so that raw material can be efficiently brought in and products brought out; for information infrastructure such as telephone and Internet; and for utilities such as dependable electric power and water.

- The term "technical knowledge" is not one of the classical factors of production, but must be available in the education and previous experience of the staff, and as patents and trade secrets.

When a product is bought steadily by a group of customers over a period of time, it must be due to the needs that are met by the product. The customer must be satisfied by the benefits from the products, and have sufficient purchasing power to cover the costs of the product at posted prices. A customer usually has a choice of other products on the market, and will do a cost/benefit analysis to determine that this product is worth buying either because it costs less or has better benefits than any other competing products. The needs and constraints of buyers are different for consumer products such as nylon stockings, for business products such as oil tankers, and for government products such as nuclear submarines. The entrepreneur tries to understand the needs, the budget, and the competition of each group of customers, and tailor the products and prices to suit these realities.

Throughout history, the global reach of an invention often followed trade, migration, and colonization. The traders were often the first visitors to an isolated community, and the first to expose the natives to new goods and methods. Marco Polo was credited with bringing back a number of inventions from China to the Western world, such as noodles. Peaceful migration and military conquests bring a larger influx of exposure, such as the appearance of horses and firearms in the New World. There are a number of consumer barriers to the diffusion of new technology, thus the process can take many years before a new technology is adopted. Besides cost/benefit and competition, there can be many social barriers to the market acceptance of new products, such as inertia, xenophobia, and religious and cultural taboos. The rate of adoption of a new technology such as the mp3 player can be compared to a parade led by the innovators for their own uses, followed closely by a few intrepid early adopters. There is a tendency for the innovators and early adopters to be risk takers who are younger, better educated, of higher socioeconomic status, and who take pride in being the trendsetters for their generation. They are followed by the majority, who join when they are encouraged by the satisfaction and recommendations of the early adopters. The trailing groups are the late adopters, who are more conservative, more inflexible, and suspicious. Finally there are the laggards who do not join this parade at all. The late adopters and laggards can be older skeptics who have an aversion to change, are less educated, and are socioeconomically lower.

The US Census Bureau provided data for 2007 on the relatively new technology of the Internet, which is used by more than 4/5 of the wealthier users with more than $100,000 in annual income and by the younger users less than 44 years old. On the opposite end, the Internet was used by less than 1/3 of the lower-income population who earn less than $20,000 and users older than 65.

The World Bank publication of World Development Indicators gave the following data on popular use of information technologies as an international comparison per 1000 population. The high-income nations use the television 10 times more often than low-income nations, and use personal computers 100 times more often. In this instance, wide usage is not simply a matter of need, but is driven by purchasing power and education as well.

Nation	Radio set	TV set	Personal computer	Internet
Low income	139	91	7.5	10
Lower middle	360	326	37.7	46
Upper middle	466	326	100.5	149
High income	1266	735	466.9	364
United States	2117	938	658.9	551

1.3 CHANGING THE WORLD

Inventions lead to technological developments and effects on society that can be either beneficial or harmful. Many people feel that new technology is basically beneficial and gives society more choices, even if it is sometimes hazardous, and people should be willing to adapt to the inevitable march of progress. Others view many technological developments as basically harmful, because they may fall into the hands of unscrupulous individuals and institutions who would misuse them. The steam engine and penicillin are generally viewed as beneficial inventions with a few unfavorable consequences, but weapons of mass destruction are more often viewed as primarily harmful. In the free market, inventors respond mostly to demands from people with strong purchasing power, and ignore the weaker signals from people with ethical concerns or weak purchasing power. Inventors can serve the needs of the poor and powerless only if there is a market demand created by governments and nongovernment organizations.

1.3.1   Transforming Work

When Adam and Eve were banished from the Garden of Eden, they were told that "By the sweat of your face you shall eat bread." Since the beginning of time, mankind has had to labor to produce the necessities of life. Human history took great leaps forward when outstanding inventions appeared to expand our capabilities so we could produce results previously thought to be impossible. However, most inventions have the more modest goal of making our work more efficient and productive. The productivity of our work largely determines how well we live, and how easy it is to produce the necessary and luxury goods and services for ourselves and our families. Our employment also affects our self-esteem, sense of accomplishment, purpose in life, and ability to help others.

There were a few truly outstanding inventions that so greatly expanded human capabilities that we could accomplish work that could never be done before. Several times in history, a new invention changed the way we work to make a living, and ushered in a new age in our history.

The *stone axe* was the first human invention that put us on the evolutionary path. To be human means that, simply by our ingenuity, we can supplement the gifts of nature. The first human invention appeared 2–3 million years ago in East Africa, marking the beginning of the Paleolithic Age. Humans are not the only animals that use natural objects as tools to supplement our teeth and claws, but only humans can design, manufacture, and constantly improve their numerous fabricated tools. Perhaps the first inventions of humans were clubs of wood or horn used in hunting and warfare, which have perished through the years. The stone axe is the oldest surviving man-made object, which was designed and manufactured for a function, and was continuously improved over millions of years. But the most important consequence of this invention is that humans no longer relied entirely on nature to provide useful tools, and began to supplement their insufficient inheritances by inventing their own tools. The stone axe is really the father of all inventions.

Before *fire* was tamed, humans had to live in warm climates and work only by daylight. Their diet consisted of easily digestible fruits and nuts, and they had difficulties with uncooked cereal grains and tough meat. Fire was tamed some 500,000 years ago, with the immediate benefit of providing heat to keep humans warm in winter, light to see and work at night, and a weapon to scare away ferocious animals. Fire also led later to many other wonderful inventions, including cooking for improved flavor and digestibility, making durable pottery, and creating metallurgy, which in turn led to the Bronze and Iron Ages (3300 and 1200 BCE in the Middle East).

Before the invention of *agriculture*, the primary human diet was obtained by gathering plants, hunting and fishing, and scavenging dead animals. This often meant a nomadic life with no fixed residence, as the availability of food was seasonal so people followed warmth and rainfall favorable to plant growth and animal grazing. Since wanderers have no place to store food for long periods of time, the bleak time in winter before the next growing season could mean hardship and starvation. About 10,000 years ago in the Fertile Crescent and Mesopotamia, agriculture appeared for the first time in the human world, which ushered in the Neolithic Age. The Agricultural Revolution introduced a method of food production far more abundant and dependable than hunting and gathering; it led to better nutrition and health, an explosive population increase, and it allowed people to abandon the nomadic life by settling down in fixed houses and villages. Agriculture became the principal source of wealth and power for the next 10,000 years. When farming became extensive on land far away from rivers and springs, artificial irrigation became necessary to feed the distant farmlands. A network of irrigation canals and ditches required organization and leadership, and led to hierarchy and government.

Before *writing* was invented, our knowledge about ancient people was based on their bones and the objects that they made. Writing began as a means of keeping track of inventories of grains and beer, and of contracts and promises. Recorded human history began in Sumer around 3500 BCE when people wrote down their histories and stories, their laws and religions, and their communications to each other. The written record is more accurate and detailed than the oral tradition of passing stories from one person to another. The earliest civilizations are known to us because we have the written records of Gilgamesh, the Code of Hammurabi, and the Bible. This led to the beginning of books and libraries, and the accumulation of human knowledge and technologies, so that each educated person can learn from the storehouse of experiences of the past. Isaac Newton said that he could see very far because he was riding on the shoulders of giants.

Energy and power are required for construction, for manufacturing, and for transportation. Initially, we learned to supplement human power with animal power. Much later, we harnessed wind power with windmills and water power with water wheels to grind corn. The coming of the *steam engine* in the 1780s was based on combustion and heat, which gave us almost unlimited power, and led to textile mills, steamships, and railroad locomotives. Since the Industrial Revolution, manufacturing became the principal source of wealth and power in the last two centuries, and led to mass migrations of people from the countryside to cities where the factories were located. Large-scale transportation and trade between continents became practical, for commodities as well as luxuries, and each region on earth could choose to work only in some specialized areas, and export the surplus to trade for other necessities.

The *Information Revolution* involved the convergence of many information technologies in telecommunications, computers, and the Internet. This led to a globalization of information, which made it possible to access any information from networks of libraries and databases, and to send messages to other people. In manufacturing, it is possible to unbundle a complex task into components that can be made in different parts of the world, and then transport and assemble the parts at one place. This outsourcing also applies to many service activities, such as the reading and diagnosis of X-ray photos by physicians, and the reservation of airplane tickets by clerks in remote locations such as India. It is no longer necessary for an educated person to leave her or his country to take up residence in another country in order to participate in high paying and satisfying careers. These are all profound changes brought about by inventions and technologies, affecting how we make a living, how well we live, and how our society is organized.

Years ago	Invention	Transformation
2,000,000	Stone axe	Human progress
500,000	Fire	Migration to Asia and Europe, cooking, pottery, bronze, iron
10,000	Agriculture	Secure abundant food, villages, population and density increase
3,500	Writing	Record keeping, history, documents, communication, trade, contract
250	Steam engine	Textile mills, steamship, railroad
20	Information revolution	Information access, analysis, storage, communications, globalization

For a given state of technology, a wilderness area may be regarded as barren and useless, but inventions can create new resources out of the wilderness. Early men in East Africa looked at their forests and grasslands and recognized food items in fruits, nuts, small animals, and decaying carcasses; but the rest of their world was not edible to them. Then came the stone axes which enabled them to move into the ecological niche of lions as they could now cut up meat from fresh carcasses, of hyenas as they now had the tools to crush bones to extract bone marrows, and of warthogs as they were now able to dig for roots. Fire enabled humans to cook and soften grains, so that they expanded their diet into the niche of horses and cattle for grain. Fire also enabled the conversion of clay into ceramic material for food and containers, as well as the extraction of ores into metals for tools and weapons.

Fertilizers are essential for the growth of food, and there is a limited supply in nature. It was clear by 1900 that the earth could not provide food for much more than 1.5 billion people, as there was a limit to naturally available fixed nitrogen for the growth of plants. Fritz Haber and Carl Bosch changed all that by converting water, air, and energy into synthetic ammonia, so that we now have enough fertilizers to support more than 6.5 billion people. The burning Arabian Desert is one of the most barren places on earth, but oil drilling changed it into one of the wealthiest communities on earth. Flat low land is ideal for growing rice in flooded paddies, but not the mountainous regions of Asia. New flat lands were created by the construction of terrace farms that march up the steep mountains. Holland did not have enough land to support its growing population, so they turned to the sea, and expanded agricultural territory and living space with landfill in areas of the North Sea.

Inventions in technologies for work have made labor much more productive and less onerous, and produced higher quality goods and services. When a farm tractor can plow land 10 times faster than a horse can, the farmer can cultivate much more land and become much richer, as well as have time left over for recreation and leisure. When the Watt steam engine required much less coal than the Newcomen engine, the coal that was not wasted could then be used for other valuable tasks.

Our ancestors in the Paleolithic Age lived in isolated communities, which may be comparable to the modern-day aborigines in New Guinea. Each tribe had to be totally self-sufficient, and live entirely by the local resources. Small bands of hunter–gatherers tended to be nonhierarchical and egalitarian, except for the sexual division of labor where men do most of the hunting, and women do most of the gathering. When the stone axe became widely used, some bands may have had the luxury of a part-time artisan who made and repaired stone axes for others in return for a share of successful hunting and gathering expeditions. When the superior quality of stones at distant quarries became known, the new occupations of transportation and trade would have become involved in acquiring these stones. When agriculture created sufficient food surplus, there was enough to support specialists in tool making, health care, trade, priesthood, art, war, and leadership. Since the copper and tin ores were seldom found locally, the Bronze Age brought even more specialization in the transportation of ores and the fabrication of tools and weapons. Today in villages and small towns with a low population density, there is enough demand to support specialized health care providers and doctors; in small cities, the health providers can specialize further to internal medicine and surgery; and in a great metropolis, the surgeons can specialize even further into brain surgery and plastic surgery. The diversity of occupations increases with market size; when two communities trade, the effective market size is increased, as is occupational diversity. In the modern world, we can all be specialists who make or do a few things very well, and buy everything else.

The gainful occupations of the modern world are traditionally divided into three groups:

(1) *Primary Activities.* Farming, forestry, fishing, hunting, mineral extraction.

(2) *Secondary Activities.* Manufacturing of goods, construction, public utilities.

(3) *Tertiary Activities.* Trade, transportation, information, finance, insurance, real estate, education, health, art, entertainment, recreation, restaurant, hotel, government.

In the history of more developed nations, there was a steady shift from agriculture to industry and then to services, together with enormous increases in productivity and the standard of living. This pattern changed for developing nations: when the low productivity farm labor poured into cities for industry jobs, their productivity increased as well as the economy of the entire society. The following table lists the wealth of Ethiopia, Indonesia, Brazil, and United States today, together with their labor distribution among agriculture, industry, and services. The values of gross national product per capita ($GNP/capita) are according to the official rates of currency exchange, and the $GNP/capita purchasing power parity (PPP) are based on purchasing power, which is more relevant for buying local products and services. The poor nations mostly work on low efficiency farms; it is evident that improvements in wealth involve the shift of occupation from agriculture to industry and services.

	$GNP/ capita	$GNP/capita PPP	Labor in agriculture (%)		Labor in industry (%)		Labor in services (%)	
			M	F	M	F	M	F
Ethiopia	204	630	84	76	5	8	10	18
Indonesia	1,420	3,310	43	41	20	15	37	44
Brazil	8,515	8,700	25	16	27	13	48	71
USA	44,710	44,070	2	1	30	10	68	90

Since these four nations have different climates and resources, it would be more instructive to consider the influence of occupation shifts within a single country. The relative productivity of these three sectors in China in 2007 can be seen in the following table. Since 43% of the labor in agriculture produced only 11% of the GNP, but 25% of the labor in industry produced 49% of wealth, clearly everyone gained when some of the farm labor migrated to the cities to enter industry or services. This migration will continue to be profitable for many more years to come, till the percentage of labor involved is roughly the same as the % GNP generated for all sectors.

	Agriculture	Industry	Services
% GNP generated	11	49	40
% Labor involved	43	25	32
Relative productivity	0.26	1.94	1.25

The new technologies of information and transportation are helping to narrow the gap between rich and poor nations. When world trade became more liberal, and shipping by ocean liners and air cargo planes became more affordable, the high-income nations began to outsource much of its low-tech manufacturing overseas to nations with lower wages, and to keep only the most proprietary and critical manufacturing at home. The world GNP is 48.7 trillion dollars and growing at 4.6% each year, but the world merchandise export is 12.1 trillion dollars and growing at 8.0% each year. Food, fuel, and ores take up 27% of merchandise export, and manufactured goods take up 73%. The United States is steadily losing low-tech manufacturing jobs, and Wal-Mart is piling up with imported low-tech goods. In the last decade, the economy of China was growing at an average rate of 10.6% per year, and India was growing at 5.9%, principally fueled by this shift of migrating surplus farm labor into more productive manufacturing and service occupations.

While goods can be outsourced, some services still require that the provider is in front of the client, such as for haircuts and baby-sitting. However, with the emergence of information technology, especially rapid and inexpensive telephone and Internet, it is now possible to outsource many service works to other countries, such as the preparation of tax returns, tutoring of school children, and examination of X-ray photos to India. Banks and financial investment firms traditionally divide their jobs into the "front office," where officers and agents talk to the investors and Wall Street analysts, and the "back office," where the staff collects information, does data analysis and makes reports. Fast and cheap telephone and Internet made it unnecessary for the back office support staff to be on site, so they can now live where they want to and still participate in high level professions and good pay. World export of services is now 7.8 trillion dollars, and growing at 7.8% each year. The major items of service export are transport, travel, insurance and finance, and information.

1.3.2 Transforming Lives

Humans were descended from East African tropical fruit eaters. We have already seen that the invention of tools enabled humans to eat the food of other animals and invade their ecological niches. An equally impressive expansion was their ability to live in the habitat of any animal, and to colonize the entire globe. Humans have proven that they can even live in research stations in the Antarctica and space stations, where no animal has ever lived before.

How do we measure the well-being of an individual, so that we can make comparisons among nations, and measure progress with the help of inventions through the years? It is easier to measure material well-being using economic purchasing power to acquire desired consumer goods such as food and clothing, and consumer services such as health care and travel. It is more difficult to quantify many intangible measures such as the pleasures of family, friends, community, freedom, justice, beauty, and truth. Besides purchasing power, individual well-being also depends on the rate of business investments on production machinery and plants; on transportation and utilities; and on government expenditure on schools, police, and national defense.

Some of the most dramatic episodes in human history took place when a large population migrated to new habitats, which were either uninhabited or home to indigenous people, which the newcomers then colonized permanently. The early hominids had developed a set of technologies to cope with life in the highland savannah of East Africa, then some of them decided to leave for other lands. Perhaps the reasons for migration were *Push* to escape dangers from human enemies in war or persecution, or from nature's wrath in disease or disaster; or *Pull* to benefit from the opportunity of better climate, health, resources, and land. The first great wave was the migration of *Homo erectus* from Africa to the Middle East, and then to Asia and Europe around 1.5 million years ago, and the second great wave was by the *Homo sapiens* out of Africa some 75,000 years ago. These were heroic journeys that required solving many problems encountered en route. It is particularly difficult to cross from one climate zone into another, which involved adaptations: dealing with unfamiliar climate and terrain, identifying food among new plants and animals, seeking shelter, finding material to make tools and weapons, and facing new enemies.

Each animal in nature is usually best adapted to one habitat or biome, where their bodies have the nature-provided tools to obtain resources and to avoid hazards. A Bengal tiger would not trade places with a Gobi camel, as neither one has the tools to cope with the swapped habitats. When the season changes, some animals migrate a great distance each year to favored habitats, such as the annual arctic tern flight from the Arctic to the Antarctic and back. The fable of the Hedgehog and the Fox describes two animals with different strategies: the single-minded hedgehog has one big idea to cope with any problem, by rolling up into a ball; but the versatile fox has many small ideas, by studying each situation and arriving at a tailored solution. Humans are the most versatile in adaptations, as the clever fox that can live anywhere.

A climate zone is defined by the parameters of temperature, rainfall, soil, vegetation, and animal life. Let us consider the Köppen climate classification system, the environment, resources, and people that live there now.

Climate	Environment	Resources	People
A. Tropical	Rainforest	Banana, yam, sugar, coconut, grass	Masai—Africa
	Monsoon	Zebra	Polynesian—Hawaii
	Savanna		
B. Dry	Steppe	Cactus, date palm	Tuareg—Sahara
	Desert	Lizard	Zuni—Arizona
			Australia Aboriginal
C. Template Warm	Broadleaf forest	Rice	China
	Long grass	Chicken, pigs	Italy
			Cherokee—SE US
D. Template Cold	Evergreen forest	Wheat, grass	Scandinavians
	Short grass	Sheep	Russian
E. Polar	Boreal forest	Lichen, moss	Eskimo
	Tundra	Reindeer	Laplander
	Ice sheet	Seal, fish	Chukchi—Siberia
F. Highland	Short grass	Potato, corn	Inca—Peru
	Ice and snow	Goat, yak	Tibetan

The *Homo habilis* (handy man) was inventor of the stone axe, and lived in East Africa in A. Tropical Climate. The *Homo erectus* (upright man) migrated from East Africa during the Early Pleistocene 2 million years ago. They journeyed first to the Middle East, and went through B. Dry and C. Template Warm zones. Possibly it was during a period of time when the Sahara was wet and supported vegetation. The journeys to Mediterranean Europe and to Java were in warm and tropical climates. Their journey and final settlement in northern China would seem very difficult without the invention of fire to keep them warm.

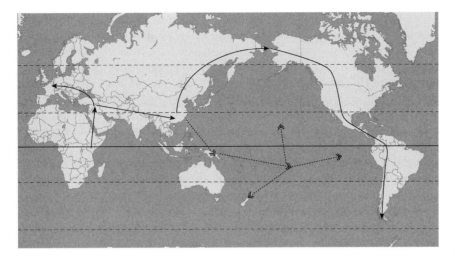

FIGURE 1.1   World human migration before history.

The modern men *Homo sapiens* originated in East Africa. That was the start of the last ice age, which lasted 60,000 years and reached its maximum 20,000 years ago. During the height of the Last Glacial Maximum, ice sheets covered North America and Western Europe, and mountain glaciers covered the Alps, Himalaya, and Andes. The frozen sheets tied up so much water that the ocean was as much as 600 ft lower than today. Land bridges were formed between Siberia and Alaska, and from Southeast Asia to Java and Borneo. The modern men traveled and colonized the Near East about 70,000 years ago, shown in Figure 1.1. These were incredible journeys of courage and resourcefulness, overcoming tremendous obstacles with very primitive tools. This audacious conquest and colonization began from the tropical grassland of Kenya. They had to cross the Red Sea either in the south near Yemen, or at the north near Suez. In either case, there were more than 1000 miles of desert before reaching the temperate grassland of the Fertile Crescent. What were their motivations to undertake such a journey, leaving comfortable grasslands to enter hostile deserts, meeting the challenge of finding food and water, shelter from the scorching sun, and avoiding new and unfamiliar dangers? How would they know that after 1200 miles of hardship, they would find temperate grassland that may be no better than their original home?

The next journey took them into the temperate forests of Europe and Asia, arriving at Southeast Asia 50,000 years ago, and Europe 40,000 years ago. Some of them turned north into the frozen taiga and tundra of Siberia before they reached the polar desert of the Bering Strait. Then they crossed over to Alaska 12,000 years ago, and went south all the way to Tierra del Fuego at the tip of South America in approximately 1500 years! In this process, they repeatedly crossed from one climate zone into another, with the need to adapt and adjust from keeping warm in the frozen lands to keeping cool in the steamy jungles, and back to frozen land again.

The conquest and colonization of Oceania by the Polynesians was completed by the year 1000, and depended totally on the inventions of the sailing canoes and celestial navigation. The Polynesians started from Southeast Asia for Fiji, and reached New Guinea (5°S), which formed their base to reach Samoa around 1600 BCE, then Tahiti (18°S) around 300 BCE, Hawaii (20°N) around 500, New Zealand (40°S) in 850, and Easter Island (27°S) between 500 and 1200. The distance from Tahiti to Hawaii is 4500 km, and they traveled in the sailing outrigger canoes that were open to wind and storms. We do not know how they navigated without charts or compasses in an alien ocean. There are not even any fixed stars in that area to serve as reference points, as Polaris is not visible south of the equator and the Southern Cross is not visible north of the equator.

These pioneers were followed by the subsequent voyages of discovery and conquest of Christopher Columbus (1492), Vasco da Gama (1497), and Magellan–Elcano (1519–1522). These later voyages were made possible by the invention and development of ships in the form of carracks and caravels that are deep ocean-going vessels with great cargo spaces to carry food and water for the journey, as well as spices to pay for the journey.

In the future, if we run out of living space on earth, there is the moon beyond as well as the planets and the cosmos. Perhaps there are life forms beyond our planet in this solar system, or out in the galaxies. There are many science fiction stories about the conquest of space, such as the "Martian Chronicles" of Isaac Asimov. Rockets and spaceships are our technology for such grand conquest and colonization, if we are desperate enough about the future of the earth to take this very risky adventure. The challenges of living on the moon or Mars would be very great indeed, but we have the tradition of 2 million years of inventions, and we are the descendants of resourceful pioneers.

What is the ideal life? The gardens of Eden and Paradise were thought of as walled orchards or gardens in simpler times, when men live off God's bounty and had no need to toil for a living. The climate would be sunny and mild, and the gardens always well watered. However, Thomas Hobbes believed that in the state of nature, human lives used to be "poor, nasty, brutish, and short." The moving description by Charles Darwin about the naked natives shivering in the frozen Tierra del Fuego would suggest that Hobbes was not far off the mark. We would like to know how our lives today compare with our ancestors 2 million years ago in East Africa. Historical comparisons are usually very qualitative, as such economic records were not a high priority among historians. A proxy is to make comparisons of the welfare among various groups of contemporary people, on their technologies and their lives.

What are the most important human needs and desires, and how do we measure the quality of life? Survival is usually the prime objective, so the most basic needs are food, clothing, and shelter, followed by health and security. When survival is no longer the issue, we can tend to more advanced and less urgent needs. How do we put together and sort out a list of economic and noneconomic needs? Abraham Maslow proposed a hierarchy that starts from the lowest level of basic needs for everyone to the highest level of self-actuation for the elite.

- *Basic Needs:* Food, drink, shelter, clothing
- *Safety:* Security, health, freedom, stability
- *Love/Belonging:* Family, friendship, community
- *Esteem:* Recognition, fame, social standing
- *Self-Actuation:* Fulfillment for prophets and saviors

Maslow suggested that after the most basic needs for survival have been satisfied, one would move up to the next level of safety and love, then to the next levels of esteem and self-actuation. In this view of sequential climbing of a pyramid, we consider love and friendship only after our stomachs are full and we are safe from wild animals. But we also know about selfless acts that put the welfare of loved ones ahead of oneself. The notion of a family budget suggests that there is actually a parallel as well as sequential distribution of attention. People with few resources would assign larger amounts on the lower levels of needs, but people with more resources could assign more to the higher levels. Inventions and technologies play a large role in satisfying the basic needs of the first level, such as food production that requires plows, fertilizers, irrigation, harvesting, and storage. As one travels up the levels, the role of technology becomes less obvious and less direct, but still very important. Fame is difficult to achieve without the printed word in books and journals, and the media of radio, television, and the Internet.

There are many methods to measure the wealth of a nation, including the GDP, the GNP, and the GNI. They are used to measure different things, but their differences are small. The official GNI of a nation in US dollars is computed by conversion of the national currency by the official currency exchange rate. An important variant is the PPP, based on the local cost of food and labor instead of the official exchange rate. The GNI is given in two different forms by *income* or how it is earned in the farms, factories, or office towers and by *expenditure* or where it is spent on food, housing, health care, and investments. The GNI per capita of the nations currently vary from a high of $68,440 in Norway to a low of $170 and $130 in Ethiopia and Congo; the PPP would compress this disparity from a high of $50,070 in Norway to a low of $630 in Ethiopia and $270 in Congo.

For an approximate idea of the relative importance of the components of the national expenditure, the percentage of US expenditure during 2000 can serve as a guide, shown in Figure 1.2. About 2/3 of US expenditures are controlled by households, 1/6 are controlled by business investments in production and administration tools and facilities, and 1/6 are controlled by government expenditures, particularly in health, security, transportation, and education.

The Human Development Index (HDI) of the United Nations is a more comprehensive measure, which rates the well-being of a group of people by three indexes: life expectancy at birth, adult literacy plus years of schooling, and standard of living. Out of 177 nations ranked in 1993, Norway was the first and Niger ranked the last, and the breakdown of a few nations in 2003 is shown in the following table.

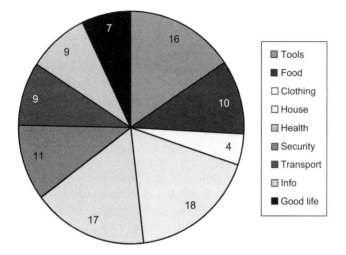

US Expenditure, 2000

- Tools
- Food
- Clothing
- House
- Health
- Security
- Transport
- Info
- Good life

FIGURE 1.2 US National Expenditures in 2000.

Nation	HDI	Life expectancy at birth	Adult literacy (%)	School enrollment (%)	GDP (US$/cap PPP)
Norway	0.963	79.4	100.0	100	37,670
US	0.944	77.4	100.0	93	37,562
Russia	0.795	65.3	99.4	90	9,230
China	0.755	71.6	90.9	69	5,003
India	0.602	63.3	61.0	90	2,892
Niger	0.281	44.4	14.4	21	835

This table shows that the rich live longer and with better education. Between the lowest Niger and the highest Norway, the standard of living is increased 45 times, the educational opportunities are increased by a factor of 5, and the life expectancy is almost doubled. Perhaps the residents of Niger have something that those of Norway do not have, but it would be difficult to find many envious Norwegians. Many factors contribute to this disparity in the quality of life; however, inventions and the use of technologies are essential.

1.3.3 Transforming Society

The enjoyment in life of average people is better correlated with the productivity of their society than with their individual abilities and efforts. An average teacher of arithmetic in Ethiopia or Haiti has a much lower life expectancy or income than his or her counterpart in the United States or Norway, even if they all have similar training and dedication. Great civilizations have dynamic cycles of growth and expansion associated with new inventions, followed by maturity and decay. As societies become more complex, hierarchy, and inequality usually follow.

Inventions have always been an aspect of the creativity of a dynamic civilization and a prosperous society. The people of central Italy slumbered for a long time, and then awakened during the Roman Republic and Empire by making creative and triumphant solutions in organizations, sciences, inventions, arts, which led to prosperity, power, influence, admiration, and imitation. After a span of time in the sun, Italy returned to slumber and stagnation till a second awakening during the Renaissance. What is behind this transition from the static YIN to the dynamic YANG, followed by a return to YIN again? What nations are the leaders on earth today, how long have they held leadership, and how long will it be before they surrender to newcomers?

Spengler and Toynbee discussed the evolutions of great civilizations in history, such as Mesopotamia, Egypt, Greco-Roman, China, India, Arabia, and Western. They all went through periods of awakening in springtime, growth in summer, maturity in autumn, and decline in winter—measured by their creativity and influence in philosophy, religion, mathematics, art, music, and politics. Did great inventions in technology play significant roles in their evolution? According to Spengler, Western civilization began around 900, when there was an awakening in the pace of technology. The first great invention of printing by movable type was in 1454 by Johannes Gutenberg, at the beginning of summer, and the other great invention of the steam engine in 1764 by James Watt in the middle of autumn. Other inventions sparked the beginning of Western civilization, and these two great inventions were instrumental in its growth and glory.

Inventions have been an integral part of civilizations. A prosperous society has the optimism and ability to support the explorations of a few creative members, which can lead to inventions that benefit the whole society. New inventions have been the engines of growth in wealth and power, and have provided resources for contributions in the arts and sciences. The scientific contributions of Pythagoras, Archimedes, Newton, and Darwin were made by prosperous people whose lives were not totally consumed by the necessity of earning a living. All of their inventions took place during the golden age of their respective civilizations, not during the early periods of struggle with subsistence living nor during the later periods of indolence with decadence.

There are no reliable estimates of the size of human population in prehistory. It was probably stable during the long period of hunter–gatherers according to the carrying capacity of the land, but the Agricultural Revolution brought about a rapid increase. Better and more dependable nutrition led to a great improvement of health and longevity, increase in fertility and decrease in death by disease and starvation. The sedentary habit of life was also more favorable for family formation, as there was no need to carry dependent children in a nomadic lifestyle, so that the spacing between babies could be shorter and birth rates could soar.

A UN report in 2004 made estimates on the world population since the Agricultural Revolution, which showed three periods of rapid growth that occurred around 10,000 BCE, 800 BCE, and since 1800. The first and third of the growth

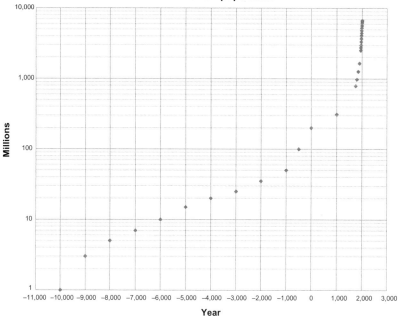

FIGURE 1.3 World population from 10,000 years ago to now.

spurts can be associated with the Agricultural and the Industrial Revolutions, shown in Figure 1.3.

The population density of the number of people per area of land is limited by the "carrying capacity" of the land. In the world today, population density varies from 2 persons/km^2 in Mongolia and Namibia, to 6500 in Hong Kong and Singapore. In the United States, population density varies from less than 0.4 persons/km^2 in Alaska, to more than 3065 in the District of Columbia. Since half the population of Alaska lives in cities such as Anchorage, the population density in the countryside is actually only 0.08 person/km^2.

In a society of hunter–gatherers, the amount of food that can be found in a given area is quite limited. Modern agricultural land in North Dakota and South Dakota currently produces enough food to feed 2800 persons/km^2. Ancient cities existed for political administration, trade, education, and culture. The Industrial Revolution accelerated the process of urbanization, so that more people relocated from the countryside to cities to do manufacturing. The highest population densities today are found in urban high-rise office towers and apartment buildings. It is impossible to live in such concentration without an efficient system of roads, mass transit, electricity, water, sewage, and elevators in high-rise towers.

Urbanization is a worldwide phenomenon. The United Nations projected that world population was 13% urban in 1900, but grew to 49% in 2005, and will increase further to 60% in 2030. The principal reason for the migration of rural populations to the cities is economic opportunity, better prospects for education, diversity of occupations, as well as diversity of consumer goods and services. When one lives on a small and isolated family farm, there are few opportunities to improve the standard of living beyond basic sustenance. Cities offer many opportunities to increase wealth, social mobility, job opportunities, education and access to health care, and quality services such as restaurants and a variety of entertainment venues. The dark side of urban life includes noise, overcrowding, crime, and social stratification. Since the prevailing wind in North America tends to blow from west to east, and the pollutants tend to move downwind, so many urban areas are divided between the good "west" neighborhoods and the bad "east" sections of town.

The advantages of free trade were explained by David Ricardo (1772–1823), due to comparative advantages in factors of production. England had a strong textile industry and India had a good supply of raw material; so when they specialized on producing what they were best suited for and traded for the rest, both nations benefited from the exchange. We are all familiar with life in small towns where the stores carry only basic necessities, and one needs to visit a bigger city in order to get specialized products, and to a metropolis for fancy luxury goods. A corollary to this theme is that trade would also increase career options and encourage job specialization. A village can barely support a single part-time health worker, and it takes a small city to support a doctor who is a general practitioner; a bigger city can support a surgeon and a clinic, and it takes a metropolis to support brain surgeons and surgery centers with the latest medical diagnostic equipment.

New inventions and technology can have long-term consequences in shifting power among groups within a society, and also among nations. In an economic activity such as sowing wheat to make bread, the intended consequence is to satisfy the hunger of the farmer and his customers; but there are also long-term unintended consequences for other parties that did not participate in this transaction, such as the society and the environment.

Some of the important tasks in our lives can be carried out by small independent units, such as gathering fruits and nuts, and growing corn and chicken. The production group may be organized around a family or a closely knit clan, where there may be little concern for rank and distribution of duties. There are other important tasks that can only be carried out by a large assembly of people, such as warfare and building an irrigation network. Such work is divided among numerous subgroups of people with specialized skills, and led by a hierarchy of leaders. Farmers on land without adequate rainfall in the growing season need irrigation from a river or oasis, which may be far away. Large-scale irrigation requires the cooperation and coordination of many people, who need organization and rulers to compel all participants to make required contributions. A group of prosperous farmers living near subsistence nomads creates the danger of invasions and pillage, such as what happened with the Mongols and the Vikings. This brought about the rise of governments to establish hydraulics and armies, and eventually the rise of kings.

An invention can have a profound impact on the relative welfare of different groups within a society. Products of a technology are sold by the producers to a group of users to enhance their lives, which can place nonusers at a relative disadvantage since they do not experience a similar enhancement. Some technologies are egalitarian, so that early users obtain a temporary advantage over later users, but equality returns after the adoption rates become substantially equal. Other technologies are exclusive, where the users have to possess special qualities that are not widely shared, so the advantage of those users may become long lasting if not permanent.

- *Exclusive technology* can be used only by the elites who have access to required resources, and have the aptitude and means to benefit from special training. Agriculture requires warmth, rainfall, and soil, which are not available for people living on high mountains and dry deserts. The invention of writing led to the formation of a class of scribes and elites, who had the necessary time and means to learn writing, and had access to books that were hand copied and kept in libraries away from the general population. This led to a division of the population into the literate lords and priests with knowledge and power, who ruled over the illiterate common people. When a technology is new, it tends to be used only by the elite to increase inequality; as the years go by and the cost keeps falling, it becomes more affordable for everyone and thereby more egalitarian. An example is the Internet use in the United States in 2009 for groups with different educational levels as shown in Figure 1.4. It shows that for those with less than a high school education, only one fourth would use the Internet; but for those with a bachelors degree or higher, 90% were users.

- *Egalitarian technology* can be used by every person as it does not require special resources, or special ability and training. The stone axe required fieldstone and fire required combustible plants, neither of which are difficult to obtain. The invention of printing made books cheaper and more easily

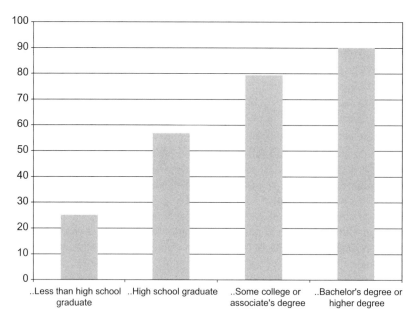

FIGURE 1.4 US Internet use by education level in 2009.

available, leading to increased literacy and decreased dominance of priests and scribes. Today, the common person can own and read books on every subject, from the most sacred scriptures and technical manuals to the tabloids. An ordinary parishioner can read the Bible without interpretation from a priest, and a student can read a classic and challenge the interpretations of a professor. Watching television also uses a technology that does not require special knowledge and education, and can be enjoyed by all.

The relative importance of the two sexes can be changed by an invention. In the days of hunter–gatherers, hunting was likely dominated by a few fit and skillful males. Meat can be more satisfying, as well as being more dangerous and difficult to acquire, which made it a more prestigious food than fruits. An invention such as the bow and arrow increased the ability and power of the hunters and helped them to be even more productive, and increased the social inequality between men and women. Conversely, gathering is more egalitarian and gives women equal status.

Lynn White proposed the theory that the stirrup for horse riding was responsible for the rise of chivalry and aristocracy in medieval history. The Greek warriors in the Iliad rode on their chariots to the sites of battles, and then jumped off to fight with spears and swords on foot. The Roman cavalry have neither stirrups on their saddles to steady their feet nor the pommel and cantle at the front and back of a saddle to prevent being thrown forward or backward, and could not deliver a lance thrust against foot soldiers without being thrown off their horses. Thus, the infantry could stand up to the cavalry.

The stirrup changed all this and kept the medieval cavalry steady on their saddles, which was augmented by their plate armor. So the mounted knights became invincible to foot soldiers, which began to diminish in importance during the time of Charlemagne. Cavalry became the main force of the army, which was expensive to maintain, and necessitated the granting of land and serfs in feudalism to ensure that the cavaliers had the means to support horses and the required armor. The castle with walls and moats also increased the power of the feudal lords over peasants, who had no means to challenge them. These inventions contributed to the rise of chivalry and feudalism, by dividing the population into aristocratic knights with stirrups and common peasants on foot.

Eventually, power shifted back to the infantry with the invention of the English longbow that decimated the French cavalry during the battles of Crecy and Poitiers during the Hundred Years' War. Then came gunpowder that made the foot soldiers with muskets the equal of the cavaliers. The cannon could rapidly throw metal and stone balls against the strongest castles to bring them down. The city walls of Constantinople withstood numerous assaults for a thousand years, but fell to the Ottoman Turkish cannons in 1493, which ended the Byzantine Empire of Eastern Rome. Gunpowder helped to end feudalism, and led to the rise of the modern age of common men.

In peacetime, a successful invention can create many winners and losers in economic and cultural wars. The first satisfied lady customers of synthetic mauve could make a fashion statement to other women, and the early users of these expensive new colorful garments included Empress Eugenie of France and Queen Victoria of England. The initial beneficiaries of this invention were the inventor, William Perkin, and his financial backers—his father and brother. The other immediate beneficiaries were the employees in his factory, the suppliers of raw material and machinery, and the textile manufacturers who had a product that their competitors did not have. The next level of winners includes the merchants who sell groceries and real estate to the primary winners, as well as the teachers of their children, the keepers of their taverns, and the tax collectors.

The immediate victims of this invention were the manufacturers of old-fashioned vegetable and mineral dyes, whose goods looked drab and unfashionable by comparison, and the envious women who did not have the means to buy mauve. As time went by, mauve became much less expensive, and other dyes were developed to displace mauve, so that almost everyone could afford colorful and inexpensive synthetic dyes. The displaced workers in old dye factories could find employment in the growing new dye industry, or move on to other occupations.

Globalization today often means moving manufacturing from developed nations to emerging nations. In the book *Winners and Losers of Globalization* by de la Dehesa (2006), and the book *The World is Flat* by Friedman (2006), it is suggested that if capital and technology can freely move, the winners and losers of globalization are as follows.

Countries	Winners	Losers
More developed countries (MDC): US, Western Europe, Japan, Australia	Consumers, capitalists, professionals, high-skilled labor	Low-skilled labor
Less developed countries (LDC): China, India	Consumers, capitalists as partners with internationals, professionals, high-skilled labor, low-skilled labor	Capitalists as independents

Consumers everywhere will benefit from a wider choice of goods and services at lower prices. The capitalists from the MDC can build factories in the LDC, sometimes in partnership with local capitalists, and employ the low-skilled labor there, which displaces the less resourceful local capitalists. The younger high-skilled labor in MDC can learn rapidly, and adjust to new job requirements with better benefits. All the low-tech manufacturing in MDC would be closed, resulting in unemployment of the MDC low skill labor. In the LDC, the most conspicuous losers are the native capitalists who remain independent. Would this trend march on without arousing contrary forces?

Jeremy Bentham concentrated on more happiness for the most people, but did not consider the important roles of fairness and justice. Is it appropriate to increase the happiness of very many and sacrifice the happiness of a few, or should they be compensated? The manufacturing process may create problems of safety and health for factory workers, and the waste streams may cause community pollution and environmental problems. Ultimately, many parameters are needed to measure the various aspects of human well-being.

The inequality of wealth or income within a society can be measured by the Gini coefficient, which measures the degree of departure of a society from perfect uniform distribution. First we rank everyone by income from the lowest to the highest. In a nation with perfect equality, the bottom 10% and the top 10% of the households both enjoy 10% of the national income; but in any real nation, the top 10% owns much more than the bottom 10%. The Lorenz curve is a graphical representation of income distribution, where the vertical axis shows the percentage (y) of total national income enjoyed by the percentage (x) of households with income equal to or up to this value, shown on the horizontal axis. For a nation with perfect equality, the Lorenz curve is the 45° line, and the Gini coefficient is 0. For a nation with perfect inequality, where one person owns all the wealth, the Gini coefficient is 100. For an actual nation, the Lorenz curve hangs below the 45° line, and the area between these two lines is proportional to the Gini coefficient. In the United States of 2000, the degree of inequality can be seen in the following table.

Families by income (%)	Wealth owned (%)
20	4.1
40	13.7
60	29.2
80	52.4
95	79.6
100	100.0

The poorest 20% of all the US families own only 4.1% of the wealth in the nation, and the next 20% own $13.7 - 4.1 = 9.6\%$ of the wealth, and so on. Figure 1.5a shows the Lorenz curve of four nations. For the United States, the area between the Lorenz curve and the 45° line is 39% of the area of the triangle, so that the Gini coefficient is 39. It is also instructive to divide the families into five groups by rising income, called the quintiles. The bottom quintile is the families with incomes in the lowest 20% and the top quintile is the families with incomes in the highest 20%. Figure 1.5b shows that in the United States, the top quintile gets 46% of all the income and the bottom quintile gets 6%. Denmark is currently the most egalitarian society with a coefficient of 24.7, and Namibia is the least equal society with a value of 70.7. Even for egalitarian Denmark, the top 10% has an income 8.2 times that of the bottom 10%; in the United States the ratio is higher at 15.7 times, and in Namibia the ratio is much higher at 129 times.

Among nations on earth, there is a strong correlation between high income and low Gini coefficient, indicating that all things considered, the more developed nations tend to be more equal. Denmark has a mostly homogeneous population of race and educational attainment; Namibia has a small white minority who owns most of the material wealth, and a big black majority that lives by subsistence farming in very dry soil. However, increasing wealth does not always bring more

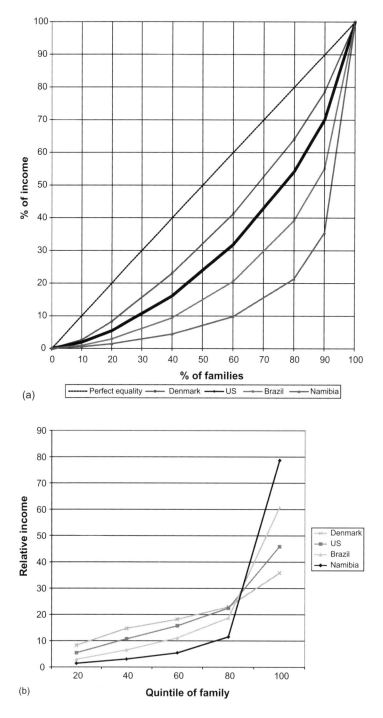

(a)

(b)

FIGURE 1.5 Gini coefficient of income inequality in four nations.

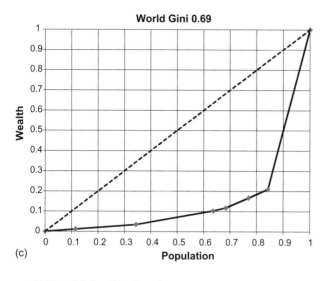

FIGURE 1.5 (*Continued*)

equality: the income of the United States has been steadily rising in the past decade, yet the value of the US Gini coefficient has also been increasing, so the United States is becoming more unequal as she becomes richer.

Besides these inequalities among families within a single nation, it is also possible to compile an international inequality comparison among nations or regions. We use the World Bank classification to divide the world into seven regions:

High Income: Australia, Germany, Japan, UK, US, Hong Kong, Saudi Arabia, and so on

Europe and Central Asia: Hungary, Russia, Turkey

Latin America: Argentina, Brazil, Mexico, Venezuela

Middle East and North Africa: Egypt, Iran, Lebanon, Libya

East Asia and Pacific: China, Indonesia, Philippines, Thailand

South Africa: India, Bangladesh, Pakistan

Sub-Sahara Africa: Congo, Ethiopia, Kenya, Namibia, South Africa

In 2006, the world population was 6,438 million, and the GNI was 45,135 billion dollars, for an average of $7,011 per person. When we prepare a Lorenz curve using these world regions as units, the result is shown in Figure 1.5c. This curve can be used to calculate the world Gini coefficient by these world regions to have a value of 69. Of course, if we break down the world regions into separate nations, and even further into individual families, the Gini coefficient would increase to much higher values.

Societies with the most extreme inequality are those who practice colonialism, imperialism, and slavery. The power of one people over another is greatly increased when one group has more powerful technologies than the other. The advent of nautical travel and navigation led to easier access to many foreign lands from Western Europe, and gunpowder plus cannons added unprecedented military power to conquer and enslave the less developed people of the world. Western Europeans created colonial empires in North and South America, Africa, India, and Australia. They deprived the Native Americans of their lands, and they captured and shipped Africans to be their slaves. In 1462, the Russians expanded their land all the way from Muscovy to Siberia and Alaska, and later into Central Asia, as they had the technology to easily overcome the tundra and nomadic people.

The conquering nation acquired a monopoly on political and military power, controlled the best jobs with high income and prestige, and retained the best living conditions. The subject people provided the raw material for manufacturing, the low-skilled labor for menial jobs, and lived in subsistence conditions and even slavery. This dominance can be relatively benign, but can also be very cruel, such as the Spartans who harassed and killed their subject people without cause.

There are many factors that lead to winning wars, particularly leadership and organization, but technology has often played a critical role. In ancient Mesopotamia and Egypt, the horse-drawn chariot was one of the most depicted images in paintings. Such chariots were superb technology that provided the top commanders with speed, the ability to observe the battle conditions, and a high platform from which to shoot arrows and throw javelins at enemies. The rapid mobility of horses often gave the less numerous nomadic archers a decided advantage in warfare over the plentiful farmers as foot soldiers. This is exemplified in the Mongol conquest from China to Eastern Europe, as they overwhelmed many nations that had much larger populations and longer histories of intellectual achievements. The Manchu conquest of China placed a small group of nomads in charge of a huge and civilized population. The Viking conquest of the more civilized states of Normandy and Sicily can be attributed to the same rapid mobility, bestowed by their dragon ships, which could be navigated through narrow rivers and sailed in broad seas. The conquistador Cortez had only 257 soldiers when he entered Mexico, but he defeated the Aztec Emperor Atahualpa's army of several hundred thousand. Cortez's superior technology consisted of guns, steel armor, and horses. The British Empire attained colonies with superior technology in rifles and cannons, transported over the ocean by ships that could enter harbors and point cannons at the palaces of rulers. The Russian conquest of Siberia and Central Asia depended principally on the Trans-Siberian railroad of 1891–1902.

It is very fortunate that there is a general world consensus that such invasion and enslavement is immoral and will not be tolerated in the world today. This enlightened view is strongest among the most advanced nations that practiced imperialism and colonialism in the past. Today the superpowers have far more terrible weapons of mass destruction, of biological warfare with deadly viruses, and of nuclear warfare with atomic and thermonuclear bombs. But as a consequence of this modern humane view, the superpowers have lost the ability to dictate their terms to the smaller and less powerful nations, as well as to pirates and terrorist groups that are not represented by governments.

1.3.4 Transforming Environment

When the human population was low and equipped with primitive tools, there was very little disturbance to nature and the environment, as people were content to live off the land and had neither the motive nor the power to do harm. As they grew in number and in power, they began to transform the environment in profound and sometimes disturbing ways. Fire gave people the power to clear large tracts of weeds and forests. Agriculture propagated the plants that are economically useful, displacing those that were less useful to humans. It became feasible to create new land by draining swamps, filling in the shallow shores of a lake or sea, and building terraces on steep hillsides, to the detriment of native plants and animals. The need for water also led to irrigation with ditches and canals, and building dams on rivers. The parallel development in animal husbandry also led to the cultivation and multiplication of economically useful species such as dogs, pigs, chickens, and cattle, and their needs displaced the natural habit of wild or unnecessary animal species—especially predators that posed a danger to humans and their possessions.

The coming of industrialization bestowed even more changes to the environment, as there were now demands for minerals as sources of metals, ceramics, and fuel. Some of the most obvious changes on earth are due to the cities which house the vastly increased human population at much higher density. Urbanization and trade also led to transportation infrastructure of roads and other pavements for urban commerce and residential needs. Often it takes many years and decades after a technology has been introduced and widely used, before serious environmental consequences are recognized, such as the ozone hole from refrigerants and global warming from burning fossil fuels. This has made people more suspicious of new inventions that may lead to unpredictable and sometimes irreversible consequences.

The human activities that have the greatest environmental impact are the heavy physical projects that manufacture goods and move them around such as agriculture, mining, construction, manufacturing, and transportation. Most of the knowledge-based service industries involve office work, and have less environmental impact per dollar produced. Consider for example the emission of CO_2 from the use of fossil fuels in generating heat and power. Low-income countries use very little energy and have low emissions per capita; many middle-income countries play a catch up game in rapid industrialization, and are very heavy in emissions per capita; and the high-income countries have moved on to service industries and outsourced a lot of manufacturing, and have returned to low emissions per capita. This rainbow curve of emission per $GDP versus income per capita is sometimes call the Kuznets curve, shown in Figure 1.6. It can be seen that low-income countries are relatively clean; but when they industrialize and become middle income, they become heavy emitters; and the high-income countries become cleaner again. It

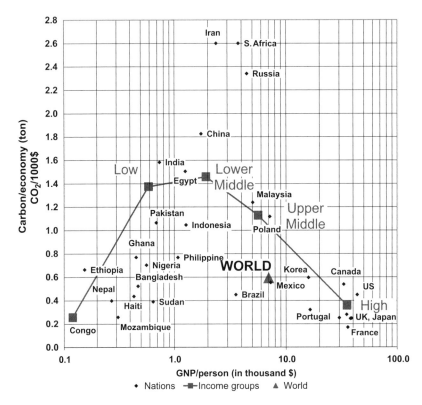

FIGURE 1.6   National carbon/GNP versus GNP/capita.

should also be remembered that the middle-income nations are the workshops for the world, and emit not only for their own consumption but also for world consumption.

Household consumption has also had a significant impact. In the opinion of the European Environmental Agency, the aspects of modern household consumption that cause the greatest problems are as follows:

- *Food and Drink.* Consumers upgrade from basic grains and vegetables to choice meats and milk.
- *Housing.* From basic shelters to spacious and power-hungry residences with heating, air conditioning, and electric appliances.
- *Personal Travel.* Commuting to work, school, leisure, shopping, visits.
- *Tourism.* Long distance travel.

Some of the greatest impacts fall on

- *Air Quality.* Urban and regional air pollution, dust, carbon monoxide, and hydrocarbons.

- *Water Quality.* Domestic water discharges, industrial and mining discharges, pollution in ponds, rivers and lakes, estuaries, drinking and groundwater contaminations.
- *Water Scarcity.* Decline in mid-latitude and in semiarid low altitude water supply, increased drought, glacial melt in the Himalayas and Andes.
- *Agricultural Productivity.* Decrease due to drought and variations in rainfall.
- *Health Risks.* Malaria in Sub-Saharan Africa, malnutrition, diarrhea, infectious diseases.
- *Ecosystems.* Harm to coral systems, biodiversity, deforestation, desertification.
- *Biodiversity.* Changing climate and disappearance of species that are unable to adapt; the practice of monoculture in agriculture incurs the risk from disease attack, which was illustrated by the Irish potato famine of 1846.
- *Nature Disaster.* Floods and droughts.
- *Climate Change.* Global warming, ozone hole, shift in rainfall pattern, ocean level rising.

An example of a society that has collapsed due to environmental degradation concerns the remote and isolated Easter Island in the Pacific Ocean, which is 3700 km from Chile and 2090 km from the Pitcairn Island. A group of Polynesians arrived around 900, and found a forest of tall trees. They divided into tribes, and the population rose to some 30,000 people. They cut down the forest to create agricultural land, and built their famous stone statues in competition with other tribes. When Dutch ships arrived in 1722, they found no trees and a much smaller and impoverished population. Anthropologists surmised that they had no wood left to sustain their civilization or to build canoes and escape from the island.

For any given invention, we cannot predict all the consequences, and we do not have a consensus on how to assess the impact, and how to balance the negative and the positive consequences. Some changes are irreversible, such as the loss of the carrier pigeons and the dodo birds. The prevailing attitude is that nature before the industrial revolution was perfect, and that any change in the environment by human intervention in any direction is considered harmful by definition. Global warming would have beneficiaries who remain silent, but would harm others who protest and try to influence the political process. A decrease in temperature would have the opposite effect, but the silent and the protesting parties would be reversed. The unattainable goal, or nirvana, would be to launch a new invention to the satisfaction of all the producers, users, and nonusers, but to leave the environment untouched, just as it was before.

REFERENCES

Adams, R. M. "Paths of Fire", Princeton University Press, Princeton, New Jersey, 1996.

Alexander, J. "Economic Geography", Prentice-Hall, Englewood Cliff, New Jersey, 1963.

Asimov, I. "Asimov's Chronology of Science & Discovery", HarperCollins Publishers, New York, 1989.

Ausubel, J. H. and H. E. Sladovich, editors, "Technology and Environment", National Academy Press, Washington DC, 1989.

Basalla, G. "The Evolution of Technology", Cambridge University Press, Cambridge, 1988.

"Beowulf", translated by William Alfred in "Medieval Epics", Modern Library, New York, 1963.

Billington, D. P. "The Innovators: The Engineering Pioneers who Made America Modern", John Wiley, New York, 1996.

Boesch, H. "A Geography of World Economy", John Wiley, New York, 1974.

Bogucki, P. "The Origins of Human Society", Blackwell Publishing, Malden, Massachusetts, 1999.

Boorstin, D. J. "The Republic of Technology", Harper & Row, New York, 1978.

Bugliarello, G. and D. B. Doner, editors, "The History and Philosophy of Technology", University of Illinois Press, Urbana, 1979.

Cardwell, D. "Fontana History of Technology", Fontana Press, London, 1994.

Carlisle, R. "Inventions and Discoveries: Scientific American", John Wiley, New York, 2004.

Constable, G. and B. Somerville, editors, "A Century of Innovation", Joseph Henry Press, Washington DC, 2003.

Cossons, N. editor, "Making of the Modern World", John Murray, London, 1992.

Darwin, C. "The Voyage of the Beagle", Doubleday Anchor, New York, 1962.

Daumas, M. editor, "A History of Technology and Invention", Presses Universitaires de France 1962, translated by E. Hennessy, Crown Publisher, New York, 1969.
  Vol. I: The Origins of Technological Civilization.
  Vol. II: The First Stages of Mechanization.
  Vol. III: The Expansion of Mechanization, 1725–1960.

Sprague de Camp, L. "The Ancient Engineers: Technology and Invention from the Earliest Times to the Renaissance", Barnes & Noble Books, New York, 1960.

de la Dehesa, G. "Winners and Losers in Globalization", Blackwell Publishing, Malden, Massachusetts, 2006.

Derry, T. K. and T. I. Williams. "A Short History of Technology: From the Earliest Times to AD 1900", Dover Publications, New York, 1960.

Diamond, J. "Guns, Germs, and Steel: The Fate of Human Societies", W. W. Norton, New York, 1997.

Diamond, J. "Collapse: How Societies Choose to Fail or Succeed", Viking, New York, 2005.

Eco, U. and G. B. Zorzoli. "Picture History of Inventions: From Plough to Polaris", Macmillan, New York, 1963.

Ehrlich, P. R., A. H. Ehrlich, and J. P. Holdren. "Ecoscience: Population, Resources, Environment", W. E. Freeman and Company, San Francisco, 1977.

Evans, L. T. "Feeding the Ten Billion: Plants and Population Growth", Cambridge University Press, Cambridge, UK, 1998.

Fagles, R. Translation of "Iliad" and "Odyssey", Viking, New York, 1990, 1996.

Fellmann, J., A. Getis, and J. Getis. "Human Geography", William C. Brown, Dubuque, Iowa, 1990.

Finniston, M. editor, "Oxford Illustrated Encyclopedia of Inventions and Technology", Oxford University Press, Oxford, 1992.

Friedman, T. "The World is Flat", Farrar, Strauss and Giroux, New York, 2006.

Goudsblon, J. "Mappae Mundi: Humans and Their Habitats in a Long-Term Socio-Ecological Perspective, Myths, Maps and Models", Amsterdam University Press, Holland, 2002.

Grant, P. R. and B. R. Grant. "Evolution of Character Displacement in Darwin's Finches". *Science* 313, 224–226, 2006.

Grun, B. "The Timetables of History", Simon and Schuster, New York, 1963.

Hagen, E. E. "The Economics of Development", Richard Irwin, Homewood, Illinois, 1968.

Heilbroner, R. L. "The Worldly Philosophers: The Lives, Times, and Ideas of the Great Economic Thinkers", Simon & Schuster, New York, 1992.

Highsmith, R. M. Jr. and R. M. Northam. "World Economic Activities: A Geographic Analysis", Harcourt, Brace & World, New York, 1968.

Hodges, H. "Technology in the Ancient World", Barnes and Noble Books, New York, 1970.

Hounshell, D. A. and J. K. Smith. "Science and Corporate Strategy", Cambridge University Press, New York, 1988.

Hughes, T. F. "Human-Built World: How to Think About Technology and Culture", University of Chicago Press, Chicago, 2004.

"Human Development Report 2005", United Nations Development Program, New York, 2005.

James, I. "Remarkable Mathematicians: From Euler to von Neumann", Cambridge University Press, Cambridge, 2006.

James, P. and T. Nick. "Ancient Inventions", Ballantine Books, New York, 1995.

Kahn, H. "World Economic Development", Morrow Quill Paperbacks, New York, 1979.

Knauer, K. "Great Inventions: Geniuses and Gizmos, Innovation in Our Time", Time Inc. New York, 2003.

Kranzberg, M. and W. H. Davenport, editors, "Technology and Culture: An Anthology", Schocken Books, New York, 1972.

Leakey, R. E. "The Origin of Mankind", Basic Books, New York, 1994.

Lewin, R. "In the Age of Mankind: A Smithsonian Book of Human Evolution", Smithsonian Books, Washington DC, 1988.

Li, N. "Shi Jing 诗经" (in Chinese) Da Xian Publishing, Taipei, Taiwan, 1994.

Lonnrot, E. "Kalevala", translated by Keith Bosley, Oxford University Press, Oxford, 1989.

McClellan J. E. and H. Dorn. "Science and Technology in World History", Johns Hopkins University Press, Baltimore, 1999.

McNeil, I. editor, "An Encyclopedia of the History of Technology", Routledge, London, 1990.

McQuarrie, D. A. and P. A. Rock. "General Chemistry", W. H. Freeman, New York, 1987.

Mellinger, A. D., J. D. Sachs, and J. L. Gallup. "Climate, coastal proximity, and development", in "The Oxford Handbook of Economic Geography", G. L. Clark, M. P. Feldman, and M. S. Gertler, editors, Oxford University Press, Oxford, 2000.

Messadie, G. "Great Inventions through History", Chambers, Edinburgh, 1988.

Misa, T. "From Leonardo to the Internet", Johns Hopkins University Press, Baltimore, 2004.

Mitchell, S. "Gilgamesh: A New English Translation", Free Press, New York, 2004.

Needham, J. "Science and Civilisation in China", Cambridge University Press, 1954.

Vol. I: Introductory Orientations.

Vol. II: History of Scientific Thoughts.

Vol. III: Mathematics and the Sciences of Heaven and Earth.

Vol. IV: Physics and Physical Technology, 3 parts.

Vol. V: Chemistry and Chemical Technology, 13 parts.

Vol. VI: Biology and Biological Technology, 6 parts.

Vol. VII: Language and Logic, 2 parts.

Nicolaou, K. C. and T. Montagnon. "Molecules that Changed the World", Wiley-VCH, Weinheim, Germany, 2008.

OECD Factbook 2009. "Economic, Environmental and Social Statistics", OECD Publishing, 2009. www.oecd.org/publishing

Ostlick, V. J. and D. J. Bord. "Inquiry into Physics", West Publishing, Minneapolis, 1995.

Pacey, A. "Technology in World Civilization: A Thousand-Year History", Basil Blackwell, Oxford, UK, 1990.

Peel, M. C., B. L. Finlayson, and T. A. McMahon. "Updated world map of the Koppen-Geiger climate classification". *Hydrology and Earth System Sciences* 11, 1633–1644, 2007.

Perpillou, A. V. "Human Geography", translated by E. D. Laborde. Longmans, London, 1966.

Polenske, K. R. editor, "The Economic Geography of Innovation", Cambridge University Press, 2007.

Schon, D. A. "Technology and Change: The Impact of Invention and Innovation on American Social and Economic Development", Dell Publishing, New York, 1967.

Schumpeter, J. A. "The Theory of Economic Development", Harvard University Press, Cambridge, Massachusetts, 1934.

Singer, C. et al., editors, "A History of Technology", Clarendon Press, 1954.

Vol. 1: From Earliest Times to Fall of Ancient Empires.

Vol. 2: Mediterranean Civilizations, Middle Ages, 700 BC to 1500 AD.

Vol. 3: From the Renaissance to the Industrial Revolution, 1500 to 1750.

Vol. 4: The Industrial Revolution, 1750 to 1850.

Vol. 5: The Late Nineteenth Century, 1850 to 1900.

Vols. 6–7: The Twentieth Century, 1900 to 1950.

Song, Y.-X. "Tian Gong Kai Wu: Chinese Technology in the Seventeenth Century", translated by E-tu Zen Sun and Shiou-chuan Sun. Pennsylvania State University Press, University Park and London, 1966.

Spengler, O. "The Decline of the West", Vol. 1: Form and Actuality, Vol. 2: Perspectives of World-History, 1922, translated by C. F. Atkinson, Alfred A. Knopf, New York, 1928.

"Statistical Abstract of the US", US Census Bureau, Washington DC, 2007.

Strahler, A. N. and A. H. Strahler. "Elements of Physical Geology", John Wiley, New York, 1989.

Thornton, A. and K. McAuliffe, "Teaching in World Meerkats". *Science* 313, 227–229, 2006.

Tignor, R., J. Adelman, S. Aron, S. Kotkin, S. Marchand, G. Prakash, and M. Tsin. "Worlds Together, Worlds Apart", W. W. Norton, New York, 2002.

Toynbee, A. "A Study of History", Oxford University Press, London and New York, 1947.

Travers, Bridget, editor, "World of Invention: history's most significant inventions and the people behind them", Gale Research, Detroit, 1994.

Tybout, R. A. editor, "Economics of Research and Development", Ohio State University Press, Columbus, 1965.

United Nations Development Programme, "Human Development Report", New York, 2007.

Waley, A., translator "Shih Ching: The Book of Songs", Grove Press, New York, 1996.

White, A., P. Handler, and E. L. Smith. "Principles of Biochemistry", McGraw-Hill, New York, 1964.

White, L. T. "Medieval Technology and Social Change", Oxford University Press, London and New York, 1963.

Wiener, N. "Invention: The Care and Feeding of Ideas", MIT Press, Cambridge, Massachusetts, 1993.

Williams, T. I. "A History of Invention: From Stone Axes to Silicon Chips", Checkmark Books, New York, 2000.

"World Development Indicators", The World Bank, Washington, DC, 2008.

Web Sources
Nobel Foundation, http:\\www.nobelmuseum.org

INVENTIONS FOR WORK

We work to produce goods and services for our own needs and enjoyment, and for sale to customers. Work also gives us an opportunity to demonstrate our knowledge and skill, our organization and diligence, and thus to earn esteem and glory. The accumulation of inventions and technologies over millions of years gives us the ability to create ever more wealth for less effort. In the future, we are counting on new inventions to make life even better.

Most animals are born with specialized tools as part of their bodies to do the necessary work of their lives, such as the teeth and claws of predators. Darwin described a wide variety of finches of the Galapagos Islands, shown in Figure 2.1, that have different beaks to eat different types of seeds and nuts prevalent on the island: large beaks are suitable for cracking heavy nuts, which are plentiful during rainy years; and agile small beaks are suitable for gathering small nuts, which are plentiful during dry years. During a prolonged drought, birds with small beaks thrive, but those with large beaks suffer. Nature-given bodily tools are not versatile under changing conditions, and are severely limited in use. If you have only a good hammer, you are ready to hammer nails, but what if you need to saw wood? Humans lack such highly specialized tools on their bodies specifically designed for limited purposes. As a compensation for the parsimony of nature, we have invented numerous tools for different purposes. We continuously add new inventions to the toolboxes we have inherited from our predecessors, and collectively, these inventions enable us to invade and live in the niches of other animals and create new ones not found in nature. These are the roles of the "invented tools" of humanity.

Besides the tools used in individual homes, farms, and factories, the proper functioning of our civilization also depends on our infrastructure of public works. For transportation, we need roads, bridges, harbors, waterways, and airports. Our homes and factories need to be connected to the utilities of electricity, water, sewage and garbage removal; they also need telecommunication linkages of cables, satellites, and the Internet.

Our standard of living depends on our productivity. Economists understand productivity as the amount of wealth that can be created per hour worked. Early economists such as Adam Smith considered the factors of productivity to be land, labor, and capital; but modern economists place more emphasis on the role of inventions and innovations. Joseph Schumpeter, in particular, distinguished between an invention—an important discovery— and innovation—the introduction of a new good or method of production, or the opening of a new market, a source of new raw material, or a new industrial organization. Robert Solow calculated that about

Great Inventions that Changed the World, First Edition. James Wei.
© 2012 John Wiley & Sons, Inc. Published 2012 by John Wiley & Sons, Inc.

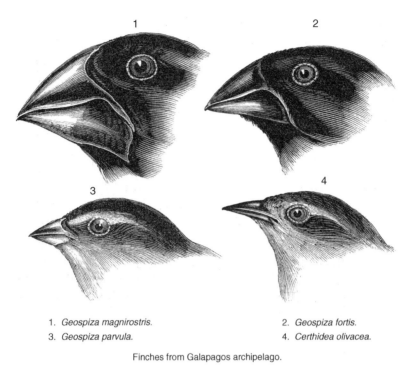

1. *Geospiza magnirostris.*
2. *Geospiza fortis.*
3. *Geospiza parvula.*
4. *Certhidea olivacea.*

Finches from Galapagos archipelago.

FIGURE 2.1 Darwin's finches and beaks. Reprinted with permission Mary Evans Picture Library/Alamy.

four-fifths of the growth in the United States output per worker was attributable to technical progress.

Every nation on earth embraces the stated goal of increasing labor productivity as a means to increase the standard of living, to become influential in world affairs, and to be masters of our destiny. This goal assumes a reliance on progress, so if there is a problem that cannot be addressed by current technologies, we try to learn and copy from our neighbors, or try to invent a new technology to solve the problem. Progress is not an effortless entitlement, but a result of investment in education and research, and of giving up some current enjoyment for the brighter potential of future generations. The greatest invention of all is the culture of inventions—to solve problems and to expand capabilities.

2.1 TOOLS AND METHODS

Our teeth and fingers are limited tools for survival and making a living. The Paleolithic stone axe was mankind's first invention, which began the glorious tradition of using tools to shape and manipulate objects to our desire. Humans are not the only animals that use tools: a species of Galapagos finch uses twigs to search for insects in tree holes, and sea otters smash clams with stones to reach the edible flesh inside.

Humans, however, are the only animals that design and manufacture a large variety of tools for specific uses, store them for future use, and improve upon the designs they inherited. We use hand tools to obtain food, fabricate clothes, and construct shelters and buildings; we also use tools to upgrade and increase the value of resources obtained from nature.

A *hand tool* is a simple device to change material into a desired shape: by separating a piece into several pieces, by joining pieces together, and by bending and cutting the pieces. To get work done, the carpenter has the axe, saw, drill bit, nail, and screw; the stone mason has the hammer, chisel, straight edge, and trowel; and the blacksmith has the furnace, hammer, anvil, tong, and cutter. A *machine* is a more complex device, and often has several parts that can move separately. Machines make it possible for us to do work that is too heavy or tedious with precision and speed, and without getting tired and endangering our safety. Our productivity, or how much wealth we can create per hour of work, is very much dependent on the tools and methods at our disposal. We invented such simple machines as the lever and the inclined plane to lift heavy loads and the pulley to change the direction of forces. We also use the wheel and axle to reduce friction in transportation, and the shaduf and screw pump to raise water.

When the work involves many steps and operations, to be carried out by a team of different workers, it benefits from an *organization* to coordinate their efforts. An underground mine involves digging vertical shafts for access to greater depths, ladders and pulleys to raise and lower people and objects, radiating horizontal galleries to reach veins of minerals, props to prevent the collapse of tunnels, and water removal to prevent flooding. The manufacture of a wood bench may involve a single carpenter who saws wood into pieces and hammers them together. The manufacture of an automobile involves many different workers to make the frame, the body, the engine, and the wheels; and then another set of workers to assemble them together on a moving line.

2.1.1 Stone Axe

In the Paleolithic Age 2.5 million years ago, *Homo habilis* (handy man) in East Africa began experimenting with stone tools. They had small brains, prominent jutting jaws, and walked on two feet—thereby freeing their hands for manipulating and carrying objects. Bands of 10–20 made their living as hunter–gatherers and as scavengers. They foraged for soft plants and nuts as other tree apes, and they learned to broaden their diets to include meat from animals. The carcasses of larger animals represented bounties, but they were difficult to handle. How would they rip open the tough hide and tear off a limb from the carcass, suitable to carry up the nearest tree before a lion or a leopard would show up to interrupt their meal, and to bring home to feed the family? Bone marrow is a rich source of fatty food, but accessing it requires crushing the bone to extract the contents. How would they flay the hides to make clothing? They had seen how lions and leopards used sharp teeth and claws, but nature had given them neither. Some of them had the inspiration and creativity to find a fieldstone with a broken edge that could be used as a cutting tool. It is likely that there were even older tools made from wood or horn, but they did not

FIGURE 2.2 Paleolithic Oldowan chopper.

survive—thus, we have only a huge collection of stone hand axes, and have come to regard them as mankind's first invention.

The stone axe is not a single invention that occurred only once in history, but a sequence of inventions that gradually improved over millions of years by different species of the human animal, forming a relay team of inventors and inventions. The first stone tools of the *H. habilis*, known as Oldowan choppers (Fig. 2.2) after the Olduvai Gorge in northern Tanzania where a large number of them were found, are crude products that could be easily mistaken for naturally broken pebbles. They can be made by using a hammer stone to strike a fist-sized pebble at a sharp angle, to chip off a thin flake from the core. The core with the sharp edge can be used as a cutting or scraping tool for dismembering carcasses; digging for tubers, roots, and insects; and stripping bark from trees to access gum and insects. The blunt end can be used for hammering open bones, and for pounding and cutting nuts and fruits. With this invention, humans could eat the food of lions, hyenas, and hogs.

The best stone for this tool is a fist-sized pebble that is fine grained and uniform, so that it can be chipped and shaped into any desired form. Rocks with large grains such as granite, and rocks with layers such as slate, are not easy to shape into cutting edges; rocks with minute grains such as flint, and rocks that are glassy like obsidian, are the best. The rock should also be hard and heavy, so that the sharp edge created would remain sharp for a long time. *H. habilis* might have found pebbles on river banks with gravel bars of cobbles. It is more difficult to hammer out a piece from solid bedrock. At this stage of development, there were no standardized sizes, shapes and forms, and a tool of any shape with some sharp edges would serve a purpose.

Beginning 1.5–1.7 million years ago, a strikingly different and refined version of the stone axe first appeared among the *Homo erectus*, a new tribe with larger

FIGURE 2.3 Acheulian hand-axe. Reprinted courtesy of Library of Congress.

brains and smaller jaws. This tool, called the Acheulian hand-axe (see Fig. 2.3) was no longer a crude tool with a random shape, but was built according to a standard design that remained constant for more than 1 million years, and spread from Spain in the west to India in the east. The Acheulian hand-axe has an approximate length of 6 in. (150 mm) and a weight of 1.3 pounds (600 g), with a graceful tear-drop or pear shape. Its creators were more like sculptors who had a final form or template in mind. At one end, there is a point for piercing, two symmetrical cutting edges on the sides, and a blunt end for grasping and perhaps also for bashing bones and nuts. A hard stone hammer may have been used for the primary flaking to get the axe roughly into the desired approximate shape, then soft hammers of antlers and bones for the secondary flaking for better control of the finer work of trimming and detaching numerous small flakes. The best stones were European flint that was mined from chalk and chert limestone, as well as volcanic obsidian glass, but these valuable stones were available only in special locations and had to be obtained by long distance transportation and trade.

Homo sapiens, who made their first appearance around 200,000 years ago in Ethiopia, found many sophisticated uses for the stone flakes as tools, in addition to

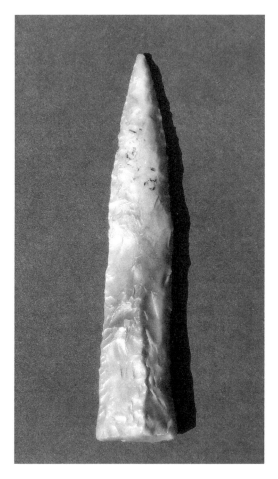

FIGURE 2.4 Neolithic stone tools. Reprinted courtesy of Library of Congress.

the cores. Modern humans appeared around 35,000 years ago, and created a great number of new stone tools. In the Neolithic Age (from 10,000 BCE), even more sophisticated stone implements (Fig. 2.4) were made with much finer control of the shape, sharper edges, and pleasing shapes. The flakes were used to flay hide and to pierce holes for making garments. The new tools were made with the technique of pressure flaking, which ingeniously substituted the hard-to-control hammer blows with steady pressure using bone, ivory, or hard wood against a stone edge, giving better control and producing smaller flakes. These microflakes could then be fastened to wooden shafts to make spears, arrows, and scythes for farming.

STRENGTH OF MATERIAL

Paleolithic inventors were primarily concerned with whether their tools worked, but scientists ask *how* they worked. Such knowledge leads to scientific understanding, and sources of future inventions. A scientific and quantitative description of the mechanical behavior of a material considers its response or strain under stress. A rod with an original

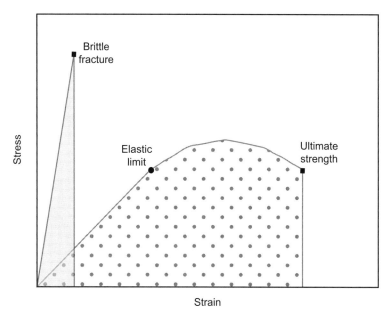

FIGURE 2.5   Stress–strain graph of brittle and ductile materials.

length and cross-section area will become longer when stretched. The force per area is called the "stress," and the resulting "strain" is the fractional change in the length of the rod. Figure 2.5 shows a stress–strain graph where the strain is plotted in the horizontal x axis against the stress in the vertical y axis. A brittle matter such as stone is shown in the solid curve. When the stress is small inside an elastic region, the deformation or strain is elastic and reversible, so that when the stress is released, the deformation will return to zero. When the material is stretched beyond the elastic region, the material suddenly breaks into two or more pieces at the point of ultimate strength. The amount of energy needed to break the material, the toughness, is measured by the triangular area underneath the solid curve.

The behavior for an elastic material, such as copper, is shown in the dotted curve. The elastic region is exceeded at the point of yield strength when the material begins to undergo irreversible or permanent plastic deformation; it eventually breaks at the point of ultimate strength. In comparison with brittle stone, copper needs less stress to deform, but more energy to break it.

Stone axes are made of hard materials, which are used to shape and manipulate softer materials, such as hide and bones. The word "hard" has two separate meanings, referring to either a material's resistance to bending and scratching, or its resistance to breaking. Diamond, for instance, is the hardest substance known that cannot be scratched, but it is brittle and can be shattered with a moderate force; on the opposite end is steel which is more flexible and easier to scratch, but is tougher and can sustain much heavier blows. Stone tools are extremely hard, and therefore make effective tools for cutting through softer materials. Their brittleness, however, means that they are easily shattered if struck against an object of superior toughness. While stone tools certainly represented an advance over fingers and teeth, they could not be used where flexibility and toughness were required, such as for making a sword or a lever. Even so, stone was the principal material for tools and weapons till the advent of the Bronze Age, only 5500 years ago (or 3500 BCE). Flint knapping is the name given to making stone axes and other implements in modern times, which is done by experimental archeologists to understand the past. It is often taught to outdoorsmen for wilderness survival, and used to make flintlock firearms as a hobby.

The subsequent development and specialization of hand tools continued with metallic tools. In the middle ages, each village needed three craftsmen: a blacksmith to work on metal objects and horseshoes, a carpenter to make furniture and wheels, and a mason to construct walls and houses. Each of them needed a set of tools to carry on their trades. In the modern world, we have many specialized industrial machines and tools, manmade sources of energy and power, and stronger materials to increase labor productivity and the quality of life.

2.1.2 Machines

When you tie a stone axe to a wooden handle, you have created a simple machine, which is a mechanical device that changes the direction or magnitude of a force, and can use leverage or mechanical advantage to multiply the force. There are six classical simple machines: the wheel and axle, pulley, screw, inclined plane, wedge, and lever. The first three involve rotary motion around a central axis, and the other three provide mechanical advantages in linear motion.

Inventions involving wheels and other forms of continuous rotary motion began to appear around 3000 BCE. Road vehicles with wheels are undoubtedly among the greatest inventions in history, as they made possible the easy transportation of people and heavy goods, as well as the waging of wars. Men and women could carry small items on their backs or heads, and larger items could be hung from poles carried between the shoulders of two people. A sledge with a flat bottom or runners that could be dragged across grass, mud, or snow was an improvement, especially when the friction could be lowered by lubrication with water. Another method, important in the construction of the Egyptian pyramids, was to move heavy stones over rolling logs. None of these methods could compare to the convenience of a wheel with a hollow hub rotating around a fixed axle, especially with the application of lubricating oil or grease in the hub. An obvious early use for carts and wagons was to carry heavy harvests from fields on the farm to living quarters, and to carry farm implements and manure in the opposite direction. Wall paintings in Egyptian tombs show families traveling in ox-carts with solid wheels. Even before 3000 BCE, wheeled vehicles and chariots were used in warfare. War chariots greatly increased the speed of travel and the quantity of weapons that could be carried, in comparison with the infantry, and were such prized possessions that they were buried with powerful kings in royal tombs.

Wheels enable people to make things as well as move them. Some of the most important early uses of wheels include the horizontal grinding stone to mill wheat and corn into flour, and the crane with vertical pulley to lift stones for construction. The potter's wheel first appeared in Sumer around 3000 BCE. The centrifugal force of a spinning potter's wheel throws wet clay outward into a perfectly symmetrical shape that can be guided by a finger or tool. The wheel disk was probably first marked by tying a string to the center and rotating the loose end with a point to trace out a circle. A better design is to connect the wheel supporting the pottery to a co-rotating base wheel with a vertical post, so that a seated potter can rotate the base with his or her feet. The potter's wheel diffused from the Middle East to the rest of the world, and became the first machine that was able to mass produce consumer goods.

Fibers from linen and wool must be twisted together to form yarns of greater strength and constant thickness, which can then be woven into fabric. The weaver's spindle was developed around the same time as the potter's wheel, and used the wheel to improve productivity. The spindle is a slender rod spun by hand to twist the fibers from a ball of fibers held by the distaff. A small weight called a whorl can be attached to the spindle, making it easier to spin at a constant speed by hand. Since this work was done by the women in the ancient world, the distaff became a symbol of womanhood and home life. A heavier spinning wheel could be activated by a foot treadle, or by water power.

The carpenter's screw uses a spiral groove along an axis to combine rotary motion with the mechanical advantage of a wedge, which makes it a stronger fastener than a nail. Archimedes introduced the screw to raise water from a lower level to a higher level with moderate force.

BIOGRAPHY: ARCHIMEDES

Archimedes of Syracuse in Sicily (287–212 BC) was the most famous inventor and scientist in the ancient world. He was related to King Hiero II, the ruler of Syracuse on the island of Sicily, and was a wealthy aristocrat with great influence. He was said to have traveled to Egypt and Alexandria to study. He worked on numerous inventions, including every day tools and weapons used against the Romans. The most famous anecdote about Archimedes was reported by Vitruvius about how he discovered the method to determine the density of an object with an irregular shape, such as a crown in the shape of a laurel wreath of King Hiero. Archimedes was to determine whether the goldsmith had added silver to solid gold, which is a heavier substance. While immersing himself in the bath, he noticed that the level of the water in the tub rose, which can be used to measure the volume of a submerged object and thus, its density. In excitement of his discovery, he took to the streets naked, and shouted "Eureka!" or "I have found it!"

It is said that during the Second Punic War, he produced many war machines that kept the Roman army at bay for 2 years. He made ballistic weapons shooting missiles and stones, he had a crane that can lift Roman ships out of the water, and he built burning mirrors that set ships on fire. However, all his ingenious military engineering was not enough to keep the Romans from breaking into the city walls of Syracuse. He was working on his mathematical problems at home, when a Roman soldier under General Marcellus came to his room and interrupted his work. Archimedes said "Do not disturb my work," but the enraged soldier killed him.

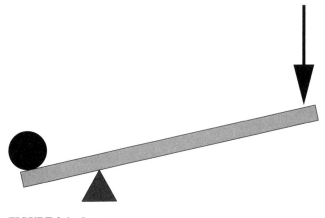

FIGURE 2.6 Lever.

The lever was used by Egyptians to move heavy objects, but Archimedes was the first person to write about its role to lift weights too heavy for the strength of unassisted men. A lever is shown in Figure 2.6 to have a fulcrum or pivot that divides a rod into a longer force arm and a shorter weight arm. If your lever is 10 times longer on the force arm than the weight arm, a force of 1 kg on the long arm can lift a weight of 10 kg on the short arm. This leverage is not magic, as the product of force by distance is fixed. When the force of 1 kg is pushed down by 10 cm, the weight of 10 kg rises by only 1 cm. Archimedes famously boasted, "Give me a lever and a place to stand on, and I will move the earth." But this is only a figure of speech, as the mass of earth is 100,000 billion billion times greater than that of Archimedes. Where would he find a sufficiently long and strong lever, and where would he stand? The screw, wedge, and inclined plane all provide mechanical advantages to amplify a small applied force onto a large resulting force. A ramp with a length of 10 and a height of 1 can be used to raise 10 kg of weight by the force of 1 kg. Strong forces are needed to squeeze out all the juice from an olive to produce oil and from grapes to produce wine, which are provided by the screws on the top of olive and wine presses.

The pulley can be used to change the direction of a force, so that you can lift up a weight by pulling down on a rope with your body weight, shown in Figure 2.7. The upper wheel is fixed and the lower wheel is movable; when the rope is pulled down by 2 cm with the force of W, the lower wheel moves up by 1 cm to pull up with the force of $2W$. A compound pulley with several blocks and tackles, set on top of a fixed wall or a movable crane, can also create more mechanical advantages. In the ancient Greek theater, when human affairs became hopelessly tangled, a god would descend from above, suspended by a pulley, to solve their problems. This plot device is called *deus ex machina*, or "god out of the machine." Sailors can use compound pulleys to move heavy ships, which is perhaps the source of the legend that Archimedes made war machines that could pull Roman ships out of the harbor and dash them on the rocks.

More complex ancient machines involved simple machines as components, such as air and water pumps and treadmills. Egyptian wall paintings from 1500 BCE show scenes of metal working, where the fire was enhanced by

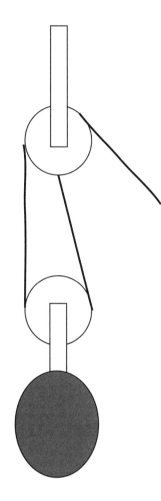

FIGURE 2.7   Pulley. Reprinted with permission
© National Portrait Gallery, London.

air bellows operated by the feet of slaves. The bellow had an air intake controlled by a valve so that air could only enter but not leave, and an output
nozzle also controlled by a valve so that air could leave but not enter. Hephaestus or Vulcan was the smith of the gods, and forged the thunderbolt for Zeus as
well as the splendid armor for Achilles; book XVIII of the Iliad describes his
equipment, which included bellows. Hero of Alexandria described a hydraulic
organ for music, where air is propelled by the flow of water.

Rotary motion was needed for a mill to grind corn into flour, and for a hydraulic
station to raise water to a higher place. These devices can be powered by human
muscle, animal muscle, or wind and water. Our knowledge of ancient machinery comes
from a number of distinguished sources. Marcus Vitruvius of Rome (80–15 BCE)
wrote 10 books in *De Architectura* on architectural constructions, on machines used in
structures such as hoists, cranes, and pulleys, and on war machines such as catapults
and siege engines. Hero of Alexandria (10–70) wrote about the aeolipile—the earliest
steam engine—and about a wind wheel that harnessed wind energy.

2.1.3 Production Organizations

Mining is almost as old as agriculture. The earliest mines were open-pit mines where the miners worked to extract stones and metals from ores exposed to the sky. When these easy pickings were exhausted and miners started looking for precious metals, gems and salt, they turned to underground mines with vertical tunnels for descent and horizontal galleries that required ventilation and illumination by torches.

The earliest known mine may be the "Lion Cave" in Switzerland from 43,000 years ago, which was the source of a red ochre pigment known as hematite. Later came the flint mines in northern France and southern England. The Grimes Graves in Norfolk, England (4000 BCE) consist of a pit some 12 m deep and additional pits connected by radiating tunnels. This was a tremendous industry for producing flint that must have employed many miners over a long period of time. Presumably the miners were full-time specialists who traded their flint for food and other daily necessities with buyers who may have come from far away.

A variety of tools and machinery were used in the ancient mines. In the Laurium silver mines near Athens (approximately 1100 BCE), workers used hammers, picks with wood handles, and lamps. Workers accessed the mine by ladders and used pulley wheels to haul up ores. Underground mines also required ventilation to bring fresh air to miners. Sometimes, this was accomplished by giant bellows, but more advanced mines included separate shafts to introduce fresh air and to evacuate stale air. An underground mine was often flooded by rainwater from above, or from groundwater, which had to be removed quickly, often by slaves and convicted criminals. All of this had to be organized, which required the large-scale planning and coordination of different functions, and an investment over many years. Much of our knowledge of ancient mining and metallurgy came from the Roman Pliny the Elder who wrote *Historia Naturalis* in the year 77, from the German Agricola who wrote *De Re Metallica* in 1556, and from the Chinese Song Yingxing who wrote *Tian Gong Kai Wu* in 1637.

It is somewhat easy to dig soft clay and sedimentary rocks with metallic picks and shovels, but removing hard igneous rocks is much more difficult. The process of splitting hard rock was made easier by the technique of "fire-setting," in which rocks are heated with fire and then drenched with cold water. The sudden temperature change causes the rocks to split, revealing the stones inside. The greatest advance in mining technology came with the use of explosives to break down rocks and to create tunnels. Black gunpowder was used in this way for many centuries until the invention of modern high explosives such as nitroglycerine and dynamite.

A complex product, such as a revolver, has many parts that are designed to fit together smoothly with very small tolerance for error. In the traditional craft method, a single master smith will make all the parts, and fit them together with last minute adjustments at the end. When the demand is for hundreds or thousands of revolvers for a coming war, the craft method is not adequate as there may not be enough master smiths on hand, and not enough time to accomplish the task. There is a long history to the concept of interchangeable parts through exact specifications and standards, and an assembly line to put them together.

One of the earliest assembly lines was put together around 215 BCE to produce 8000 life-sized terra-cotta soldiers and horses for the funeral of the First Emperor Qin Shi Huangdi in the City of Xi'an in China. The separate body parts were interchangeable, assigned to different workshops to mold and bake the clay, and later assembled in a central facility. Each workshop inscribed its name on the part it manufactured to add traceability for quality control. However, each head was meant to represent a specific individual, and had to be hand crafted by different master sculptors. The result was an army of life-like individuals with distinct heads but interchangeable bodies. In the 1500s, when the Venetian navy needed many new ships for a coming war, the Venetian Arsenal employed about 16,000 people who were able to produce nearly one war galley each day, on an assembly line with standardized parts.

A muzzle loading musket or rifle receives loose gunpowder from the muzzle, to be followed by a bullet of soft lead. Making these firearms was a slow and expensive process, and two such guns did not need to have the same barrel diameter, as the soft lead bullet and loose gunpowder could be made to fit a range of diameters. In the traditional craft method, a single master gunsmith made the barrel, the revolving cylinder and the firing hammer, and then assembled them together. The craft method is fine for a small market such as an individual buyer or even a small group of buyers, but is not adequate for an army that needs thousands of revolvers for a coming war—there might not be enough skilled gunsmiths, or enough time for rapid and economic production.

The speed of fire is incomparably increased with a revolver, using a copper-clad cartridge of a standard fixed diameter. The Colt 45 has a barrel diameter of 11.4 mm, which must be precise. The revolving cylinder with bullets is designed to be aligned with the barrel on one end, and with the firing hammer on the other end. It is critical to ensure that the hammer will strike the cap squarely at the back of the bullet, which will then fly out of the barrel. Samuel Colt set up his revolver factory in Patterson, New Jersey in 1836, and advanced the practice of interchangeable parts. This modern method of mass production evolved as a solution, which divides up the work. Each worker was responsible for producing only one part repeatedly to exact dimensions, and the results were later assembled on a line. The bullets also had to be made to the same dimensions, so they would fit any gun barrel.

At the beginning of motorcars in the United States, Ransom Olds patented the modern assembly line concept, which he put to work in his Olds Motor Vehicle Company factory in 1901. The same concept was observed in a Chicago slaughterhouse, which can more accurately be described as a "dis-assembly line," where animals were butchered as they moved along a conveyor, and each worker removed only one organ or part repeatedly with great efficiency.

This concept was reported to the Ford Motor Company, which was adopted by Henry Ford with immense enthusiasm. Ensuring a perfect fit between all the separate pieces required accurate specifications and strict discipline on making them, but it was very effective. Before the assembly line, an assembly job that took 12.5 man-hours was reduced to only 1.5 man-hours, so productivity had increased by a factor of eight. The method was so successful that the drying of paint on automobile parts became a bottleneck. Only Japan Black would dry fast enough, forcing the company to make only black Model T cars.

Ford also assigned each worker to a specific location, instead of allowing them to roam about, which resulted in a reduced rate of injury. Ford's high efficiency made possible his offer of higher wages as well as lower car prices, which was copied by other car companies around the world, and eventually to other major industries. The other car manufacturers could either embrace the assembly line and produce cars at a lower cost, or go out of business. For its critics, the assembly line is a monotonous and dehumanizing system, and restricts employee freedom; but for its promoters, the assembly line represents one of the greatest industrial inventions of all times.

2.2   ENERGY AND POWER

Everything that we do involves converting energy from one form to another. Life is sustained by plants converting sunlight energy into the chemical energy of food by photosynthesis, and then food is burned in the body to create the metabolic energy of life. We keep warm and cook supper with heat energy, see in the dark with light energy, travel and work with kinetic energy, hear and play music with sound energy, obtain information from the Internet and the computer with electrical energy.

The total energy in a closed system remains constant amidst changes, while one form of energy can be converted into another form. When we say that something "consumes" energy, we are referring to how much energy goes into a process—as an equal amount has to come out, usually in some combination of "useful" energy and "wasted" energy. If you operate a car to travel, you consume chemical energy in gasoline to produce useful mechanical motion as well as emitting waste heat through the tailpipe. We use many different units for measuring energy, from the familiar British Thermal Units (BTU), to the metric calorie, and the scientific Joule. The term "power" refers to the rapidity of the rate of energy consumed over time.

The ability to convert energy intentionally from one form to another is part of the unique human history. From fire to the steam engine, humans have long experimented with techniques for turning natural energy into warmth, light, mechanical power, information, and entertainment.

2.2.1   Fire

Around 800,000 BCE, some of the *H. erectus* in East Africa decided to leave for Asia and Europe. They alone among the great apes had lost the long body hair that covered all the mammals, so they had no insulation against the cold winds, rain, and snow. They left tropical forests and the savannah in East Africa and crossed desert regions to enter the temperate Near East, and then later turned to the frozen north of Asia and Europe in the middle of the Ice Age. Keeping warm with clothes and shelter was not important in tropical East Africa, but became critical in the frozen subarctic north. They may have huddled together to keep warm, and perhaps relied on the moon and stars to see at night.

The mastery of fire was not the invention of a new object that did not exist before, but the invention of methods to control it for our benefit. We have found evidence for *H. erectus* and fire in Zhoukoudian near Beijing (40°N) which is thought to be from 460,000 years ago. The conquest of fire probably began with natural fires started by lightning or natural sparks. A person who learned to keep a reasonable distance from fire could use it to keep warm in a cold night, or to roast animal flesh to make it softer and easier to eat. The next achievement of mastery was to maintain fire at a steady level without extinguishing, by feeding it with wood or dung, sheltering it from wind and rain, and giving it oxygen by blowing on it. A fire also needed to be contained by barriers, and primitive humans learned to extinguish their fires with water or earth.

Most plant materials are primarily made of carbon and therefore make reasonably good fuels, as long as they are dried and have plenty of surface area to contact with air. Air primarily consists of nitrogen (78%), which does not burn, but the less abundant oxygen (21%) does. When you burn a twig or a piece of straw, you are witnessing the following reaction, which produces carbon dioxide:

$$C + O_2 \Rightarrow CO_2 + \text{energy}$$

In this reaction, C is carbon, O is oxygen, and CO_2 is carbon dioxide. We see this energy in the form of radiating heat and light, hear the crackling sound, and feel the warmth.

The first fuels had to be locally available and dry—presumably grass, brushwood, or branches. Massive logs are difficult to burn, as there is not much surface area and the interior is not exposed to oxygen. Dried animal dung from cattle, sheep, and camels contain undigested fibers and make good fuel. A longer burning torch can be made by soaking branches in animal fat, vegetable oil, or mineral pitch that seeped through the ground. Eventually, humans learned how to make charcoal by burning wood with insufficient air to drive off the volatile material and smoke, leaving only a core of concentrated carbon that would burn to produce a lot of heat but little smoke.

The next advancement in the mastery of fire was the ability to start a fire anywhere at any time, so that one is not totally dependent on finding natural fire. The fire-starter needed a method to create a local temperature that was higher than the autoignition temperature of the fuel—about 230°C for dry grass or thin pieces of wood. Sulfur and yellow phosphorous have lower autoignition temperatures, which is why they are used in black gunpowder and in friction matches. The friction method of starting fires, familiar to Boy Scouts, involves rubbing two pieces of wood together till there is enough heat to raise the temperature and cause a fire to start. The spark-ignition method uses percussion to create a spark by hitting a piece of stone against another stone or metal. Iron pyrite is easy to find, but flint is even better. The spark is then directed to a pile of wood shavings or sawdust, and the small fire is encouraged by blowing. The modern cigarette lighter has a steel wheel against a flint to create a spark, and a wick soaked with butane to catch fire. Friction matches, based on powdered glass, sulfur, and phosphorus, were invented in the 1800s and served as miniature and portable fire-starters.

Combustion consumes oxygen, so a vigorous flame depresses the local oxygen level in the air. A good draft is required both to provide oxygen in fresh air and to remove smoke and combustion gases. Without enough fresh air, the chemical reaction changes to partial combustion, releasing less than a third of the heat of full combustion, and producing poisonous carbon monoxide. Moving a fire indoors provides shelter from the wind and rain, but requires a chimney for necessary ventilation.

The conquest of fire was one of the most important milestones of human progress and has been celebrated in the mythology of many races, usually as the magical gift of gods. In ancient Greece, the Titan Prometheus stole fire from the gods as a gift to men, and was punished for that crime; in Persia, Zoroaster represented the forces of good associated with heaven, light, and fire; in India, the fire god Agni presided over sacrifices. The warmth of fire was critical for the human conquest of the frozen north during the height of the ice age.

Fire also provided a gathering place where family members or tribesmen could socialize and eat in comfort, protected from the cold and dark. Cooking greatly expanded the range of available food beyond soft leaves, fruits, and soft choice meat. Rice and wheat are difficult to digest without cooking in water, and roasting softens tough meats and makes them chewable. Heat also destroys many germs and other disease causing microbes. Fire opened the way for humans to advance beyond the Stone Age, by making new materials of ceramics and metals, and manufacturing tools and weapons. Much later, when we mastered the steam engine and the internal combustion engine, fire and heat led to new sources of energy and power and brought us to the Industrial Revolution.

Many forests have been cut down for fuel around major cities, and around mining and manufacturing centers. Natural biomass has been unable to keep up with the demand from an expanding and more prosperous population. Because of this, the fossil fuels of coal, oil, and natural gas have come to be the dominant fuel since the Industrial Revolution. Outcrop coal was used in Britain during Roman times, but did not play an important role till the Middle Ages as long as the supply was limited to surface coal. Underground mining of coal featured prominently in the *Tiangong Kaiwu* published in 1637 in China. The Industrial Revolution in Britain depended critically on coal mining.

Herodotus mentioned the use of asphalt in the construction of the walls and towers of Babylon, which came from oil seeping from underground. Naphtha or liquid oil was used in the Greek fire for warfare during the Byzantine Empire. *Tiangong Kaiwu* also mentioned the use of bamboo poles to conduct natural gas from wells, to create fire for the evaporation of brine to produce salt. Petroleum became an important source of energy after the development of drilled oil wells, which ensured a steady and growing supply. This was particularly spurred by the 1858 drilling of an oil well by Edwin Drake in Pennsylvania, which reached a depth of 21 m.

2.2.2 Steam Engine

The operation of machines to grind corn and to lift stones requires power, which was initially provided by human muscle, and later by animals such as horses and oxen. The water mill, that creates power from moving streams, was introduced in China and Greece around 100 BCE to grind wheat into flour. It has a horizontal millstone that is connected by a vertical shaft to a second horizontal wheel of paddles at the bottom driven by a stream of water. The Roman engineer Vitruvius introduced a far more efficient water mill with a vertical wheel mounted on a horizontal axis, which can be efficiently moved by either a rapidly moving stream at the bottom of the wheel, or falling water on top of the wheel. This rotation of the vertical wheel must be transferred by a gear to rotate the horizontal millstone.

Wind power makes sense in windy places that are without rapidly flowing water, and apparently began in Islamic Persia and spread to Europe around 1200. It used a horizontal wheel around a vertical shaft, with vertical funnels on arms that are operated by winds from any direction. However, on one side the arm collects wind and retreats to generate power, but on the opposite side the arm spills wind and advances to consume some of the power generated. The classic modern windmills of Greece and Holland have vertical wheels with arms covered with sails, and must be turned to face the wind.

Water mills and windmills must be located near sources of running water or wind. On the other hand, heat engines that depend on hot gases generated in combustion can be located anywhere. The simple steam engines built by Hero of Alexandria (62 AD) had a metal sphere on an axis, with two jets on opposite sides. When steam filled the metal sphere, it escaped from the two jets and caused the sphere to spin. It was used more as a fanciful toy than practical machinery. For steam engines to be useful, the inventors had to find an economical way to translate steam power into mechanical energy, and harness it for a practical problem.

Sixteen centuries after Hero, Thomas Savery of Scotland built a working steam engine to pump water out of flooded mines. In his engine and the subsequent improvements created by Thomas Newcomen in 1712, the reciprocal or back-and-forth motion created in a steam-filled piston was transferred through a lever to produce a lifting action on the other end. The first industrial steam engines operated by a lever built as a long rocking beam with a pivot in the middle. Newcomen placed on one arm a vertical steam cylinder with a piston, while the other arm was connected to a hoist in a pool of mine water. The cylinder was first filled with steam to make the piston rise, so the other arm of the beam would dip into the mine water. He then introduced a spray of cold water into the steam cylinder, where the sudden change in temperature condensed the steam and created a vacuum that sucked the piston down. The force of the downward motion caused the other arm to rise and pump water out of the mine. It was a great improvement over human and animal strength, but Newcomen's engine was slow and inefficient. Only the down stroke was a power stroke, so half of the time and motion was not productive; moreover, the repeated heating and cooling of the cylinder consumed a lot of fuel. Finally, it was a low-pressure system that needed a very large engine to do a modest amount of work—the cylinder was pushed down by no more than one atmospheric pressure, or only 760 mm of mercury.

BIOGRAPHY: JAMES WATT

James Watt was born in 1736 in Scotland. His father was a ship builder and owner, and his mother was well educated. He was home schooled by his mother, he had good manual dexterity, and enjoyed mathematics and Scottish legends, but he disliked studying Greek and Latin. When he was 18, his mother died and his father's health began to fail. He traveled to London to study instrument-making, but returned to Glasgow after a year without serving the full 7-year apprenticeship, and was thus not accepted by the Guild.

FIGURE 2.8 James Watt.

Instead, he went to work at the University of Glasgow with the physicist Joseph Black, who became his friend.

Watt was a very skilled mechanic and ingenious inventor, and always looked for more refinement in his engines. He was not a good businessman, and disliked bargaining and negotiating. Despite his lack of formal education, he was good at society and conversations in scientific societies. He retired in 1800 when his patent and partnership expired. He continued to invent, became a fellow of the Royal Society and a foreign associate of the French Academy of Science. He died in 1819 at his home in Birmingham at the age of 83. A colossal statue was placed in the Westminster Abbey, which was later moved to Scotland Fig. 2.8 shows an engraving of Watt as a prosperous gentleman.

James Watt had a machine shop in Scotland. Professor John Robison of the University of Glasgow called the attention of Watt to the Newcomen steam engine, and he began to experiment with it. The University owned a model Newcomen engine that was in London for repairs, and Watt had it returned and repaired in 1763. He demonstrated that 80% of the steam heat was wasted from the repeated heating and cooling of the cylinder. In 1765, he invented a second condensation chamber that could be kept cool (20°C), which made it possible to keep the first power cylinder hot (100°C).

The next stage of making a full-scale steam engine required much more capital, and Watt also needed precision machining of the piston and the cylinder. It was not enough to come up with a great invention, as he required sources of finance for the long and expensive path of development, manufacturing, and marketing. In addition, the system of patent protection for inventors was in the early days, and required an Act of Parliament. His early backer, John Roebuck, went bankrupt forcing Watt to find other employments to pay his bills. He finally found another partner in Matthew Boulton, and they formed the partnership of Boulton & Watt that lasted 25 years.

In the year 1776, his first commercial engines were installed and worked for pumps, mostly in Cornwall for pumping water out of mines. The market for his engines was greatly increased when Watt converted the reciprocating motion of the piston by a system of mechanical linkages, to produce rotation power for grinding, weaving, and transportation. He also introduced a double acting engine, shown in Figure 2.9, in which the steam entered from the left in the upper frame to push the cylinder to the right, and later from the right in the lower frame to push the cylinder

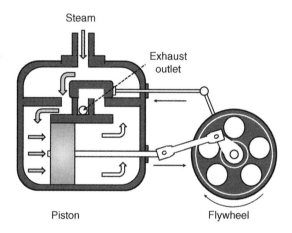

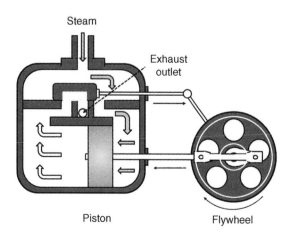

FIGURE 2.9 Watt's double-acting steam engine.

to the left. Note that there are three key advantages of this design: each stroke is a power stroke, which doubles the horsepower; the steam is the source of power and can be much higher than one atmospheric pressure; and the back-and-forth motion has been converted to a rotary motion suitable for many more applications.

Hotter steam provides higher power. If we substitute a 1 horsepower single-acting piston with steam at 100°C and 1.0 atmosphere pressure, with a double-acting piston with the same volume and steam at 150°C and a pressure of 5.4 atmospheres, it becomes a 10 horsepower engine. If we were to increase the steam temperature to 200 or 300°C, the steam pressure becomes 15.3 and 86.8 atmospheres, and the engine with the same volume becomes 30 and 173 horsepower! Of course, higher temperature and pressure also bring with them a higher danger of explosion, especially in an era before the development of high-strength steel, tight valves, and cylinders. The early Industrial Revolution witnessed a number of explosions that created a great deal of public fear of this dangerous venture into high-powered machinery.

Watt also invented a way to control his engine from running too fast by a throttle valve and a centrifugal governor. Many other improvements were made through the years, and he was most proud of the parallel motion/three-bar linkage patented in 1784. The sum of all these improvements produced an engine that was up to five times as efficient in its use of fuel and many times more powerful than the Newcomen engine.

SCIENCE AND TECHNOLOGY: CARNOT EFFICIENCY

Let the amount of heat going into a heat engine be called Q_H, and the work produced be called W. In general, W is less than hot energy Q_H, so the remaining colder energy is discharged and called Q_C. The First Law of Thermodynamics specifies that $Q_H = W + Q_C$, or that the energy is conserved. Later people would say that this law shows that "There is no free lunch."

In a heat engine, Q_H is delivered at a steam temperature of T_H, and the Q_C is discharged at a cooling water temperature of T_C. The famous Carnot rule is based on the absolute temperature on the Kelvin scale, where degrees Kelvin equal degrees Centigrade plus 273.1. The Carnot rule states that $W/Q_H \leq 1 - (T_C/T_H)$ which is to say that "You cannot break even." So if the steam has a temperature of 150°C, and the cooling water has a temperature of 20°C, then the maximum efficiency is $1 - (20 + 273)/(150 + 273) = 0.307$ or 30.7%. This is the maximum efficiency, and you usually do worse since you have leaks, heat losses, and frictions. You can increase the efficiency by making T_C lower by finding a source of colder cooling water; or you can make the steam hotter by making T_H higher, which also means a higher pressure.

In retrospect, James Watt was not the inventor of the steam engine. By a series of critical improvements, he made the steam engine more than five times as efficient as the previous versions, and made it popular with many mine operators and industrial manufacturers. He has been justly credited with starting the Industrial Revolution.

The steam engine had an enormous impact on modern life—perhaps the most profound impact since the invention of agriculture. The successful thermal engine converted heat energy from combustion into mechanical energy and gave us nearly unlimited power. The textile mills in England made tremendous improvements in manufacturing speed and efficiency, and their productivity made European and Asian handlooms obsolete. The industrialized nations became more rich and powerful being the workshop of manufactured goods for the world; the rest of the world were given the roles of customers of manufactured goods and providers of raw material, and became intimidated and dominated.

Within the island of Britain, manufacturing became a much more productive source of employment than traditional agriculture and animal stock keeping, which led to a mass migration of underemployed farm workers to the cities looking for better-paying jobs. The population of London grew rapidly, from 1.3 million in 1825 to 6.5 million by 1900. The steam engine also led to a revolution in transportation, including the navigation of rivers by steamboat and the connection of continents by railroads. A trip from New York to Philadelphia used to take 2 days by horse-drawn coach; on a train, it took only 2 h. Oceangoing steam ships cut the trip from London to New York from a month to 3 days. Rapid transit could now deliver heavy freight at a fraction of the cost of animal power, which lowered the regional differences in the cost and availability of goods, and diminished barriers in languages and cultures. The Trans-Siberian Railroad played a major role in the Russian conquest of Asia, and made possible the rapid movement of troops from Moscow to Alaska.

The following decades saw many advances in the heat engine that produced power with a much smaller volume and weight, which was critical in transportation. Instead of a boiler to generate steam, and a separate cylinder and piston to generate thrust, the internal combustion engines combined the two volumes into one, such as the gasoline and diesel engines. The invention of turbines replaced the back-and-forth motion of the cylinders with the smooth rotation of wheels with blades. Since higher temperatures are needed for higher Carnot efficiency, modern turbines use very high temperature alloys and ceramic parts to withstand up to 1400°C. The coming of nuclear power created its own bright advantages and dark menaces.

For all of its virtues, it must be said that the Industrial Revolution also devastated the lives of millions. As Charles Dickens made famous, the housing in London and Manchester was crowded and dangerous, the air was polluted by dark smoke, and the sun was seldom seen. William Blake called them the "Dark Satanic Mills." In the countryside, the demands of fuel created Welsh coal mines with appallingly low standards of safety and sanitation. Today we have a deep concern with global warming, which has been linked to the effects of burning fossil fuels and releasing carbon dioxide into our environment. The steam engine and the Industrial Revolution started a path that changed our lives and our world.

2.2.3 Electric Power

Power generated by heat has many drawbacks. The steam and internal combustion engines that powered the Industrial Revolution had to sit next to the mills and pumps that they powered. All the heat engines, from Watt's steam engine to its successors, the gasoline and jet engines, are smoky and noisy and require stores of fuel nearby. In the nineteenth century, a series of inventors recognized the need for small, quiet, and clean engines that could start and stop instantaneously, and could be powered by a remote energy source without a local coal pile or a can of gasoline. They found the answer in electric power, starting with what is sometimes called the Second Industrial Revolution.

Electricity's ability to be transmitted over long distances made it possible to have a separate location for the power generation and its users. Today, power from Niagara Falls, fossil fuel generation stations, and nuclear reactors can be transmitted to every home, office, and factory. We can generate power in remote locations far from the elegant homes and businesses in densely built cities. Electricity was originally intended to provide lights for homes and businesses, but soon took over many other chores such as heating and cooling, ventilating, cleaning, communications, and entertainment. Today we live in homes filled with devices that are powered by electricity.

An electrical current is simply the passage of free electrons through a conductor, typically a metal wire. Electricity can be generated in nature in the form of lightning, and by the rubbing of amber to create static electricity. But lightning is too unpredictable, and static electricity is too feeble for use in practical applications. The creation of electricity by chemical action was discovered by Alessandro Volta around 1800, when he invented the chemical battery, with a positive and a negative electrode immersed in a chemical solution. While effective and practical, a battery can only produce a small current, and the voltage gradually drops over time as the metal electrodes are consumed by the chemical reaction.

In 1831, the American scientist Joseph Henry used one of these batteries to generate an electrical current that could turn a wheel in a magnetic field, which yielded the first electric motor. Fortunately the process can also be reversed, and Michael Faraday invented the dynamo or electric generator, where a copper disk rotating between the poles of a magnet generated electricity by the application of a mechanical force. Werner Siemens and Charles Wheatstone independently and simultaneously announced the first practical design for a dynamo in 1867. When powered by a steam engine, a dynamo generates a steady stream of electricity that can be used for such applications as lighting and propelling vehicles.

A dynamo generates direct current or DC electricity, where the current always flows from a positive to a negative electrode. A second kind of generator, called an alternator, generates alternating current or AC electricity, where the current flows alternately between two electrodes in rapid alternations, such as 50 or 60 cycles per second. Although both AC and DC currents produce flows of electrons, they behave very differently in motors and electrical distribution. Thomas Edison, perhaps the greatest American inventor, championed DC. This would ultimately prove to be the wrong choice in what became known as "the War of the Currents," which was over whose outcome would shape the technology and the world of the future.

BIOGRAPHY: EDISON AND TESLA

Thomas A. Edison (1847–1931) is one of the best known of the American inventors, and is often called *The Wizard of Menlo Park* as he built the first industrial research laboratory in Menlo Park, New Jersey, with a large team of workers who engaged in inventions

FIGURE 2.10 Thomas A. Edison.

on a full-time basis. His picture is shown in Figure 2.10. He had 1,093 US patents in his name, particularly in telecommunications, and in electric power generation and distribution. He was born in Ohio, and grew up in Michigan. He was not a good student in school, and was homeschooled by his mother. He sold candy and newspapers on trains, and considered himself a practical man instead of a scholar. His method of discovery is often described as "Edisonian," based on numerous empirical trials instead of calculations and predictions based on scientific understanding.

He first gained fame for inventing the phonograph in 1877, involving a tinfoil wrapped around a grooved cylinder that was cut by a needle connected to a microphone. However, the recording had poor sound quality and was also difficult to mass reproduce and store. The eventual winner in the sound recording market was the gramophone disk invented by Emile Berliner popularized in the Victor Talking Machine.

After the "War of Currents" with Nikola Tesla, Edison went on to do other research, lived very comfortably at his winter retreat in Fort Myers, Florida, and became a good friend of Henry Ford of Ford Motor Company. He died at his estate in West Orange, New Jersey in 1931 at the age of 84, from diabetes.

Nikola Tesla was born in 1856 in the former Yugoslavia, of Serbian parents. His picture is shown in Figure 2.11. His father was a pastor, and his mother's father was also a pastor of the Serbian Orthodox Church. He was tall and slender, standing at 6 ft 2 in. (1.88 m), with large hands. He always tried to dress elegantly, and gave an impression of

FIGURE 2.11 Nikola Tesla.

culture and learning. He was a life-long bachelor and never showed any interest in women or family.

Tesla returned to research on many wide-ranging and ambitious projects, including: wireless communications, vibrations, radio-controlled rockets, and even a death-ray. He always lived in first rate New York hotels, dressed impeccably in tails while dining alone at the Waldorf-Astoria, and liked to feed pigeons. In 1912, it was rumored that he and Edison would share the Nobel Prize, which did not take place. It is ironic that in 1917, he was given the Edison Medal of the American Institute of Electrical Engineers. In 1943, he died impoverished and alone in his hotel.

Edison built the Menlo Park research laboratory in 1876–1881, and worked on many research projects at the same time. One of his most important projects was the invention of the incandescent electric light, and the key element was the filament that had to be heated to red-hot for many hours without burning out. He evacuated the bulb so that there was no oxygen around the filament. He experimented with thousands of substances, and in 1879 he had a carbon filament that could last 40 h. Subsequently, his most successful product was a carbonized bamboo filament, which could last 1200 h.

The electric light has many advantages over a candle or an oil lamp: it has tremendous brightness, the ability to be turned on instantly, and lacks smoke and hazard of fire. Its main disadvantage is that it requires a steady supply of electricity. Many problems hampered the lighting system that Edison installed in the Madison Avenue mansion of the financier J. Pierpont Morgan, who was one of his early backers. Edison installed a coal-powered steam boiler to power a dynamo that had to be run by an experienced engineer. The engineer would arrive at 3 p.m. to generate sufficient steam to have the lights on by 4 p.m. Guests arriving for cocktails and dinner were duly impressed by the wonderful illumination, but at 11 p.m. they had to revert to candles and kerosene lamps when the engineer left. The boiler and generator also clanked and throbbed, belted out noxious smoke and fumes, and produced vibrations felt throughout the entire neighborhood. The Morgan demonstration, meant to show off the wonders of electric lighting, made it clear instead that generating electricity would have to be done remotely and at a greater distance from homes and offices.

Edison patented a system for electricity distribution of DC current from a central generating station in 1880. In 1882, he founded the Edison Illuminating Company and built the first electric utility on Pearl Street in New York City, just half a mile from the New York Stock Exchange and Wall Street. He had a steam engine powering a dynamo to produce 100 kW of electricity, sufficient for 1200 lamps for homes and businesses. Edison chose the low voltage of 100 V, to reduce the risk of accidental fire and shock, but he then faced a distribution problem.

SCIENCE AND TECHNOLOGY: DC AND AC POWER

The power delivered by an electric cable is controlled by the equation $P = VI$, where the pressure or voltage V is measured by the unit of volt, and the current I is measured by the ampere. Thus, a 100 W light bulb can be supplied by 100 V at the low current of 1 A, or by 1 V with a high current of 100 A. It is safer to have low voltage in the house, as a high voltage can lead to a fire or electrocution. However, electric transmission along a cable suffers from line losses equal to $Q = I^2 R$, where R is the electrical resistance of the cable. The 100 V from the generating station will drop to 90 V after it travels for 1 km on its way to distant customers, who would obtain less power than nearer customers. The line loss depends only on the current I and not on the voltage V. Therefore, it would be advantageous to have 1000 V from the generating plant, since a drop to 990 V after 1 km does not cause too much of a problem. The associated advantage of high voltage is the current needed at 1000 V is only 0.1 A so that the line loss is only 1% of 100 V power.

The voltage of AC power can be boosted to a hundred thousand volts for transmission, and stepped down to 110 V just before it enters the house, by the use of a transformer. The problem with DC power is that there are no transformers, so it has to be transmitted and consumed at home at the same 110 V. Therefore, we have to use low voltage in transmitting DC power, and cannot have long distance transmission.

In DC electricity, the voltage cannot be changed easily so the power must be transmitted and consumed at the same low voltage. A low voltage therefore requires a high current and a dramatic loss in power over a relatively short distance. Edison's customers had to live within a radius of half a mile (0.8 km) from the central station. Only the central business districts of major cities have a high enough density of customers in such a small radius from the generating station. A major customer such as a big factory could afford a local dedicated generator, but the small and scattered city and suburban customers had to rely on the construction of many smaller generating stations. At this point, alternating current AC made a dramatic entry.

The principle of the AC transformer was discovered simultaneously in 1831 by Michael Faraday and Joseph Henry. A transformer has two sets of wires wrapped around an iron ring, with more turns in the primary set of wires than in the secondary set of wires. When AC current runs through the primary coil, it induces AC current in the secondary coil, and the ratio of the primary to secondary voltage is precisely the same as the ratio of the number of primary turns to the number of secondary turns. In a system of AC power, the voltage can be easily stepped up by a transformer at a generation station to several hundred thousand volts for transmission over long-distance cables. The voltage can then be stepped down again by another transformer to hundreds of volts at a local substation prior to distribution to homes and businesses. AC power is therefore safer to use as well as offering the possibility of long-distance transmission with low line loss. But there was a catch: AC motors were not yet perfected at the time. This would change when Nikola Tesla was able to develop his 1880 concept of an AC motor in Budapest, followed by his demonstration in New York in 1887.

When Nikola Tesla was studying engineering at Graz in Austria, he was shown an electric machine, which could be used as either a DC dynamo or a motor. He noticed that the commutator, which is a necessary part for DC motors, sparked inefficiently. This inspired him to think about how to design an alternating current AC motor. In 1880, he went to work in Budapest in a telephone office, and conceived of a method to generate a rotating magnetic field without any moving parts, which would form the basis for an AC motor without commutators. It worked with two or three electric coils wrapped around a ring core, supplied with the same AC electricity; however, the coils were not in phase, so that when the voltage in the first coil was at the maximum, the second coil could be at the minimum. Thus, the coils would take turns in becoming the north magnetic pole, which created a revolving magnetic field. This was an original and unprecedented idea, but he had neither the time nor the means to build and test such a motor.

In 1882, he went to Paris and worked at the Continental Edison company where he had the opportunity to build and demonstrate his first two-phase AC motor in the city of Strasbourg. At the advice of his manager Charles Batchellor, he went to the United States in 1884 and obtained a position with Thomas Edison. Tesla offered the AC technology to Edison, who rejected it because he was already totally dedicated to promoting DC power. Tesla resigned after a year with Edison, and was reduced to digging ditches to pay for his bread. In 1887, he obtained some financial support and was finally able to produce a three-phase AC motor, gave a famous lecture at the Institute of Electrical Engineers, and received a patent for this invention. The AC induction motor that he invented worked by a rotating magnetic field driven by two circuits with the same frequency but out of phase.

George Westinghouse was a famous industrialist, known for inventing and promoting air brakes for trains. Westinghouse bought the AC patents from Tesla for 1 million dollars, in addition to 1 dollar per horsepower of each motor. Westinghouse decided to enter the same electrical light and power market as Edison, but with the AC rather than the DC system. Edison fought Westinghouse for many years, primarily on the basis of safety. But gradually the users, particularly in rural areas, chose the efficiency and the long reach of the AC system. Eventually the distribution systems of different power companies were merged together into a single AC power grid from coast to coast, tying together power generated by different sources including hydroelectric dams, combustion generators with fossil fuels or biomass, nuclear reactors, wind farms, and geothermal sources.

Reliable electric power is an incredibly versatile energy source. In the modern hospital, it powers the lights as well as the myriad of diagnostic, monitoring, and life-saving machinery. Theaters and sports stadiums all depend on electricity for powerful lights that pose neither fire nor smoke hazards, and to power the sound equipment and ventilation. Today, our modern world consumes a staggering quantity of energy, most of which goes toward combustion engines and the generation of electricity. In the year 2007 alone, the sum of supplies from coal, oil, natural gas, nuclear, hydro, geothermal, solar, and renewable biomass and wastes was equivalent to 11,400 million tons of oil. When we convert these raw sources into useful forms, such as gasoline and electricity, and then transport them by trucks and electric lines, the energy available for final consumption is reduced to 7,910 million tons of oil equivalent—only 69% of the primary source energy. The biggest loss is in the generation of electricity, which is seldom more than 30% efficient. Over 80% of our global energy comes from fossil fuels (coal, oil, and natural gas) whose combustion releases carbon dioxide and other greenhouse gases, and therefore, contributes to global warming. Nuclear energy is a relatively clean, high-technology energy source, but it can have accidental radiation leaks as in Three Mile Island and Chernobyl, as well as the 2011 Japan earthquake. It also creates nuclear waste and presents the danger of bomb production in the hands of unstable states and terrorists. Natural sources such as hydroelectric, geothermal, and solar power are safer and cleaner solutions, but they are expensive and require special locations. Future generations have inherited a society utterly dependent on mechanical and electrical energy, so the inventors of tomorrow must continue to look for ways to generate cleaner and more efficient power that would increase our productivity without destroying our environment.

2.3 MATERIALS

The performance of our tools, devices, and structures depends critically on the materials they are made of. Materials have been so important in understanding human history that they have been used to identify the prehistoric eras: the Stone Age, the Bronze Age, and the Iron Age. Our ancestors first used natural materials—stone, wood, bones, horns, shells, and plant fibers—to make their tools, weapons, clothes, and homes. Much later, humans began inventing new materials with properties that were very different from, and in some ways far superior to natural materials. Ceramics and metals proved particularly useful for their durability, heat resistance, malleability, and strength. Synthetic polymers arrived less than 300 years ago, and were originally motivated by the need for low-priced materials to supplement the exhaustion of natural materials.

Leonardo da Vinci designed a human-powered flying machine, requiring strong but lightweight materials which did not yet exist. In 1979, Paul MacCready built the Gossamer Albatross to fly across the English Channel. It was constructed with a carbon fiber frame wrapped in a plastic Mylar film, and weighed only 32 kg.

Contemporary high-tech materials can be very superior, combining different substances with complementary properties, and have become essential for the success of computer displays, space capsule tiles, and artificial organs. Archeologists of the future, surveying our record of space-age materials, might justifiably call our era the Designer Materials Age.

2.3.1 Ceramics, Metals

Ceramics are probably the earliest man-made material. The biggest uses of ceramic materials are in construction materials and household goods, and more recently for high technology space age applications. Mud and clay, the precursor to ceramics, have long been used for walls and buildings, but they are vulnerable to erosion by rain and water. Bricks have a history dating from 7500 BCE, made by mixing clay with water and sand to the desired consistency, and then pressing into molds to dry in the sun. Bricks are much better than clay at resisting compression but are vulnerable to tension or shear. Their construction was important enough that even the Bible discussed the reinforcement of bricks with straw to strengthen the product.

With improvements in kilns some 4000 years later, superior kiln dried bricks were made that could withstand more weight and resist erosion from water. In construction, masons lay down the bricks in rows and bind them together with mortar, which is made from water, sand, and a binder such as lime or gypsum. Masonry, one of the oldest professions, was held in high regard in both the ancient and the medieval world. The fraternal organization of the Freemasons uses the square and the compass as their symbol, with a distinguished list of members that included George Washington.

Cement is also a ceramic material used in construction. The Romans prepared dry cement with crushed rock, volcanic ash, and burnt lime as a binder; subsequently cement was used to make mortar and concrete. Modern hydraulic cement, known as Portland cement, sets after mixing with water. The manufacturing of Portland cement begins by burning limestone (calcium carbonate) with clay or shale containing silica in a 1450°C rotary cement kiln to drive off carbon dioxide and produce calcium silicates. The rotary kiln can be 60 m long, and is set at a slight angle so that solids can slowly slide down from the high inlet end to the low outlet end, while air and fuel enter from the low end and flow in the opposite direction. The rotary kilns run 24 h a day, typically stopping only once a year for maintenance. The resulting "clinker" is discharged from the low end and then ground up with a small amount of gypsum (calcium sulfate) to make ordinary Portland cement, which is used to make mortar, grout, and concrete. Cement kilns are among the most energy-intensive processes in use today, and generate a tremendous quantity of carbon dioxide, soot, and dust.

Finer ceramics are used as kitchen utensils, to store oil and wine, and to cook and serve food. The making of fine ceramics requires a complex set of skills and tools, including the judicious blending of minerals, and the building and tending of fires. An enclosed insulated kiln was introduced to make the heating more even, to decrease heat loss and to increase the temperature; the introduction of bellows and charcoal made the fire burn even hotter. These kilns may have originated in the

Middle East around 6500 BCE, as indicated by the age of the oldest burnt clay objects that have been discovered.

As the kiln temperature increases, ceramics grow stronger. Earthenware are made of mixtures of clay, kaolin, quartz, and feldspar, and are fired in a kiln at 800–1200°C. They are less strong and have porous surfaces. Stoneware fired from 1100 to 1300°C is stronger. Surface porosity can be reduced by applying a glass-like glaze, which is made of high melting silicon and aluminum oxides, then added to a mixture of sodium and calcium oxides to act as a flux that melts at a lower temperature.

Porcelain fired at 1300–1400°C is stronger still and less porous. Early porcelain began in China around 1600 BCE, and reached a high level of excellence during the Tang Dynasty (618–906), and was a main article for trade on the Silk Route. The secret of porcelain manufacturing was discovered in Europe around 1700, which led to the Meissen and other works.

Most modern glass is based on ordinary sand and the mineral silicon oxide, which has a very high melting point of 1700°C. The melting point of glass is lowered by mixing it with soda and lime; this molten material is cooled quickly to form a noncrystalline solid. The most important glass found in common windowpanes and bottles is soda–lime glass, made with a mixture such as 16% soda and 10% lime, which lowers the melting point to a manageable 700°C. Glassmakers from ancient times to the present dope their glass mixture with other elements to create attractive tints and dazzling colors. Optical glass, such as that used for eyeglasses, binoculars, or luxury Waterford and Swarovski glassware, has as much as 35% lead oxide to create a higher refraction index that bends light more strongly into rainbows. Some wine and whiskey decanters contain so much lead that they have been accused of causing health problems such as gout.

High-temperature ceramics are used in the most demanding applications that are subject to high temperatures and erosion, such as the tips of turbine blades in jet aircrafts and electric power generation stations, the nose cones of rockets, and the heat shield panels of space shuttles. Flak jackets use ceramic plates to resist bullets, and glass fibers form the core of the global communications network in the form of fiber optics.

The Stone Age faded away with the arrival of the Bronze Age. When ancient man first turned to metals, he used what could be found in nature in elemental form: copper, silver, gold, and meteoric iron. However, elemental metals are usually found in quantities too small to make a major impact on the quality of life. In addition, although native copper flakes can be hammered together to make ornaments, it is too soft for tools and weapons.

The major advance in the technology of metalworking was the invention of smelting, which frees the metals from the much more abundant ores. Metallic ores are found in many places in the form of oxides or sulfides, such as the copper ore cuprites (copper oxide), chalcopyrite (copper iron sulfide), and the much more abundant iron ores of hematite and magnetite (forms of iron oxide). The metals are tightly bound to oxygen or sulfur in these ores, and the separation requires a lot of energy.

The earliest discovery of smelting was most likely accidental, involving lead and tin ores in wood fires sufficiently hot for these low melting metals. Evidence of smelting has been dated back as early as 6500 BCE in Catalhuyuk of Anatolia. The mastery of bronze and iron required advances in the art of making hotter fires in furnaces and crucibles that may have originated in the Middle East for the purpose of making ceramic pottery.

The Bronze Age began around 3300 BCE in the Near East, initially using copper alloyed with its natural impurity—arsenic, and much later by deliberately adding tin to copper. The production of bronze objects required careful planning, as copper and tin ores were seldom found in the same place. The bronze maker had to import these ores over great distances by trade and transport. Sometime around 3000 BCE it was discovered that mixing copper with tin over a hot kiln fire created molten bronze at 1100°C. The liquid could then be poured into molds to make tools and weapons and then hammered on a forge over a hot fire to harden and achieve the final shapes. The Bronze Age was at its zenith around 1100 BCE, around the time that *The Iliad* and *The Odyssey* were written. Bronze is an alloy of copper mixed with tin—the best bronze has about 10–15% tin. Bronze is much harder than copper, and highly prized for weapons and ornaments. Bronze has many advantages in comparison with stone, as it dents and bends after a heavy blow but does not shatter; blunted blades can be resharpened, and a broken blade can be reforged. It is excellent for making sculptures, as it expands slightly just before it sets, thus filling in the finest details of a mold. It also makes excellent ornamental and ritual vessels for food and drinks, as well as musical instruments such as bells. It oxidizes only superficially, and the thin oxide layer protects the underlying metal from further corrosion. Bronze tripods and cups were used in the Shang dynasty in China, for religious ritual and royal banquets, which were distinct from the ceramic bowls for everyday use.

Iron was known to appear in elemental form in meteors since a very early age—the Egyptians made weapons from meteoric iron as early as 3000 BCE. The Sumerian and Hittite name for iron meant "fire from heaven," and both Attila and Timur had swords from heaven which gave them a mystic aura and psychological advantages. Iron has the disadvantage of quick oxidation leading to rust, in comparison with the more stable bronze.

A more reliable source of iron is in the form of the abundant ores of hematite, magnetite, and pyrite. The mastery of iron required the separation of iron from its ores, and then melting the iron to pour into molds; this required extremely high heat of around 1500°C, which could not be achieved without insulated kilns and bellows. The Iron Age began in the Near East, Iran, and India around 1200 BCE. Despite the need for the more demanding high temperature kilns, iron eventually replaced bronze as the primary metal for weapons and tools, because iron is harder than bronze and iron ores can be found almost everywhere.

Elemental iron is quite soft and unable to keep an edge. What we think of as useful "iron" is therefore an alloy containing some percentage of other elements, typically carbon. Since iron ore is smelted over charcoal, and charcoal is nearly pure carbon, all forms of refined iron contain at least some carbon. The simplest and oldest form of iron is known as "pig iron," due to the resemblance of the molding configuration to nursing piglets.

Pig iron can be exceedingly brittle, as it contains significant (up to 5%) quantities of carbon, and undesirable impurities of silicon, sulfur, and phosphorous.

When pig iron is remelted in the presence of air, much of the phosphorous and sulfur burns off, producing the more useful product known as cast iron. When the carbon content of cast iron is reduced to a very low level of 0.5% by further heating and burning in air, then removing the silica as slag, the final product is wrought iron—soft and ductile but too soft to keep an edge. All of these alloys offer some advantages over bronze, but none have the versatility of steel.

Steel is a form of refined iron that contains about 1–2% carbon. It can be forged, hammered, tempered, and annealed to become harder and more ductile. We use steel today in everything from weapons, tools, bridges, and buildings, to pots and pans. The most prized steel swords in history were manufactured in Damascus in Syria, Toledo in Spain, Solingen in Germany, and Sheffield in England. To this day, the steel Japanese samurai sword has few equals in sharpness and flexibility.

Swords need hardness to grind into and maintain a keen edge, as well as requiring flexibility to avoid shattering upon impact. For a sword to have the necessary hardness at the surface with flexibility at the core, it needs a higher carbon content on the cutting edge and lower carbon content in the core. Swords were heated to red hot in a charcoal fire, then hammered, and then rapidly cooled by quenching in water, to become harder and more brittle in a form called martensite. The opposite manufacturing technique is tempering, which involved allowing the hot forged steel to cool slowly thus yielding a less brittle form known as austenite. Hand forging is a technology sufficient to produce a small supply of steel for aristocratic weapons, but not for producing the large volume needed for buildings and structures.

The modern industrial use of steel traces its history to an invention in 1855 of Sir Henry Bessemer, a British industrialist who created a crucible to reduce the carbon content in molten iron by a blast of hot air. This was a very fast method to make steel, which made it economical and widely used for low cost applications such as nails and beams for construction. Since then, the process has been replaced, first by the more advanced Siemens open hearth process in 1865, and then by the Basic Oxygen Process in 1952. Engineers continue to tinker with the composition of steel, sometimes adding materials such as nickel or manganese to increase its strength. The familiar stainless steel of pots and pans is an alloy with 18% chromium and 10% nickel added to inhibit corrosion. In 1977, US consumers and manufacturers used 131 million tons of steel, which averages to 397 kg of steel per person per year.

SCIENCE AND TECHNOLOGY: MATERIAL SELECTION

The most important properties of materials are usually their density, mechanical strength, melting point, and durability. The mechanical strength of material is measured by the stress–strain curve, shown in Figure 2.5. When a rod is stretched, the stress is measured by the force per area, and the strain is the percent elongation of the rod. A brittle material such as a rock would elongate very little, and would suddenly break at the fracture. A tough material such as steel would elongate much more and recover when the stress is removed, till the elastic limit when the elongation can become permanent, and then break at the ultimate strength. The areas under these two curves are the energies required to break the materials. So a material can be considered "strong" with two different meanings, one is high resistance to elongation like a rock, and the other is high energy required to break like steel.

Besides mechanical properties, materials have other properties that led to their use in critical technologies.

- *Electrical Conductivity.* Silver and copper have the highest electrical conductivity, silicon and gallium–arsenide are semiconductors used for signal, information, displays, and computers. Superconductors have essentially no electrical resistance when cooled below a critical temperature.
- *Heat Resistance.* Nickel–chromium alloys and ceramics have very high melting points, and are used for turbine blades and for heat shields for space capsules.
- *Adhesion.* Superglues are used to glue airplanes together, Teflon does not adhere to anything and is used for frying pans.
- *Magnetic Properties.* Magnetized compass is used for navigation, iron oxide is used to store information on tapes and computers, magnetized rails are used for magnetic levitation for rapid trains.
- *Optical Properties.* High refractive index glasses and polymers with optical clarity are used for lenses for eye glasses, cameras, telescopes, microscopes; optical fibers are used for transmission of information, and for light and sight in organ surgery; LED or light emitting diodes are used for computer and television displays and for lighting; solar cells are used to convert sunlight into electricity.
- *Biological Properties.* Titanium and ceramics are biocompatible with human tissue and used in biomedical devices to replace hip and knee joints, and in implanted heart pacers; biodegradable polymers are used for the controlled release of drugs, and for the surgical suture or stitches that do not have to be removed.

Subsequent developments in metallurgy continued to explore new methods to extract metals from their ores, and to convert them into alloys with useful properties. The Hall–Heroult process for making aluminum relied on electrolysis rather than high heat to remove oxygen from aluminum oxide, which was a major advance that transformed aluminum from one of the most expensive metals to one of the most common. Titanium is one of the best space age metals which is as strong as steel but much lighter. Today, in their search for new metal technologies, materials scientists are extending the legacy of the alchemists who searched, unsuccessfully, for ways to turn lead into gold.

2.3.2 Polymers, Composite

Early in human history, all materials were natural materials, with the later addition of simple ceramics and metals. Since the 1700s, inventors have introduced many more advanced materials, including composite materials that combine the advantages of different materials. In the twentieth century, we have witnessed a true materials revolution with an explosion of organic substances made from nothing but air, water, coal tar, and petroleum.

Many natural and manmade materials are polymers, whose structure consists of joining together numerous small units that are called "monomers." Plant fibers such as cotton are polymers of sugar units; proteins, such as animal tendons, are polymers of amino acids; and the genetic material of DNA is a polymer of nucleic acids.

Rubber is a natural polymer with a particularly long and colorful history. Christopher Columbus and his companions noticed that Native Americans used rubber from tree sap to form balls that they used in sports. They were produced by making cuts into the bark of the rubber trees, collecting the sticky milk-colored latex sap, and refining into rubber. These balls were adequate for games, but natural rubber was less satisfactory for industrial and everyday applications, as it became brittle in cold weather and tacky and smelly in warm weather.

In 1839, Charles Goodyear invented vulcanized rubber by adding sulfur to rubber latex and heating it in an oven, with a resulting product that stays elastic in all kinds of weather. Subsequently, enormous plantations were established in Brazil, Ceylon, Indonesia, and British Malaya to produce rubber. The most important use of rubber is for the tires of automobiles as well as for airplane landing gear. During World War II, when Japan invaded British Malaya and cut off access to the rubber plantations, a crash program was developed in the United States to produce synthetic rubber from petroleum oil and gas.

Ivory from elephant tusks was once used to make billiard balls, piano keys, and combs, but by the mid-1800s, it was becoming scarce and expensive. The path to develop a substitute for ivory took several decades. In 1846, Christian Schönbein discovered that immersing cotton into a mixture of sulfuric and nitric acids created a new product called nitrocellulose, which became an explosive also known as guncotton. It could also be dissolved in a solvent, and used to make lacquer and films. Cellulose nitrate derived from wood also had the potential to mold objects, but it was highly flammable and difficult to mold. In 1845, Alexander Parkes made the product flexible by adding camphor, but his version of cellulose nitrate burst into flames when touched by a cigarette. Finally, John Hyatt invented a better method in 1870, when he molded nitrocellulose, camphor, and alcohol, and heated the mixture under pressure to make a product that was allowed to harden under normal pressure. Hyatt's celluloid became a great success used in a wide variety of products, and slowed down the slaughter of the elephants. It also arrived in time for the mass production of films for photography by Eastman Kodak, and for motion pictures by Thomas Edison.

In contrast to celluloid, Bakelite is a totally synthetic product that is not based on an existing product, but made from raw materials of coal, petroleum, water, and air. In 1910, Leo Baekeland an American immigrant from Belgium, introduced a product he called Bakelite. Made of phenol and formaldehyde from coal tar or petroleum, Bakelite was a thermosetting polymer that did not melt once set. Enormously popular in the years immediately following its introduction, Bakelite found its way into everything from billiard balls and jewelry to transistor radios. It is still used today for electrical outlets. Bakelite, the first fully synthetic material, opened the door to a string of synthetic polymers that would transform the way we live and work: Plexiglas (Rohm and Hass, 1928); polystyrene (Dow, 1937); nylon (DuPont, 1939); polyester (DuPont, 1941); and Teflon (DuPont, 1946) are only a few of the most famous of these plastic products.

Modern composite materials combine the properties of different materials for better performance. Nature is full of composites, such as granite, which is a conglomerate of three different crystals: quartz, mica, and schist. Bricks are often composites made of clay reinforced with straw, as the clay resists compression and the fibers in straw provide resistance to tension. Similarly, concrete beams alone cannot be used to span large distances because they sag in the middle and fracture at the bottom, but they can be reinforced with steel bars at the bottom that resist tension, and enable the spanning of longer distances.

Biomaterials are one of the most valuable forms of composites being explored today. Artificial hip and knee joints are constructed with alloy stems and heads that rotate in polymer cups, and are used to replace joints damaged by arthritis. They are designed to avoid immune response, toxicity, inflammation, and rejection. Other biomaterial replacements include pacemakers for the heart, stents to open arteries clogged by atherosclerosis, artery and heart valves, and dental implants. Bio-degradable polymers are designed to dissolve gradually, and are used in surgical sutures, eliminating the need for another trip to the doctor to remove the stitches. Such polymers are also used in controlled release of drugs to produce a steady dose into the bloodstream without needing a pill or an injection every few hours.

Aerospace vehicles encounter the most extreme conditions of temperature and stress, and demand the most advanced materials. The tips of the turbine blades for GE and Rolls Royce jet engines reach the highest sustained temperatures, and are usually made of heat-resistant alloys and ceramic components. The space re-entry capsules need thermal protection systems (TPS) for short times when they first enter earth's atmosphere, and they have ablative layers that will char, melt, and sublime through pyrolysis, to divert the highest heat away from the skin of spacecraft.

There are many materials with highly specialized properties that become indispensable for special uses. Some of these materials, known as "smart materials," can even change their properties in response to external changes in environmental stress, temperature, moisture, pH, electric or magnetic fields. Chromogenic materials are used in smart sunglass lenses that are light-colored indoors, but turn dark under sunlight. Chromogenic materials are also used to make the liquid crystal displays (LCD) used in televisions and computers that change color under different electrical or magnetic signals. Non-Newtonian polymer fluids change their viscosities in response to force or pressure. Drag reduction polymers are used to reduce the drag of submarines that need a burst of speed, and for oil pipelines to reduce the cost of pumping.

A design engineer for a given application looks for an available material that has a reasonable cost, many desirable properties, few undesirable ones, and a good balance of cost and effectiveness. Materials such as metals vary widely in their price, from less than \$0.40/kg for scrap iron and steel, to \$1600/troy ounce for platinum (equivalent to \$51,000/kg).

The human propensity to use large quantities of materials for applications also makes a tremendous impact on the environment, from mining to manufacturing, and from initial use to disposal. Mining can lead to erosion and disfiguring of the landscape, and contamination of soil and water. Copper and tin mines have produced collapsing tunnels, acid discharge into lakes and rivers, and slag hills. Municipal wastes are loaded with glass, cans, and plastic items. Some of this waste can degrade in nature, but others will persist for a long time in the environment. These problems were not so significant when the human population was small, and when the rate of consumption was modest. But when the population increased and began living in ever-greater wealth, such accumulation of waste became a tremendous problem. Just as the inventors of the past century have constantly looked for new materials with superior properties in performance, the inventors of tomorrow must constantly be on the lookout for materials with biodegradability.

REFERENCES

Alexander, J. W. "Economic Geography", Prentice-Hall, Englewood Cliffs, NJ, 1963.

Ball, P. "Made to Measure: New Materials for the 21st Century", Princeton University Press, Princeton, NJ, 1997.

Billington, D. "The Innovators", John Wiley, New York, 1996.

Billmeyer, F. W. "Textbook of Polymer Science", John Wiley, New York, 1971.

Bowden, M. E. "Chemical Achievers", Chemical Heritage Foundation, Philadelphia, 1997.

Callister, W. D. "Materials Science and Engineering: An Introduction", John Wiley, New York, 1997.

Charles, J. A., F. A. A. Crane, and J. A. G. Furness. "Selection and Use of Engineering Materials", Butterworth Heinemann, Oxford, 1997.

Clagett, M. "Archimedes", in "Dictionary of Scientific Biography", C. Gillispie, editor, Charles Scribner, New York, 1970.

Daniels, C. "Master Mind: The Rise and Fall of Fritz Haber, the Nobel Laureate who Launched the Age of Chemical Warfare", Ecco, New York, 2005.

El-Wakil, M. M. "Powerplant Technology", McGraw-Hill, New York, 1984.

Dijksterhuis, E. J. "Archimedes", Princeton University Press, Princeton, NJ, 1987.

Goran, M. "The Story of Fritz Haber", University of Oklahoma Press, Norman, Oklahoma, 1967.

Grant, P. R. and B. R. Grant. "Evolution of Character, Displacement in Darwin's Finches". *Science* 313, 224–226, 2006.

Haber, L. F. "The Poisonous Cloud: Chemical Warfare in the First World War", Oxford University Press, New York, 1986.

Hermes, M. E. "Enough for One Lifetime: Wallace Carothers", American Chemical Society, Washington DC, 1996.

Hounshell, D. A. "From the American System to Mass Production 1800–1932: The Development of Manufacturing Technology in the United States", Johns Hopkins University Press, Baltimore, 1984.

Hounshell, D. A. and J. K. Smith. "Science and Corporate Strategy: DuPont R&D, 1902–1980", Cambridge University Press, Cambridge, 1988.

Judson, S. and S. M. Richardson. "Earth: An Introduction to Geologic Change", Prentice-Hall, Englewood Cliffs, NJ, 1995.

Kobe, K. "Inorganic Process Industries", Macmillan, New York, 1948.

Kooyman, B. P. "Understanding Stone Tools and Archaeological Sites", University of Calgary Press, Alberta, Canada, 2000.

McGrayne, S. B. "Prometheus in the Lab", McGraw-Hill, New York, 2001.

Morris, P. J. T. "Polymer Pioneers", Center for History of Chemistry, Philadelphia, 1986.

Netz, R. and W. Noel. "The Archimedes Codex", Da Cappo Press, Philadelphia, 2007.

Pickover, C. A. "From Archimedes to Hawkins: Laws of Science and the Great Minds Behind Them", Oxford University Press, London and New York, 2008.

Shackelford, J. F. "Introduction of Materials Science for Engineers", MacMillan, New York, 1988.

Schick, K. D. and N. Toth. "Making Silent Stones Speak: Human Evolution and the Dawn of Technology", Simon & Schuster, New York, 1993.

Smits, A. J. "A Physical Introduction to Fluid Mechanics", John Wiley, New York, 2000.

Simmons, J. "The Scientific 100: A Ranking of the Most Influential Scientists Past and Present", Citadel Press, New York, 1996.

Smith, W. F. "Principle of Material Science and Engineering", McGraw-Hill, New York, 1986.

Sparke, P. "The Plastics Age", Overlook Press, Woodstock, NY, 1993.

Stein, S. "Archimedes: What did he do Besides cry Eureka?", Mathematical Association of America, Washington DC, 1999.

Thornton, A. and K. McAuliffe. "Teaching in Wild Meerkats". *Science* 313, 227-229, 2006.

Timoshenko, S. P. and D. H. Young. "Theory of Structures", McGraw-Hill, New York, 1945.

Wagner, D. B. "Iron and Steel in Ancient China", E. J. Brill, Leiden, Holland, 1993.

Walker, J. "Discovery of the Germ", Totem Books, New York, 2002.

Whittaker, J. C. "Flintknapping: Making and Understanding Stone Tools", University of Texas Press, Austin, Texas, 1994.

Wrangham, R. "Catching Fire: How Cooking Made us Human", Basic Books, New York, 2009.

CHAPTER 3

DOMESTIC LIFE: FOOD, CLOTHES, AND HOUSE

Early inventions and tremendous progress in human affairs were driven by the necessity to meet the basic needs of a domestic life: food, clothing, and shelter. We have come a long way in human evolution. Consider this moving report in the *Voyage of the Beagle*, written by Charles Darwin, on the primitive Fuegians he saw at the windy and freezing Tierra del Fuego, 55° South on the southern tip of South America:

> . . . we pulled alongside a canoe with six Fuegians. These were the most abject and miserable creatures I anywhere beheld. . . . these Fuegians in the canoe were quite naked, and even one full-grown woman was absolutely so. It was raining heavily, and the fresh water, together with the spray, trickled down her body. In another harbour not far distant, a woman, who was suckling a recently-born child, came one day alongside the vessel, and remained there out of mere curiosity, whilst the sleet fell and thawed on her naked bosom, and on the skin of her naked baby! . . . At night five or six human beings, naked and scarcely protected from the wind and rain of this tempestuous climate, sleep on the wet ground coiled up like animals . . . They often suffer from famine . . . When pressed in winter by hunger they kill and devour their old women before they kill their dogs . . . we were told that they often run away into the mountains, but that they are pursued by the men and brought back to the slaughter-house . . . "

The first priority of all animals is to find a steady supply of food and drink, to sustain their lives and health, and to feed their young. Food in the wild can be found year round in some wet tropical regions where the temperature and rainfall are evenly distributed throughout the year, as well as in some well-watered oases and river valleys. The food supply is seasonal in most other regions that have summers and winters, as well as wet and dry seasons. For the winter and the dry seasons when fresh plant food is not available, some people go hungry, others learn to store food, and some migrate to follow their food supply. Humans also encounter many items that are not suitable as food in their natural state, but can be improved by suitable preparations, such as by grinding, baking, or boiling. Decaying flesh may contain many bacteria and toxins, but can be detoxified by cooking. Olives are bitter and must be fermented or cured with brine to make them more palatable; potatoes and taro are inedible when raw due to the toxic calcium oxalate and alkaloids, which are minimized after cooking. The greatest invention for food abundance is to

Great Inventions that Changed the World, First Edition. James Wei.
© 2012 John Wiley & Sons, Inc. Published 2012 by John Wiley & Sons, Inc.

supplement the meagerness of nature with agriculture, irrigation, tools, fertilizers, and genetics.

Humans long ago lost their fur and had only their skin to protect themselves against the sun and rain, but they invented clothing and housing that enabled them to expand their habitat beyond the tropical highlands in Africa to conquer the cold north and live in high mountains. Modern *Homo sapiens* arrived in Africa about 200,000 years ago and lived in the tropical savannah where there was a moderate seasonal difference in temperature. When they moved out of Africa, around 70,000 years ago, into the Fertile Crescent and then to Asia and Europe, they had to adapt to harsh living conditions. They invented clothing and shelter against inclement weather. A house also became a nest to nurture the young and a center of family activities, as well as a secure place to avoid prying eyes and to store valuables.

3.1 FOOD

Food is basic to life and survival, to health and fitness in facing challenges, and to reproduction and care for the young. Our daily nutrition requirement depends on a number of factors, including age and gender, but is approximately 2000–2600 kcalories of energy, 50–60 g of protein, and 2–3 L of water per day. Poor nutrition leads to shorter lifespan, more health problems, higher infant mortality, and reduced fertility.

In the Paleolithic Age, people obtained food by gathering wild plants, hunting animals, and scavenging carcasses. Green plants manufacture food by photosynthesis, which requires sunlight, water, mineral nutrients, and warmth. The productivity of the land depends on the natural ecosystem, and can vary from no plant growth at the polar icecap to abundant growth in a tropical rainforest. Many trees and grasses are not suitable for human food, as they consist of indigestible bark, wood, and other tough fibers. The "carrying capacity of the land" is the number of people that can be supported by food obtained from a piece of land, such as fruits and seeds. When you depend on nature's bounty of wild plants and animals, you may not have to travel far if you live in a lush tropical forest with uniform rainfall, and plenty of bananas and coconuts. But you may have to migrate far as nomads when you live in a climate with wet and dry seasons and shifting rainfalls. You will face seasonal shortages when you live in a climate with summers and winters, so one solution would be to invent ways to store food that was grown in seasons of plenty to be consumed in times of shortage, in the manner of squirrels storing nuts for the winter.

The modern-day chimpanzees and bonobos eat vegetation such as fruits, leaves, seeds, gums, and stalks, supplemented with insects, eggs, and small animals. Such herbivores have enormous jaws and grinding teeth, as well as long alimentary canals, to process grasses that are fibrous, more difficult to digest, and have a relatively lower protein content. About 2.5 million years ago, animal foods began to occupy an increasingly prominent place in the diet of our ancestors. Comparing the modern *Homo sapiens* with the ancient *Australopithecus*, the brains grew much bigger and the jaws grew much smaller. The change in the digestive organs included decreased molar size, a less robust mandible, reduced gut length, and alterations in incisor shape, suggesting a diet requiring less grinding and more tearing. The increase in brain size may be related to the diet with more fat or lipids.

Subsistence gathering is still practiced today in the rainforests of the Congo, Amazon, and New Guinea and the tundra of Alaska, Northern Canada, and Siberia. Primitive animal herding is still practiced in the Sahara, Arabia, Gobi, and Lapland. The Eskimos live on fish and seals found in coastal waters. Most of the waking hours of the hunter–gatherers are heavily invested in finding food, which are often inadequate to the needs; consequently, there is little time for other activities, not enough surplus food to store for the winter or lean years, and no surplus to support specialists such as weapon makers and medicine men.

Ancient men were surrounded by items that they could not easily chew and digest, and many natural foods contain toxins harmful to humans and animals. Cooking is essential to convert tough and indigestible material, such as rice and wheat, into chewable and nutritious food. When cooking was invented, the list of food that we could eat was enormously expanded to include hard items that could be softened and toxic items that could be detoxified. The use of cereal in diet by humans is unparalleled as no other free-living primates can routinely consume cereal grains, unless they are cooked by boiling in water or by baking in an oven. Cooking is essential for the conversion of complex carbohydrates into simpler digestible forms, to soften and expand to allow easy contact with soluble enzymes; thereafter the starch can be split by the pancreatic enzyme into sugars such as lactose, sucrose, and maltose.

Cooking has also turned many harmful roots and nuts into edible food by destroying microbes and toxic material. Some plants contain natural toxins designed as part of their natural defense to discourage predators and avoid being eaten. Potatoes contain toxic alkaloids that must be destroyed by cooking. Taro is a tropical root vegetable, which is the basis of the paste called *poi* in the Hawaiian Islands; in its natural state, taro contains poisonous crystals of calcium oxalate, which can be destroyed by cooking. Peanuts and soybeans have to be roasted or boiled. Wild almonds contain compounds that release cyanides when crushed or chewed, and must be roasted to destroy these compounds. When a plant or an animal is left open after its death, a process of rotting and decay begins attracting bacteria such as *Salmonella* and other microbes, which leads to the production of a foul smell and toxins. These toxins are likewise destroyed by cooking.

There is no agreement among experts on when cooking began, except that it must have been after the invention of fire. In archaeological excavations, it is difficult to determine whether a fire was used to keep warm and drive away predators or was used for cooking. Charles and Mary Lamb wrote a charming tale about the invention of cooking, which involved an accidental fire in a barn with a pig inside, creating a delicious barbecue for a family. Many anthropologists believe that cooking began 250,000 years ago when hearths, ovens, and burnt animal bones began to appear across Europe and the Middle East. Richard Wrangham believes that cooking allowed us to reduce the work of chewing and digesting. The energy that our bodies saved by outsourcing the chore of digestion was redirected toward our brains, therefore cooking is what made us human.

In the modern world, most of our foods have been processed prior to consumption. A major purpose is to increase the pleasure of eating with the blending of ingredients for better taste and aroma. Common food processing techniques include mechanical operations such as peeling potatoes, chopping carrots, and squeezing fruit for juice; mixing operations such as adding salt, spice, sugar, and smoking; chemical operations such as fermenting beer and wine and pickling; heating operations such as cooking, boiling, frying, steaming, grilling, baking, and pasteurizing; and drying operations such as preserving meat, fish, and fruits.

3.1.1 Agriculture

A major warming in the climate took place 12,000 years ago in the Holocene period when the last glacier retreated, trees began to return, and many parts of the world became much wetter. Agriculture was a timely invention, as this was a time for plants to grow in abundance to produce a steady food supply, with enough surpluses from harvests to store for winters and lean years. Having adequate nutrition led to longer and healthier lives, and more surviving children. Farmers could live in settled villages and in permanent homes, in comparison to the hunter–gatherers and nomads who had to keep moving to follow the food supply and could not live in a fixed place all year round.

The invention of agriculture, and the numerous improvements over the centuries, made possible an incredible increase in the productivity of the land, so that an acre of land could support many more mouths. The carrying capacity of the land highly depends on the inventions and technologies available. In 1891, Ravenstein estimated that the average cultivated fertile land could support $80 \, people/km^2$, the short grass steppes could support $4/km^2$, and deserts with oases could support $0.4/km^2$. This old estimate was based on the technology of 1891, and should be compared with the modern American farm that can support $3000/km^2$, with the current population density in Mongolia and Namibia that can feed $2/km^2$, and with the herder population of Lapland in Finland that supports around $0.01/km^2$. The productivity of the farm land very much depends on the inventions of farming technology: farm tools and machinery, seeds and breeding, irrigation, fertilizer, and pest control. The carrying capacity also depends on the consumption pattern: a calorie of animal diet can cost up to 10 plant calories to feed the animals. We lose up to one-third of our food by rats, by throwing it away, and by diverting corn to make biofuels.

The Agricultural Revolution began in the Fertile Crescent, from Palestine north to Syria around 10,000 BCE, and then east to the Mesopotamia and south to the Persian Gulf. To supplement the gathering of wild plants, it was a giant step to learn how to take care of these plants and to make them flourish and multiply. The most important food plants were the cereals, specifically wheat and barley in Mesopotamia where it all began, as they have the great advantages of high nutrition value and are easy to store. The domestication of animals, such as sheep and cattle, also took place at about the same time, which took advantage of the more abundant grazing opportunities. Agriculture requires many natural conditions, including climate, temperature, and rainfall; and its efficiency depends on human inventions, such as domesticated plants and animals, farming implements, weed and pest removal, irrigation, fertilization, harvesting, and preservation.

We need plants and animals with desirable characteristics: readily available, easily grown in the local climate, and having a high content of nutrients or usable fibers. The most important species were often not available locally and had to be imported from another continent. Selective breeding over many years also led to improved species. Potatoes and tobacco were among the most valuable exports from the Americas to Europe and Asia after the voyages of Columbus, and the return imports included sugarcane and horses. The most important food species that we eat today originated from many parts of the globe.

After thousands of years of cultivation and plant breeding, by selecting plants with more desirable characteristics for propagation and by more complex genetic and molecular techniques, the current farm crops look very different from their ancestral wild plants. The domesticated species were gradually improved and differentiated from the wild species by becoming larger and more resistant to heat and insects, producing larger seeds, and requiring less irrigation. One example is the diversity of different breeds of dogs today in comparison to wild dogs or wolves. The introductions of corn and potatoes from America to Europe, of rice and soya from Asia to America, and of sheep from Europe to Australia have changed the fortunes of regions from relative poverty to wealth. The largest cultivated areas today have a favorable climate, abundant rainfall, and are found in East Asia including China and Japan, South Asia from India to Malaysia, Central Europe, and Eastern North America.

Seeds fall to the ground in nature, but only a small fraction are lucky enough to fall on receptive soil and sprout to form roots and leaves, as most of them are consumed by animals or fail to sprout. Successful farmers may be able to find or create suitable fields with soil that is loose and moist and contains organic matter and minerals from the decayed remains of plants. It is important to loosen heavy soil so that the seeds can fall into holes and crevices and then be covered by soil and out of reach of hungry animals. The implements for preparing the soil include digging sticks, hoes, and ploughs that are pulled by humans or draft animals.

Plants have green leaves that grow by photosynthesis with sunlight and warmth, water from the soil, and carbon dioxide from the air to make glucose. The plant must also be supplied with essential minerals from the soil, and the most important are nitrogen, phosphorous, and potassium, which are known as NPK. Natural fertilizers include human and animal urine and dung, as well as rotting organic plant matter. The valuable plants must be protected from being crowded out by weeds, which also rob the soil of nutrients, and from being eaten by insects. After the plants mature and produce edible seeds, they must be harvested, which often begins by cutting with sickles. Then the seed is separated from the stalk by thrashing and winnowing.

Modern agricultural mechanization is particularly valuable in the increase of productivity and in the decrease in the need of farm laborers. In 1831, Cyrus McCormick built the first mechanical reaper, which was followed by other inventions of mechanized tractors and harvesters. At the end of the nineteenth century, it took 35–40 h of planting and harvesting to produce 100 bushels of corn. A hundred years later, the same amount of corn would require only 2 h and 45 min. Today, farmers can ride in air-conditioned tractors and listen to music while they work.

At the time of the American Revolution, it was estimated that more than 70% of the labor force were engaged in farming. The US agriculture today employs only 2.5% or 3.4 million farm workers, who feed 300 million Americans with plenty of surplus for export. In a hunter–gatherer society, all the able bodied people are essentially engaged daily in finding food, but with technology, one farmer can feed many people, which frees them to specialize on doing other useful work such as highly skilled tool makers and warriors.

CHEMISTRY OF FOOD

The main chemicals of the human body are proteins, carbohydrates, fats, and nucleic acids. All organic compounds contain the atoms of hydrogen, carbon, and oxygen. Proteins such as muscles also require nitrogen and sulfur; the genetic DNA and RNA also require phosphorous. Numerous minor elements are required for special functions, such as calcium for bones, iron for blood, potassium and sodium for body fluids, and iodine for thyroid. These minerals must be in the soil for the plants to grow and then the plants with their minerals are consumed by animals and humans.

The principal step in the production of food and the sustaining of all life forms is photosynthesis, which takes place in green leaves by converting sunlight energy into the chemical energy of sugar. Green plants have the green pigment chlorophyll, which is closely related to the hemoglobin in the blood of mammals; it is the site for the chemical reaction

$$6CO_2 + 6H_2O + light \rightarrow C_6H_{12}O_6 + 6O_2$$

This equation shows that six carbon dioxide molecules from the air are combined with six water molecules from the soil under sunlight, to produce one molecule of the sugar glucose and to release six molecules of oxygen. Glucose is the building block and the fuel to make all the chemicals of the body. Oxygen is necessary for breathing and burning sugar to power the body chemistry of maintenance and growth.

People and animals need water to drink, and plants need water for growth. The primary source of water is rainfall, which is distributed unevenly on earth and can be highly seasonal. A secondary source of water may come from local sources such as wells, springs, lakes, and rivers in addition to being transported from a long distance by irrigation with canals and aqueducts.

When it is necessary to lift the water from the ground for domestic use or irrigation, the lifting devices include leather bags, clay vessels, the shaduf on a pole with a balancing weight, and the water wheel. Aqueducts were used by the Babylonians and Egyptians. As early as the seventh century BCE, the Assyrians built a limestone aqueduct 9 m high and 275 m long to carry water across a valley to their capital city of Nineveh. The full length of the aqueduct ran for 80 km.

The science and technology of hydraulics was advanced by many illustrious inventors and writers: Archimedes (287–212 BCE) of Syracuse in Sicily, Marcus Vitruvius Polio of Rome (80–25 BCE), and Hero of Alexandria (10–70). Archimedes wrote about the screw to lift water and about buoyancy. In

De Architectura, Vitruvius wrote extensively about water and how to find and transport it and about aqueducts. Roman aqueducts were built all over the Roman Empire, from Germany to Africa, and especially in the city of Rome itself, where they totaled over 400 km. The aqueducts were important for supplying large cities across the empire, and the Romans set a high standard of engineering that was not surpassed for more than a thousand years.

The great city of Pergamum in present-day Turkey had a cistern or a water source that was 375 m above sea level, while the Pergamum citadel was 332 m high; the aqueduct descended from the mountain and crossed two valleys that were 172 and 195 m above sea level. They probably used the siphon effect that would have required tightly sealed pipes of bronze or lead, but none of that remains today. When the aqueduct is full of water and the cistern on the mountain is at 1 atm in pressure, the valley positions are under the tremendous pressures of 20 and 17 atm, and the citadel is at the more modest pressure of 4 atm. The most famous surviving Roman aqueducts today are the Pont du Gard in southern France, built around 19 BCE, and the Segovia Aqueduct of Spain. In the new world, the Aztec capital of Tenochtitlan was watered by two aqueducts in the middle of the second millennium.

The Neolithic Agricultural Revolution led to one of the most profound changes in human history, which provided a more secure and bountiful way of producing food compared to the traditional hunter–gatherers. Agriculture also increased the carrying capacity of the land by producing more food per acre, as well as by increasing the productivity of labor by producing more food per worker. These factors led to many profound impacts on society: higher populations, settlements in villages leading to higher population density, work specialization, and sociopolitical organizations. It has been estimated that the human population before the revolution was 6 million, and the average carrying capacity of the land surface was 0.05 person/km^2; at the current time, the population is 6 billion and the carrying capacity is 54 persons/km^2. Thus, both the population and the carrying capacity have increased by a factor of a thousand!

Food security and abundance meant better nutrition and health, higher birth rates, less infant and child mortality, and longer life expectancy. This led to a large population increase, perhaps from less than 5 million to 100 million in the ancient world. Hunters and gatherers can seldom support more than one person per 10 km^2 of forest or grassland, but even primitive agriculture can support 100 persons on a farm of the same area. This led to the gradual abandonment of nomadic life since farmers could live permanently near their field and crops and settle down. They could also store the surplus food for the winter or for times of poor crops, and they could begin to accumulate tools and household possessions.

Large-scale irrigation for farms cannot be done solely on the efforts of the individual farmer, but requires a central organization with the authority to make plans and to impose those plans on the individual farmers. This has been called the hydraulic theory of government, which suggests that irrigation was the primary driving force toward strong government and social stratification between the elite and the common people.

There are two ways to measure the increase in food productivity: per area of land cultivated and per farmer engaged. When the amount of food per acre or square kilometer of land is increased, we can either increase the number of people supported by the same acre, support the same number of people in a more lavish style, or convert some of the unneeded farmland into recreation areas. When the amount of food per farmer is increased, we can rechannel the surplus farmhands into other valuable professions including manufacturing, medicine, and security.

Agriculture has also transformed a large portion of the earth, as many forests have been cut down and wetlands drained to create more farmlands. This has inevitably pushed the native species of animals and plants out of the way, some even to extinction. Elephants and tigers used to roam many Asian forests and bison used to occupy much of the American prairie, but their habitats have been very much reduced in modern times.

Modern agriculture is often practiced in the form of "monocropping," where the same selected plant species occupies a large area of many acres, to the exclusion of all other species. This is in stark contrast to nature where different plants are mingled and different breeds of a plant cross-pollinate with each other. Monocropping creates vulnerability under an environmental challenge, such as a microbe or a pest attack that can wipe out all the plants that are identical to each other, which was the case in the Irish Potato Famine. It is more difficult for the same microbe or pest to simultaneously wipe out all the different genes of many plant species, so it is important to keep some wild genes around in reserve.

3.1.2 Ammonia Synthesis: Green Revolution

The air in the atmosphere is 78% nitrogen, but free nitrogen N_2 cannot be utilized by plants and animals. N_2 has to be "fixed" or converted with a great deal of energy into a water-soluble chemical compound, such as nitrogen oxide NO or ammonia NH_3. Fixed nitrogen needed by plants is provided by nature from the energy of lightning, which fixes atmospheric nitrogen in the form of nitric oxide, as well as from the legume nodule bacteria that fix nitrogen from air by consuming energy from sugar.

The recommended daily protein need per person is $0.8\,g/kg$ or $6\,g$ of nitrogen contained in $56\,g$ of protein for a $70\,kg$ person. The world population today is 6.5 billion, so we would need 14 million tons of fixed nitrogen per year to supply enough plant food for human consumption. The actual requirement is much higher because of meat's higher nitrogen requirement: when we feed soya bean to cattle, 1 g of beef nitrogen requires the consumption of $10\,g$ of soya bean nitrogen.

The ancient society with a static population and locally grown food was a self-sustained society: the plants obtained their nitrogen and phosphorous from the soil, the minerals were consumed by humans and animals through the plants, and then the minerals were returned to fertilize the soil in the form of urine and dung. Modern society poses a new set of problems, leading to a society that is not self-sustained in food minerals: (i) a growing world population requires increased crop size, and (ii) dung from cities is treated in sewage plants and discharged into rivers and oceans, but not returned to the farms. Thus, there is a need for supplemental inorganic fertilizers and nitrogen, or else there would be insufficient food to feed the growing population.

Nitrate deposits were discovered in South America in 1809, which resulted from the millions of years of guano deposited by sea birds that had fed on the fish from the cold rich waters of the Humboldt Current. The location of these deposits is called the Atacama Desert, which is one of the most arid regions of the world with almost no rainfall each year; so this rich deposit was not washed away to the sea. The deposits are in the form of saltpeter or $NaNO_3$.

This supply of nitrate deposits was so important that the War of the Pacific (1879–1883) was fought between Chile and the joint forces of Peru and Bolivia to control these mines. These nitrate mines were a source of enormous profit to Chile for many decades, as British firms would ship the saltpeter to Europe to fertilize farms and to make gunpowder. But these mines were depleting the age-old deposits at a rate of one half million ton per year, which could not provide unlimited quantities and ultimately the mines would be exhausted.

This situation was compounded by even greater concerns. Fixed nitrogen is also the basis of explosives such as black gunpowder, nitroglycerine, and TNT, which were used in excavation and in warfare. In the early twentieth century, Germany was dependent on a fleet of ships traveling every year from the mines in northern Chile, sailing south past the Cape of Horn, and steaming up past Africa and England to reach the German port of Hamburg to deliver the precious nitrate. In case of a war with England, this route would be cut off by the superior British navy, so that Germany would be brought to its knees without its fertilizer or bombs.

Thomas Malthus said in 1800 that mankind is always trying to find balance in the race between a growing population that requires more food and improving methods to grow more food. In 1893, Sir William Crookes predicted the ultimate starvation of a large portion of the world population due to the exhaustion of these Chilean nitrate deposits. This prediction provoked discussions and research attempts to fix the atmospheric nitrogen.

In 1900, the world was struggling to feed 1.5 billion people, and there were only a few inadequate processes for fixing nitrogen. The electric arc process could fix nitrogen in imitation of atmospheric lightning, but it formed nitric oxide at a yield of less than 1%. It proceeded by the chemical reaction of $N_2 + O_2 = 2NO$. The process was in commercial operation in Norway by 1905, thanks to Norway's large and inexpensive hydroelectric power. Another nitrogen fixation process was the cyanamide process, which was also in operation in Italy at that time. Both of these processes were inefficient and not economical, as they required high-temperature reactions at enormous costs of power. Many scientists began to work on nitrogen fixation in their laboratories, including two German Nobel Prize winning chemists, Wilhelm Ostwald and Walter Nernst.

A future Nobel Prize winner from Germany, Fritz Haber was born in Breslau in 1868. Haber was the son of a Jewish chemical wholesale merchant in natural dyes, paints, lacquers, and other chemicals and drugs. He showed an early interest in languages, music, theater, and literature. Haber studied at the University of Berlin, then at Heidelberg under Professor Bunsen, earning a doctor of philosophy in 1891. In 1892, he converted to Christianity, which was a time of few opportunities for Jews. Haber received an appointment at the Karlsruhe Technical Institute as an assistant in Chemical and Fuel Technology, where he taught dye and textile printing.

In 1901, he met and married Clara Immerwahr who was also from a Jewish family, and who also had earned a doctorate in chemistry. The following year, he

FIGURE 3.1 Fritz Haber. Reprinted with permission The Image Works.

toured the United States to visit universities and chemical plants and to report on the state of the American chemical industry. He went to Niagara Falls and visited the Atmospheric Products Company that was investigating the electric arc process for nitric acid synthesis. His picture is shown in Figure 3.1 as a successful professor and a person welcomed in society.

In 1904, a Vienna chemical company asked him to do research on ammonia synthesis. He put together a team, which included an English collaborator Robert Le Rossignol, who excelled in high-pressure equipment. Haber began his work by combining hydrogen with nitrogen at the normal pressure of 1 atm and 1000°C, but obtained a negligible amount of ammonia. The chemical reaction involved in this synthesis is $N_2 + 3H_2 = 2NH_3$ where one molecule of nitrogen reacts with three molecules of hydrogen to make two molecules of ammonia. Since this reaction converts four volumes of gases into two volumes of gases, the conversion is more favorable under high pressure according to Le Chatelier's principle.

Haber decided to increase the pressure, which was also the approach of Walther Nernst of the University of Berlin. Haber and his team created a high-pressure reactor, and studied the equilibrium of hydrogen and nitrogen with ammonia under a variety of temperatures and pressures. They determined that a significant yield depended on increasing the pressure to above 200 atm. Low temperature favors higher equilibrium conversion at a slow reaction rate, while high temperature leads to negligible yield but a faster reaction. Therefore, they settled on a compromise to operate at a medium temperature, but with a suitable catalyst to speed up the reaction, and settled on using the precious metals osmium and uranium. The exit from the reactor was connected to a glass flask surrounded by dry ice at very low temperatures, so that the ammonia produced could condense into a liquid.

He gave a demonstration in 1909 to two members of the firm Badissche Aniline und Soda Fabrik (BASF), Carl Bosch and Alwin Mittasch. BASF was very impressed and signed a contract with Haber and worked with him to make the invention practical. Alwin Mittasch conducted 10,000 experiments to test 4000 catalysts, and discovered that iron oxide worked very well when contaminated or "promoted" with potassium.

BASF gave the job of building the plant for manufacturing ammonia to Carl Bosch, a 34-year-old metallurgical engineer. Bosch had the responsibility to design an apparatus to work with hydrogen under high pressures and temperatures, with an unprecedented danger of explosion and fire. By 1913, in Oppau, Germany, the manufacturing plant for chemical ammonia synthesis was operating to produce 36,000 tons of ammonia per year. The World War broke out a year later, and the ammonia produced was eagerly fed to the German ammunitions industry when the supply from Chile was cut off by the British navy. Some historians think that Kaiser Wilhelm II even delayed the war till the Haber method was ready and in production.

SYNTHESIS OF AMMONIA

The synthesis of ammonia involves the chemical reaction of

$$N_2 + 3H_2 \rightleftarrows 2NH_3 + heat$$

Building an apparatus that operates at high temperatures and pressures and is filled with highly flammable hydrogen creates a serious problem of explosion and safety. We are reminded of the disasters of the hydrogen-filled Hindenburg dirigibles, which was not even under high pressure and temperature. It is prudent to locate ammonia synthesis plants far away from centers of population, but close to transportation nodes for the supply of raw material and the delivery of products.

The raw material nitrogen comes from air, by separating from oxygen with a cryogenic process of cooling air to very low temperatures. Air is 78% nitrogen, 21% oxygen, and 1% argon. Nitrogen boils at $-195.8°C$ and oxygen boils at $-183°C$. So when the temperature drops, oxygen will liquefy first and be removed. Hydrogen is produced from a fuel such as coal or methane. The process begins with the combustion of methane to generate heat, and then the methane is reacted with water. The overall reaction produces hydrogen and carbon dioxide:

$$CH_4 + 2H_2O \rightarrow 4H_2 + CO_2$$

The fertilizer ammonia can be directly injected as a gas into the soil, which acts very quickly and requires repeated applications. The preferred method is to change ammonia into a water-soluble solid such as urea or ammonia nitrate, which makes them easier to ship and to apply, and they release slowly over a period of time.

After the ammonia synthesis success, Haber was offered a new position in 1911 to be head of a new Kaiser Wilhelm Institute in Berlin. He was a gregarious and highly sociable man, who took pleasure in cultural events, lectures, and performances. He was broadly educated, spoke many languages, and was well versed in classical literature and art. In his profession, he was admired and remembered with respect, and students came from around the world to study under him. He had a high social position, became a supporter and a close friend to Albert Einstein, and helped to negotiate Einstein's divorce from his first wife Mileva. But in family life, he was autocratic, and his wife Clara became increasingly isolated from him and his social positions.

When World War I began, Haber was looking for a new weapon to break the stalemate of the trenches in the Western Front. The Hague Peace Conference of 1899 had banned the use of unusual new weapons that might cause unnecessary suffering, which the United States did not sign. Haber organized the first poison gas attack at Ypres in 1915, with 6000 cylinders of liquid chlorine along a line of more than 2 km. When the wind blew in the direction of the French, the Germans opened all the canisters and released 150 tons of chlorine. A white cloud turned yellow-green and moved to the French trenches, which resulted in violent vomiting and coughing. According to the English staff, 7000 people were poisoned and 350 died. Germans declared that only 700 German soldiers were injured and only a dozen killed. But the Allies followed suit and retaliated with their own gas weapon 5 months later.

By 1917, Haber had 1500 people working under his command and a huge budget. Poison gas accounted for 1.3 million casualties, less than 6% of the total of 21 million casualties. Haber maintained with perfect logic that being killed by poison gas was no more abhorrent than being mangled by splinters of steel shrapnel. His son, who became a historian, wrote, "He was a Prussian, with an uncritical acceptance of the State's wisdom, as interpreted by bureaucrats, many of them intellectually his inferiors." Haber's relationship with Einstein became strained. His wife Clara pleaded with him to forsake his role in this barbaric and immoral work, but to no avail. Clara shot herself in the heart with her husband's army revolver, leaving their 13-year-old son Hermann to discover her body. Soon after, Haber remarried.

After the end of the war in 1918, Haber was given the Nobel Prize in Chemistry, with the announcement: "For his method of synthesizing ammonia from its elements, nitrogen and hydrogen. A solution to this problem has been repeatedly attempted before, but you were the first to provide the industrial solution and thus to create an exceedingly important means of improving the standards of agriculture and the well-being of mankind. We congratulate you on this triumph in the service of your country and the whole of humanity." There was no mention of gas warfare. Carl Bosch received his Nobel Prize in Chemistry 13 years later in 1931, with the citation: "It appeared impossible to find a material which, at the pressure and the temperature mentioned, would stand up to the mixture of gas involved for any appreciable time. The brilliant idea was a double wall with an inner cylindrical tube enclosed by another outer tube. The cold compressed mixture of hydrogen gas and nitrogen gas would be introduced into the space between the two tubes, so that the inner tube was exposed to high temperature and low pressure, but the outer tube would be exposed to high pressure but low temperature. You made possible for nitrogen to be made available to mankind in inexhaustible quantities, in a form suitable for agriculture and low price."

Haber was a patriot who tried to find a way for Germany to pay the war debt of 132 million marks, which was equivalent to 50,000 tons of gold. He thought that seawater contained 6 mg of gold per metric ton, and thus sought a way to extract gold from seawater. After some investigation, he realized that the earlier estimate was too high by a factor of a thousand, and he had to give up that endeavor. The rise of Hitler to power meant that Jews were no longer honored and welcomed as before, even for one who had converted to Christianity and had made such great contributions to Germany. His health was also failing, and Carl Bosch alone continued to support him. In 1933, he moved to Cambridge, England after being forced to leave Germany. The following year, at the age of 65, he suffered a fatal heart attack and died in a hotel in Basel, Switzerland.

Today, the world has learned to grow food at a rate that is four times higher than it was in the year 1900 and is able to support a population of 6.5 billion. There are many factors that led to this magnificent rate of improvement, the foremost being the availability of inexpensive synthetic fertilizers and the improvement in the breeding of crop plants that is commonly called the Green Revolution.

The science of plant breeding was greatly improved with the discoveries of Charles Darwin on evolution, Gregor Mendel on genetics, and Watson and Crick on the structure of DNA. The Green Revolution was led by Norman Borlaug (1914–2009), who received the Nobel Peace Prize in 1970. His photo is shown in Figure 3.2, where he

FIGURE 3.2 Norman Borlaug. Reprinted courtesy of Norman Borlaug Institute for International Agriculture.

was taking notes in a field of wheat. It is said that Borlaug saved more lives than any other man in history, perhaps as many as a billion people. He was born on a farm in Iowa, and worked on his grandparents' farm till he was 19. He enrolled at the University of Minnesota in 1933 where he studied forestry and was on the varsity wrestling team. He learned that special plant breeding methods could create plants that were resistant to parasites. In 1942, he received his doctorate in plant pathology and genetics.

In 1944, he began working as a geneticist and plant pathologist in Mexico with support from the Rockefeller Foundation. His work involved breeding high-yield and disease-resistant wheat. Over 25 years, he was spectacularly successful in creating a high-yield, short-stalk and disease-resistant wheat. His success was built on the extensive use of four technologies: irrigation, mechanization, fertilizers, and improved hybrid crop plants. The first two factors had already been in extensive use in developed nations for a long time, and the use of fertilizers was also a century old. Borlaug's invention was developing improved breeds of corn, wheat, and rice that are known as HYV or "high-yielding varieties."

Taller wheat grasses are better at competing for sunlight, but tend to collapse under the weight of the extra grain, which comes from high application of nitrogen fertilizers. To prevent this, Borlaug bred wheat with multiple favorable properties, which included shorter and stronger stalks that could support larger seed heads. In 1953, he acquired a Japanese dwarf variety of wheat that he crossed with a high-yield American wheat. He also crossbred the semidwarf wheat with his disease-resistant wheat to produce his new variety that combined many desirable properties.

Borlaug had the humanitarian goal to put the new cereal strains into extensive production, in order to feed the hungry people of the world. He said, "There are many breeds of wheat with long-blade, which gave increased yields but snapped when they were given more than a certain amount of artificial fertilizer. The new dwarf varieties were able to stand two or three times more artificial fertilizer, and to provide an increase of yield per decare ($1000\,m^2$) from the previous maximum of 450 kg to as much as 800 kg per decare. These varieties can be used in various parts of the world because they are not affected by varying lengths of daylight. They are better than all other kinds in both fertilized and nonfertilized soil, and with and without artificial irrigation. In addition they are highly resistant to the worst enemy of wheat, rust fungus or oromyces." By 1963, 95% of Mexico's wheat crop used the semidwarf varieties developed by Borlaug. The results were more spectacular than anyone had anticipated. Mexico had imported half of its wheat in 1943, but became self-sufficient by 1956, and even exported half a million ton of wheat by 1964.

Borlaug went to Pakistan in 1959, and to India in 1963, to introduce the wheat that had been developed in Mexico. He began another program of plant breeding, irrigation, agrochemicals, and financing. India adopted a semidwarf rice variety developed by the International Rice Research Institute, which had been established in the Philippines in 1960, that could produce more grains of rice per plant when grown with fertilizers and irrigation. In the 1960s, rice yield in India was about 2 tons per hectare, but rose to 6 tons per hectare by mid-1990s. In the 1970s, rice cost about $550/ton, but it had fallen to under $200/ton by 2001. India became a major rice exporter, shipping nearly 4.5 million tons in 2006.

Thirty years after receiving his Nobel Prize, Norman Borlaug gave a speech in Oslo. He gave tribute to Haber and Bosch, saying that 40% of the 6 billion people living at that time owed their lives to the Haber–Bosch process of ammonia synthesis. He further stated that in 1940, the US farmers produced 56 million tons of maize on 31 million hectares, at an average yield of 1.8 tons/hectare. But in 1999, the US farmers produced 240 million tons of maize on 29 million hectares, at an average yield of 8.4 tons/hectare. He said that this more than fourfold yield increase is due to the impact of modern hybrid seed–fertilizer–weed control technology! He described how developing Asia (including China and India) was able to increase cereal production from 248 to 809 million tons in four decades (1960–2000). When Borlaug won the Congressional Gold Medal in 2007, he said, "The battle to ensure food security for hundreds of millions of miserably poor people is far from won. World peace will not be built on empty stomachs or human misery." He died at the age of 95 of lymphoma at his home in Dallas, Texas on September 12, 2009.

It can be said that "Chemistry came to the aid of farmers, and changed stones into bread." The world was having difficulty feeding 1.5 billion people in 1900, but by 2010, the world could feed 6.5 billion people without difficulty. The world production of cereal increased in this period, following an increase in irrigation, fertilizers, and tractors.

Year	Cereal (million tons)	Irrigation (million hectare)	Fertilizer (million tons)	Tractor (million)
1961	248	87	2	0.2
1970	372	106	10	0.5
1999	809	179	70	4.6

The agricultural expert Lester Brown said in an article published in *Foreign Affairs*, "The new breed of grain in our age will have the same impact on the Agricultural Revolution in Asia that the steam engine had on the Industrial Revolution in Europe in the eighteenth century." This increase in food supply did not come from using more land, as the global use of land for agriculture has remained stable and may even be decreasing due to higher productivity.

The comparison of agricultural productivity among nations at different stages of development can be seen in the following table. These nations also differ in the natural factors of warmth, soil quality, sunshine, and rainfall, as well as in the capital investments and technologies used. The cereal yield of Western Europe is twice that of South Asia and 15 times that of Sub-Sahara Africa. The power of technology to make life more abundant and secure is dramatically clear.

Nation	Cereal yield (kg/hectare)	Fertilizer use (kg/hectare)
Belgium	8680	355
United Kingdom	7169	325
France	7099	237
Indonesia	4354	145

Bangladesh	3648	209
Philippine	3074	156
Sudan	663	4
Namibia	403	2
Botswana	363	2

The Green Revolution led to grain surpluses and lower prices, creating a sense of complacency about agriculture and hunger. The food supply grew faster than population from 1970 to 1990, as the Green Revolution took hold. But this trend has been reversed in recent years, and the growth rate of food production has now fallen below population growth. Part of the explanation is in the growing practice of converting food products to biofuel for the automobile. The food shortage is particularly dire in sub-Saharan Africa where irrigation is relatively scarce. More people are hungrier than ever, with contributions from political and social factors.

There are two main approaches to manage the Malthusian race between population and food supply: grow ever more food with the Green Revolution, and reduce the population growth rate. The Green Revolution appeals to people who believe in immediate feeding of the hungry people now and the power of inventions; but there are many critics who say that it cannot be sustained because it uses too much fossil fuel and creates numerous environmental problems such as changes in the ecosystems, global warming, and biodiversity. Population control appeals to people who believe that it is more sustainable in the long run; but there are also critics of birth control, abortion, and government coercion.

It may seem surprising that an attempt to feed the hungry has so many critics. The global human population grew in the last century due to medical and agricultural advances. The growth rate peaked in 1962 at 2.20% per year, but fell steadily to 1.19% by the year 2007. The growth rate remains high in the Middle East, Sub-Saharan Africa, South Asia, and Latin America. On the other hand, some countries have negative population growth, especially in Eastern Europe.

The demographic transition model (DTM), proposed by Warren Thompson in 1929, suggested that many nations go through a four-stage process of decreasing death and birth rates with time. Stage 1 is the preindustrial society when both death and birth rates are around 4% per year and roughly in balance, so the population size remains stable. Stage 2 is the current developing countries such as Sub-Saharan Africa, with death rates declining to around 1% due to better food supply and sanitation, but without a compensating declining birth rate, so the population soars. Stage 3 is more mature societies such as Mexico and India, where the birth rates fall due to a changing lifestyle from improving wealth, so population growth rates begin to decline. Stage 4 is the developed countries such as the United States and China, where the death and birth rates are again in balance, so that the population becomes stable. Some theorists think that there is an emerging Stage 5 when the fertility is subreplacement, so that population will decline and the average age will increase, such as in Europe and Japan. This often leads to a shortage of younger people to enter the labor force and the need to increase immigration from poorer nations with different cultures and ethnicities leading to potential conflicts.

3.2 CLOTHES

Most mammals have fur that insulates them against the cold and shields them from rain and snow, and many also have spring and fall shedding that may be due to seasonal changes in sunlight as well as temperature variations. But humans have much less hair and have no natural protection from the elements. In the tropics, clothes are less important, but in the colder temperatures of arctic zones and high elevations, people need clothes to avoid the freezing plight that Charles Darwin saw among the natives of Tierra del Fuego.

The first clothes were made of animal skins or hides. A more refined garment needs a knife to cut the hides into the right sizes and shapes, bone needles to sew the pieces together, and leather strips or sinews as the thread. Sewing needles of bone have been found dating from 15,000 years ago. However, natural leather turns stiff and becomes perishable in damp weather and needs to be processed by tanning to remain soft and durable. Tanning involves soaking the hide in pits that contain tanning agents such as oak gall, alum and fat, and urine and feces. Tanneries and tanners were often smelly and considered unclean by other people.

Clothing can also be fabricated from fibers, such as wool and linen. This usually involves a three-step process: spinning fibers into yarns, weaving yarns into fabrics, and finally cutting and sewing fabrics into clothes. The first step of spinning involves twisting several strands of fibers together. The most prominent fibers were wool, flax, cotton, and silk. We distinguish between coarse fibers such as flax and wool, which are more suitable for outer garments, and fine fibers such as cotton and silk, which are more suitable for wearing next to the skin.

Wool is obtained by shearing sheep, and is a strong and insulating fiber for cold weather. There were recent discoveries in Georgia of flax fibers that date from 30,000 years ago, which were spun and dyed in a variety of colors. Flax was also documented from 5000 BCE in Fayum, Egypt. To separate the linen fiber from the flax plant, a process of retting is necessary, which involves soaking the plant in a pond till the perishable organic matters have rotted away. Cotton was first introduced in India. Of the prominent fibers, silk is the premium fiber for clothes. Silk is a protein produced by the silkworm *Bombyx mori*, which has been used in China since 3000 BCE. Its origin from a worm was kept secret for many years, until it was carried over the Silk Road from Xi'an in China to Damascus and the West, where it had no rival for luxury garments for the wealthy.

3.2.1 Cotton

Cotton is a soft staple fiber grown as a boll around the seeds of the cotton plant, which is a shrub native to tropical and subtropical regions. The earliest cultivation of cotton may have been in Mexico and Peru some 8000 years ago. In the ancient world, it was first cultivated in the Indus Valley around 4000 BCE, and then spread to Egypt, Persia, China, and the Mediterranean during the Middle Ages. The growth of cotton needs long frost-free periods, plenty of sunshine, and moderate rainfall. The cultivation of cotton was very labor intensive for harvesting, ginning, spinning, and weaving. Many steps were involved in preparing cotton into a garment, and many inventions were made to make cotton more valuable as a general-purpose textile material.

Cotton is a perennial plant that lives for many years, and its harvesting involves picking the attached seeds and fibers but leaving the plants alive to grow more fibers for the harvest next year. Picking cotton is a very labor-intensive method of harvesting, which required walking from plant to plant and selectively picking out the seeds and fibers while leaving the plant undamaged. The process has low productivity, and clumsy people are said to have "cotton picking hands."

The next process of separating cotton fibers from seeds is called ginning, which is depicted in the fifth century paintings in the Ajanta Caves in India. Egyptian cotton has the longest fibers, and is used to make premium fabrics. In the American South, the higher quality long-staple cotton can be grown only in a narrow band along the Carolina and Georgia coast, and is called "Sea Island" cotton. The interior areas of the American South can only grow "Upland" short-staple cotton. The shorter length fibers have reduced yarn and cloth quality, and they have "fuzzy" seeds because the fibers are tightly attached to the entire seed surface. This strong attachment of fibers to seeds created difficulty in removing the fibers without damage, which made it expensive as it was time consuming and labor intensive. These low-productivity tasks were done by slaves in the American southern states during the 1800s.

The modern cotton gin was invented by the American Eli Whitney (1765–1825), who graduated from Yale in 1789 and is known for two great inventions: the cotton gin, which transformed the Southern economy; and interchangeable parts in machinery, which led to the mass production method of manufacturing. When he graduated, he was short of funds and went to Georgia to seek his fortune in plantations. He was introduced to the problem of finding a better way to gin the Upland short-fiber cotton, which led him to invent the cotton gin in 1794. Whitney's cotton gin made the mechanical separation of fiber from seeds much more efficient. His machine had spiked saw teeth mounted on a revolving cylinder in a box, which is turned by a crank. The teeth pulled the cotton fiber through small slotted openings that are too small for the seeds to pass; thus, the lint was separated from the seeds, followed by a rotating brush to remove the fibrous lint from the teeth, ending with the seeds falling into a hopper. Whitney's machine could separate up to 50 pounds of cleaned cotton daily, making cotton production profitable for the southern states.

Unfortunately for Whitney, his cotton gin was extremely simple and easy to copy, so he failed to profit from his invention. Soon after, imitations of his machine appeared, and his 1794 patent for the cotton gin could not be upheld in court until 1807. Instead, he decided to make money by going into the ginning business, and manufactured and installed cotton gins through Georgia and the southern states. However, his ginning plants were resented by the planters who thought that he charged too much, and hence they found ways to circumvent his patent.

In later years, Eli Whitney cultivated social and political connections through his status as a Yale alumnus and through his marriage. Although he had never made a gun in his life, through those connections he won a contract from the War Department in 1798 to make and deliver 10,000 muskets for the army in 1800. Treasury Secretary Wolcott sent him a "foreign pamphlet on arms manufacturing techniques," and Whitney began to talk about interchangeable parts. This concept was a departure from the usual method of having an individual master machinist to make each part of the musket and fit them together, which is slow and painstaking. Whitney's method was to have many less skilled machinists specialize in making only one part per person to highly specified dimensions, so that the parts would be interchangeable and could then be assembled. This is the heart of the mass production method later made more famous by Henry Ford in the manufacture of the Model-T automobile. Whitney ultimately delivered on the army contract in 1809, and spent the rest of his life promoting interchangeability.

Another step in processing cotton is spinning, which involves twisting the relatively short fibers from plants or animals into a yarn or thread. The traditional tool in manual spinning is called the distaff, which has a feminine connotation as spinning was traditionally done by women. In 1769, Richard Arkwright introduced the spinning frame to enable British weavers to produce cotton yarn by twisting fibers to form the warps, substituting metal cylinders for human fingers. This machinery could work tirelessly and at much higher productivity, and was powered by animal or water power.

Weaving involves two sets of threads called the warp and the weft, which are interlaced with each other to form a fabric or cloth. The warp consists of parallel threads fixed in the long direction of the fabric, and the weft is made by moving spindles holding threads that alternately pass above and below the warp threads in the perpendicular or short direction. In an automatic loom, the machine raises the vertical odd warp yarns and lowers the even warp yarns, so the spindle can easily go through the horizontal opening to the right; at the next moment when the spindle is ready to return to the left, the frame lowers the odd and raises the even warp yarns. The power loom is a much more productive method, and led to the closing of cottage weavers and the migration of dispersed weavers in villages into centralized textile mills in cities, first in England and later to the rest of the world. The Jacquard loom of 1801 was the first method of computer-controlled weaving that was capable of producing special patterns.

The cotton gin brought enormous wealth to many planters as the yield of raw cotton doubled each decade after 1800. Demand for cotton was also pushed up by other inventions such as spinning and weaving machines and also steamboats that could transport cotton to weavers around the world. By the midcentury, the American South was producing three-quarters of the world's supply of cotton, with most of it being shipped to England or New England to be made into cloth at textile mills. The British textile merchants bought cotton fibers from colonial plantations and processed them into cotton cloth at enormous mills in Manchester. The cloth was sold in Europe, Africa, India, and China, which reduced the need for local cottage spinners and weavers in those countries. The rapid growth of American raw cotton in bales went from 3135 in 1790 to 3,837,402 in 1860, which is a factor of more than a thousand in 60 years!

The growing popularity of cotton led to the expansion of land needed for plantations, as well as an increased demand for slaves to pick cotton. More land was also in demand for white settlers from Europe to move inland, resulting in the forced removal of Creek Indians from Georgia in 1810 and of the Cherokees by the Indian Removal Act of 1830 in a process that would be later called the "Trail of

Tears." "King Cotton," the basis for the economy of the South, arrived and Georgia's cotton production increased by a factor of a thousand. The process of separating the fiber from the seeds had become far more efficient, but the process of picking cotton from plants had not improved, so there was a greater need for cheap labor. Cotton was intimately associated with slavery, since the success of this crop was tied to the ability of a landowner to procure low-cost workers, and white Georgians bought black slaves in record numbers to do their backbreaking work of "cotton picking." The massive growth of cotton production shaped the economy and social order of the antebellum South, and was contributing to the cause of the American Civil War of 1861–1865. The Manchester textile industry also ruined the cottage spinners and weavers of India, as well as those in the rest of the less developed nations. Mahatma Gandhi tried to reduce India's dependence on western manufacturing with a return to home spinning, even introducing a flag with a spinning wheel in the center. The flag of India today has a chakra or 24-spoke wheel in the center, and has to be made with hand-spun cotton cloth.

3.2.2   Man-Made Fibers: Nylon

The first man-made fibers were regenerated cellulose from plants, which were called semisynthetic fiber. In 1878, Count Hilaire de Chardonnet worked on developing an artificial fiber and obtained a patent for making this fiber. He treated cellulose from cotton or wood fibers in a solution of nitric acid and then extruded the solution of partially nitrated cellulose through a small hole, which created a filament that is somewhat flammable. In 1891, he opened a factory in Besancon to produce the world's first man-made fiber that he called rayon. The following year in England, Charles Cross invented viscose rayon by digesting cellulose from cotton lint or wood pulp by sodium hydroxide, and further treating it with carbon disulfide to create a viscous fluid called cellulose xanthate. This viscous fluid is forced under high pressure through numerous holes in a spinneret to form fibers and spun in a bath of sulfuric acid to remove the alkali. The same viscous solution could be used to make transparent films that were called cellophane. Viscose rayon did not have the strength or glamour of silk and is used today mainly as the smooth and slippery linings for coats and sleeves.

Nylon was the first fully synthetic fiber, where the starting materials were air, water, coal tar, and petroleum. The invention of nylon involved an intellectual discovery provided by Wallace Carothers and his associates and required a sustained development effort by a large team of scientists and engineers at the DuPont Corporation. It was one of the most impressive inventions that ushered in an age of man-made materials through chemistry.

Wallace Hume Carothers was born in Iowa in 1896, and attended Capital City Commercial College in Des Moines, Iowa, in a program of accountancy and secretarial administration. He struggled to find the funds for his education, but went on to the 4-year Tarkio College in Missouri to complete a bachelor's degree in chemistry and then taught chemistry for a year at the University of South Dakota. His thyroid gland malfunctioned and he suffered spells of melancholy and pessimism. He was also interested in music and poetry, and was a person of culture and charm.

His prospects greatly improved when he received encouragement from a teacher at Tarkio, and thus proceeded to the University of Illinois for graduate school where he earned his doctorate in 1924. Thereafter, he joined Harvard University as an instructor pursuing research in polymers. At that time, the nature of substances such as silk and rubber was still in doubt. Herman Staudinger proposed the polymer theory that they were long chains of small units called "monomers" held together by ordinary chemical bonds, but this idea had many skeptics.

The DuPont Company of Delaware made and sold gunpowder for the United States and its allies in several wars, which was very profitable. In peacetime, it began seeking diversification into many other chemical technologies, including rayon for inexpensive clothing, nitrocellulose for colorful automobile lacquer, and tetraethyl lead for automobile gasoline. In 1926, Charles Stine of DuPont proposed the new concept of a corporate research unit to discover or establish new scientific facts, as a supplement to the traditional pursuit of applying known scientific facts to practical problems. He was encouraged by several university consultants, as well as by the research directors of the older research programs at Bell Telephone and General Electric. Stine recruited Carothers from Harvard in 1928, despite Carothers' interest in pure research and lack of interest in financial profit. But Stine assured him that he could choose his own research problems at DuPont.

Carothers began several programs to study complex molecules, and one of them produced practical results rather quickly. Julius Nieuwland of Notre Dame University produced divinyl acetylene, which forms an elastic jelly. Carothers assigned an assistant Collins the task to purify divinyl acetylene by distillation. In 1930, Collins left the distillation results over the weekend and by Monday he found that it had solidified into a cauliflower-shaped mass. He fished out a few cubic centimeters of the substance, which felt strong and elastic. Collins threw the mass against his laboratory bench, and it bounced like a golf ball. In a remarkably short time, Carothers' research led to the development of neoprene in 1930, which became the first mass-produced synthetic rubber. Neoprene is chemically inert and fire resistant, and is used for industrial applications such as hoses and coatings. Carothers is shown in his laboratory pulling on a piece of material in Figure 3.3.

Wallace Carothers believed in the Herman Staudinger theory that polymers are made of long chains of monomer units, and he tried to make an experiment to demonstrate the theory. He proposed building long-chain molecules one step at a time, by carrying out well-understood chemical reactions. His first scheme was to react dihydric alcohol with —OH groups on both ends, together with diacid molecules with acid groups COOH— on both ends, to make a polyester of potentially endless length. The reaction turns two monomers into a dimer and water.

$$HO - R - OH + HOOC - R' - COOH \Rightarrow H_2O + HO - R - OCO - R' - COOH$$

He and his associates were able to make chains with a molecular weight of 5000–6000, and he named them "condensation polymers." Water was a by-product of the reaction, and needed to be removed so that the equilibrium would proceed to ever-increasing molecular weights. Carothers had heard about a molecular still that operated under a vacuum to remove water, and he employed one in his research. He

FIGURE 3.3 Wallace Carothers. Reprinted with permission Hagley Museum and Library.

used a 16-carbon diacid with a 2-carbon dialcohol, resulting in a product with a molecular weight of 12,000 that could be pulled into a fiber, but it melted at less than 100°C. DuPont had experience making and selling rayon from wood chips, and the management made the decision to produce a polymer to replace women's silk stockings as a pair needed only 10 g of resin, but a wool sweater consumed much more resin and would require much higher plant investments. However, stockings at the time needed to be ironed, so the product melting point had to be much higher.

These goals created a formidable challenge to solve numerous problems never encountered before, including the design of a product that would suit the needs of the stocking market for ladies, the creation of the manufacturing technology to make the chemical raw material and the fiber, the subsequent spinning and weaving in numerous small traditional textile mills, and the advertising and marketing of a new and unfamiliar product.

Carothers decided to switch from dialcohol to diamine, as the product would have a higher melting point.

$$H_2NO - R - NH_2 + HOOC - R' - COOH$$
$$\Rightarrow H_2O + H_2N - R - HNCO - R' - COOH$$

In 1934, they tried a 10-carbon diamine with a 5-carbon dialcohol and produced a polymer called 10-5 with a melting point as high as 190°C. It was difficult to obtain large quantities of 10- and 5-carbon raw material, and soon they switched their work to a 6-carbon material that could be based on the abundant compound benzene. By 1935, they had prepared polymer 6-6 that melted at over 250°C, so finally they had a product that was worth making. The teams of DuPont then developed a process of melt spinning of the nylon polymer at a temperature of 260°C and built a small plant to produce enough fibers for a knitting test at a small mill in Maryland.

In 1936, Carothers married Helen Sweetman from the DuPont patent office. Unfortunately, Carothers' mental state was deteriorating despite his recent marriage and election to the National Academy of Sciences, and he suffered another nervous breakdown. In 1937, at the age of 41, Wallace Carothers committed suicide with cyanide at a hotel in Philadelphia shortly after leaving work. Several months later, his daughter Jane was born.

In December 1937, a mill in New Jersey finally turned out "full fashioned hosiery with excellent appearance and free of defects," and 56 pairs were distributed to the wives of the men on the nylon project. In 1938, nylon was announced at the New York World's Fair to a women's club. The announcement said:

> This textile fiber is the first man-made organic textile fiber prepared wholly from new material from the mineral kingdom . . . though wholly fabricated from such common raw material as coal, water, and air, nylon can be fashioned into filaments as strong as steel, as fine as spider's web, and yet more elastic than any of the common natural fibers.

DuPont allocated $8.5 million to build a plant that could produce 4 million pounds of nylon per year in Seaford, Delaware, and began operations in January 1940. Ten years had passed from initial discovery to full commercialization. During World War II, all nylon produced was diverted into war use, particularly for parachutes, airplane tire cords, and glider towropes that required tremendous strength. After the war, the sale of nylon to the public began, which created a marketing sensation.

Many other synthetic fibers followed the appearance of nylon, the most famous families of fibers and films being polyesters, such as wrinkle-resistant Dacron and Mylar; acrylics, such as wool-like Orlon; Spandex such as the high-elasticity Lycra; and Aramids such as Kevlar for high-strength armor and sports equipment. The polymers are made by the connection of numerous small units m, called monomers. The connection can be in a linear chain m–m–m–m,

or in a branched chain

Since 1976, the United States has made a larger volume of polymers than the sum of steel, aluminum, and copper. Currently, we use 300 pounds of polymers per year per person, which saves trees and elephants. We use polymers for such things as photographic and movie film, storage containers, and credit cards. In the 1967 motion picture "The Graduate," Benjamin the protagonist was uncertain about post-graduation plans, and he was advised by his father's friend that his future is in a single word: plastics. Plastics are inexpensive, and encourage the throwaway culture.

3.3 HOUSE

The house is a shelter from the weather, providing a barrier from the sun and heat, wind and rain, and snow and cold weather. The house also provides protection from predators, privacy for sleeping and nurturing babies, and secure storage of food and tools and belongings. A natural cave is a wonderful shelter, and can be found in the side of a hill, such as the famous Mesa Verde in Arizona. Many animals build their own shelters such as beehives, bird nests, rabbit and groundhog holes, and beaver lodges. Some of the oldest human shelters were dug into the ground, and often had covers supported by timber or bones. For an average family, the house is usually their most expensive possession.

The operations of a household benefit from furniture for sleeping and eating, equipment for storage and preparation of food, lighting, heating, and cooling, and washing and elimination. Pieces of stone furniture were found in Neolithic villages in Scotland that date back to 3100 BCE, including cupboards, dressers, beds, and seats. Knossos in Crete was built around 1700 BCE, and had storerooms with large clay vases that held oil, grains, dried fish, beans, and olives. Bathing indoors was not feasible in primitive societies, and was probably done in clean water at a pool or a creek nearby. However, Mohenjo-Daro, which was inhabited around 2800 BCE, had houses with rooms that appear to have been set aside for bathing and the palace of Knossos had bathtubs. In the Aegean island of Thera or Santorini, alabaster tubs were found with twin plumbing systems to transport hot and cold water separately. The hot water came from the geothermal hot springs on this volcanic island. Also, in Mohenjo-Daro, archaeologists found lavatories built into the outer walls of houses, which were made of bricks with wooden seats on top.

Communities must establish public methods of garbage collection and sewage removal. The Cloaca Maxima of Rome is one of the earliest sewage systems, which was used to carry the effluents from individual houses to the River Tiber around 600 BCE. In the modern world, human waste is removed from the house and treated locally in cesspools and septic tanks in rural communities or discharged into municipal wastewater treatment systems. In more elaborate societies, public utility is needed to provide a steady supply of water for drinking and washing and energy for light and power.

3.3.1 House Structure

A shelter or house should be a barrier from the weather, with barrier surfaces that are supported by a structure, strong enough to withstand anticipated forces. Some shelters have slanted surfaces that serve as both roof and wall, such as a tent by tying together a number of slanted poles rising to a single peak and covering the sides with hide or cloth. In order to cover a larger area with more headroom, most shelter structures are designed to be a set of rectangles that join to form cubic or elongated boxes with vertical walls and with horizontal or pitched roofs. The familiar basic structure is horizontal beams resting on vertical posts, such as the trilithon structure from Stonehenge that dates from 2500 BCE. Let us consider a typical structure made of two posts or pillars supporting a beam or lintel, with a slanted roof with trusses or rafters. The roof can be left flat in very dry regions, such as the Mesopotamia and Egypt, or pitched at an angle to shed rain and snow in northern climates. The roof also needs to be covered by a material that is water resistant.

A house is usually built from abundant local material, which would be clay in Mesopotamia, stones in Egypt, and wood in China and northern Europe. Vertical posts are expected to support a great deal of weight, including the structure and possible snow load. Many natural materials are good at resisting compression, such as clay, brick, and stone. The usual building material in ancient Mesopotamia was clay, which was mixed with water, straw, and dung and then formed into rectangular bricks and left to dry. These sun-dried mud bricks were somewhat durable in semi-arid places. It was expensive to produce the more durable kiln-dried bricks that required burning fuel in kilns and were reserved for palaces, temples, and elite homes.

The vertical post is subject to compression forces in holding up the weight of the structure. Figure 3.4 shows that in addition to compression, the horizontal beam is also subject to a bending force so that the center will sag under; the top surface would be under compression and the bottom surface will be under tension that can cause the beam to fracture and fail. When the length is doubled, the amount of sagging is not simply doubled but increased by the cube of 2, or 8 times! The slanted roof rafter also experiences bending, especially if the roof is heavy and the pitch angle is low. The materials clay, brick and stone are not good at resisting bending. So the Egyptian solution for horizontal spanning was either to make the stone beam extremely thick, which was expensive, or to decrease the distance between the posts, which decreased the clear space under the roof. Limestone

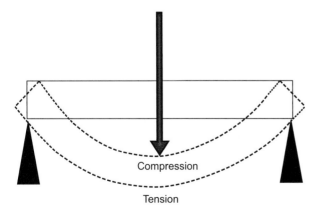

FIGURE 3.4 Bending of beam with compression on top and stretching at bottom.

beams were used to span no more than 3 m, and the stronger sandstone sometimes spanned as much as 10 m. Another solution was to use wood beams, particularly cedar from wooded Lebanon, which has strong and flexible fibers that are much better at resisting bending.

It is relatively easy to find trees that yield useful beams as long as 8 m, and there are some English cathedrals that have oak beams that are 17 m long, with a cross section of 0.7 × 1.0 m. The ancient Chinese temples and palaces had posts, beams, and rafters made of wood, but they collapsed from flammability and rotting after many centuries. That is why modern tourists can see numerous Egyptian and Greek temples with standing columns and no roofs, but the original ancient Chinese palaces and temples are no longer standing.

The problem of spanning a roof over large spaces for palaces and temples was solved later with the invention of arches and domes, where only the bricks and stones were subject to compression. Even later came the invention of steel beams that are much better at resisting bending. When a steel beam is suspended from a cable, it can span extraordinary lengths, such as the Verrazano Narrows Bridge with a span of 1300 m and the Akashi Kaikyo Bridge with a span of 2000 m.

The house also needs a number of larger openings for entry and exit and smaller openings for light and air as well as for the escape of smoke from fireplaces. For the comfort of the occupants, the openings of a house for light and air benefit from windows that can be closed against the wind and rain; for the security of the occupants, the entry needs fastening and locking to avoid the intrusion of humans or animals. Primitive windows were simply holes in the wall, but later they were shuttered with hide, cloth, or wood. The ancient Chinese used translucent paper windows, which admitted light but were not clear enough to see images. The Romans were among the first to use transparent glass windows, which admit both light and sight.

3.3.2 Light

During daylight hours, people can see indoors by sunlight through the windows. After the sun goes down on moonless nights, an alternative source of light is required for activities during the night, which is particularly long in northern winters. Fire was the oldest source of alternative illumination so that people could continue to do many useful things at night. Portable lights include torches, ceramic oil lamps, and wax candles. To shield the flame from the wind and rain, the lamp with enclosing glass housing was introduced. All such illuminations involved the risk of uncontrolled fire, especially when used in proximity to flammable substances such as curtains and Christmas trees. There have been many disastrous fires in history, such as the Great Fire of London in 1666 and the Great Chicago Fire of 1871.

When a solid body is heated to elevated temperatures, it emits a broad range of electromagnetic radiations with wavelengths from the long infrared to the short ultraviolet. This radiation is barely bright enough to be visible at a temperature of 400°C. This "black body radiation" has a peak wavelength that becomes shorter or more energetic when the temperature is raised. For instance, the surface of the sun is at 5778K (or 5505°C) and the peak radiation of 502 nm is an orange color approximately at the center of the human visual spectrum. However, the much cooler wood fire at 1200°C would peak at 2000 nm, which is mostly in the invisible infrared heat range, thus giving more heat than light. Thus, if your desire a more efficient conversion of energy into light, you need a hotter body.

Much higher temperatures can be created by electricity, which is more efficient at producing illumination. Static electricity can be created by rubbing amber with silk, which was first observed by Thales of Miletus who suggested the word "electricity" from the Greek word for amber. Upon rubbing, the substances that become positively charged include skin, leather, glass, nylon, and wool; the substances that become negatively charged include rubber, Teflon, and amber. Dynamic electricity can be created chemically by batteries or mechanically by moving magnets near electrical coils.

In 1791, Luigi Galvani published a report on "animal electricity" by using a frog's leg connected to two different metals as electrodes. Based on this observation, Alessandro Volta invented the Volta battery by replacing the moist frog tissues with a cardboard soaked in salt water. He also connected many voltaic cells in series and achieved a voltage of 50 V with a 32-cell pile.

Michael Faraday showed that electricity was created by chemical reactions of cations (positively charged) and anions (negatively charged) in a salt solution with metals. For instance, the familiar modern zinc–carbon battery has a central carbon rod as the positive terminal, surrounded by the electrolyte that is a paste of zinc chloride and manganese oxide dissolved in water, and the zinc casing as the negative terminal. Electricity is generated when zinc is oxidized to become positively charged zinc ion and releases electrons, which can migrate by a copper wire to the carbon rod; the electron is then absorbed by ammonia and manganese oxide.

In 1802, Humphry Davy had a very powerful electrical battery, and he successfully created an incandescent light by passing the electric current through a thin strip of platinum, which has an extremely high melting point of 2041K. However, this light was not very bright, and did not last long due to oxidation in the air. He also created an arc lamp in 1809, by passing electricity through two carbon charcoal electric rods (melting point 4300K) connected to a 2000-cell pile. Many subsequent attempts were made to produce a partially evacuated enclosure for the electric rods, so as to reduce oxidation and increase longevity. Joseph Swan worked with carbonized paper filaments in an evacuated glass bulb and made a demonstration in 1860, but he had difficulty producing a good vacuum and an adequate supply of electricity.

The first practical incandescent light bulb was introduced by Thomas A. Edison (1847–1931) in 1878. He was the foremost American inventor and was called "The Wizard of Menlo Park," which is the location of the first industrial research laboratory and held more than a thousand US patents in Edison's name. He also built the first electric-power central station in New York City, as well as a distribution system to reach homes, business, and factories, which has been described in Chapter 2. Many other inventors preceded him in producing electric light bulbs, but they all had flaws such as an extremely short life, high cost, and requirement of high electric currents. None of the other inventors had the entrepreneurial persistence and skill to overcome the flaws and to push their inventions into widespread use in society. Edison did not invent the first electric light bulb as is commonly thought, but he did invent the first commercially successful incandescent light that required the simultaneous distribution of electricity from power stations to homes.

When electricity flows through resistance, the rate of power consumption is proportional to power (in watts) = voltage (in volts) × current (in amperes). So a bulb of 10 W can be obtained by sending 100 V at 0.1 ampere or by sending 10 V at 1 A. Edison decided to use the relatively low voltage of 100 V, which is safer but less energy efficient. His goal was to invent a lamp bulb to contain a filament, evacuated in very high vacuum to remove oxygen, which could burn for hundreds of hours.

From 1878 to 1880, Edison and his associates tested thousands of materials for their suitability as lamp filament. He declared, "I tested no fewer than 6,000 vegetable growths, and ransacked the world for the most suitable filament material." He carbonized materials including bay wood, boxwood, hickory, cedar, flax, and bamboo. He contacted biologists and obtained plant fibers from the tropics. He even sparked a contest among his workers to contribute their beard whiskers for the experiment. After experimenting with platinum and other metal filaments, he returned to a carbon filament.

The first successful test was in 1879 for a carbonized cotton thread filament that lasted 13.5 h. His subsequent carbonized bamboo filament could last over 1200 h. This method of exhausting trial-and-error tests is called "Edisonian," in contrast to having a theoretical understanding of and guidance for systematic planning. Now that he had a practical incandescent light, the buyers of his electric lamps had to be ensured that there would be a convenient and inexpensive supply of electricity. Edison concluded that he needed to solve a far more ambitious goal: to create a complete distribution system, from generating electricity in a central station to distributing it to numerous homes and businesses by cables.

The electric incandescent light bulb had a lot of advantages over other forms of illumination, and the most important competitors at that time were gaslight and the electric arc. Compared to gaslight, the enclosed incandescent electric light did not produce soot and smoke and greatly reduced the hazards of fire. The electric light was also much easier and quicker to turn on or off and involved no maintenance. But the biggest attraction was its much greater brightness over candles and lamps, especially in public settings like theaters, sports stadiums, and major boulevards. The challenge was to persuade every home and business owner that it was worth the trouble to convert to this untried new technology, which Edison eventually won.

Throughout his career, Edison enjoyed a great deal of business success and founded many companies, including Consolidated Edison and Commonwealth Edison for electric power and the General Electric Company. He was named one of the most important people in the last thousand years. In 1983, the US Congress designated February 11, Edison's birthday, as National Inventor's Day. He died in 1931 at his home in West Orange, NJ.

Once people had electricity only for lighting, but other applications soon followed for numerous domestic uses, such as cooking, refrigeration, heating and air conditioning, cleaning, and entertainment. These labor-saving devices played a major role in the emancipation of women from long hours of daily chores in housekeeping, giving them precious time to combine domestic duties with participation in meaningful careers. Without a reliable electric supply, we could not have electric motors for elevators; without elevators, skyscrapers and the dramatic skylines of the world's major metropolises would not exist. In a modern office and factory, all the machinery operates by electricity.

Incandescent lights will eventually be phased out, as they are only 2% efficient in converting input electrical energy into output visible light, while the rest of the energy goes into heat and infrared radiation. Higher efficiency comes at a higher temperature. But even at the 7000K found at the surface of the sun (although there is no known material that remains solid at that temperature), the overall efficiency is no more than 14%. The way to produce light at higher energy efficiency is to avoid the method of black body radiation that generates a very broad wavelength distribution and to find ways to produce light only in the visible range. One current alternative is the compact fluorescent lamps (CFL), which have a rated lifespan 10 times higher and use only 20–33% of the power to produce the same brightness as incandescent lamps. In the CFL, electric current flows through the magnetic or electronic ballast into a tube filled with mercury vapor, causing it to emit ultraviolet light. The ultraviolet light then excites a phosphor coating on the inside of the tube, which emits visible light. Another alternative is the light emitting diode (LED) lamps, which are semiconductor diodes where electrons combine with holes to release energy in the form of light in an effect called electroluminescence. They are available in numerous colors and have found applications in electronic display panels for equipment, calculators, and watches. LEDs are very energy efficient, but the light output is low and the manufacturing cost is more than 10 times higher.

3.3.3 Refrigeration

The storage of food involves shielding from rats and flies, as well as from rotting and decay. A major problem is the growth of bacteria, which occurs faster in a humid environment, with an optimum growth rate between 20°C and 45°C. Bacteria do not grow well under conditions of dryness, high salinity, radiation, and extreme temperatures. Accordingly, food preservation methods involve treatments to make the food less favorable for bacteria growth, such as drying fish and meat in the sun and air; increasing salinity by salting, sugar curing, smoking, and pickling in acid or alkali; radiating with ultraviolet light or gamma rays; killing with high temperature; and suspending growth at low temperature with ice or storing in cool underground caves.

A closely related need for lower temperatures is human comfort from heat and humidity in summer and in tropical conditions. We learned how to keep warm with fire a million years ago, which played a major role in the conquest of the frozen north. But even an emperor could not cool a palace in the hot tropics and had to move to cooler mountain resorts in the summer. The Moguls in India were said to have had couriers on fast horses to relay snow from the Himalayas for their sherbet. Colonial America had underground icehouses that were insulated with straw to store ice that had been taken from frozen lakes in winter for summer consumption. The evaporation of water could be used to cool a room, but it only works in a dry climate. It is particularly difficult to be comfortable and productive in crowded indoor places such as hospitals, restaurants, lecture halls, theaters, cars, airplanes, and automobiles.

In 1748, William Cullen in Glasgow, Scotland, developed the first vapor compression refrigerator with ether as the refrigerant. The evaporation of ether cooled the remaining liquid and the vessel, but the evaporated ether could not be vented to the room and had to be collected and used again later. In 1859, Ferdinand Carre in France used ammonia as the refrigerant in a refrigerator. It was reported that in 1877, frozen mutton was shipped from Argentina to Le Havre in France, a journey of 6 months. This was a very profitable invention as frozen mutton was highly prized in Paris in the winter, and the Argentina sheep farmers became very wealthy. The practice of refrigeration spread, but the refrigerant remained hazardous. The refrigerants in use during 1930s included the following flammable and toxic chemicals: butane, ammonia, sulfur dioxide, chloroform, nitrous oxide, and methylamine.

A refrigerator has a set of evaporating cooling coils inside, connected to a set of condensing coils outside, all filled with these refrigerants. They are designed to operate under high pressures so that contaminants from the room will not leak into the coils. This means that the refrigerants have a tendency to leak out into the room, but all the refrigerants at that time were either flammable (butane) or toxic (sulfur dioxide), or both. One could go to work in the morning and return in the evening to find the kitchen full of deadly fumes. By 1929, there were 2.5 million home refrigerators in America, and refrigerator leaks were frequent. Many people died in a Cleveland hospital from a refrigerator leak, and the press clamored for a ban on the "Killer Refrigerators."

There were actually two ways to have home refrigeration safely. The first, called "The Iceman Cometh," involves constructing large industrial ice-making plants in remote rural settings where a leak would not affect many inhabitants, and the ice would be delivered to the city dwellers. The second way is to have numerous small home refrigerators, requiring the invention of a refrigerant with many required properties: effective, nonflammable, nontoxic, noncorrosive, and economical. This was the challenge presented to Thomas Midgley Jr.

Midgley was born in 1889 in Pennsylvania to a family of inventors. His father held patents for automobile wheels, and his father-in-law patented toothsaws. Midgley studied mechanical engineering at Cornell, and had little formal exposure to chemistry. When he obtained his B.S. in 1911, he was recruited by the National Cash Register Company in Dayton, Ohio. In 1921, he went to work for Charles Kettering in the Dayton Engineering Laboratories, which turned out to be a lifelong association. Kettering was famous as the inventor of the electrical starter for the automobile; and also for his role in establishing the Sloane–Kettering Cancer Center in New York City with his future boss, Alfred P. Sloane, who became CEO of General Motors. The first assignment from Kettering to Midgley was to solve the knocking in gasoline engines, which caused cars to lose power when they needed power the most—going uphill, pulling a heavy load, or accelerating upon entering a freeway. Midgley's solution was the additive tetraethyl lead, which stops knocking at the rate of 3 cc per gallon (3.8 L). It was a great invention to boost the octane number of gasoline and increase engine efficiency, so that higher compression ratios became practical. Unfortunately, it could also cause lead poisoning and was finally banned when the catalytic converter was introduced. His publicity photo in Figure 3.5 shows him as a smiling and confident person.

FIGURE 3.5 Thomas Midgley Jr.

The next great moment in the life of Midgley came in 1928, when Charles Kettering called him on the telephone and asked him to develop a nontoxic and nonflammable refrigerant. Midgley was initially skeptical that a single compound could have such properties and was considering a mixture of components. Then, he remembered the Mendeleyev periodic table that helped him to find tetraethyl lead. The relevant part of the table is shown in Figure 3.6. He studied it again, and quickly ruled out the noble gases (He, Ne, etc.) on the left column as having boiling points that are too low and also ruled out the metal elements (Li, Na, K, etc.) that are solids in room temperature. This leaves a group of nonmetal elements in the upper right hand corner of the table. He noticed that the hydrogen compounds on the first row are CH_4 methane, NH_3 ammonia, H_2O water, and HF hydrogen fluoride, which form a significant trend: the molecules become less flammable as one moves from the left to the right. There is also a vertical trend that is just as striking when one moves from the bottom upward. He found the vertical sequence of arsenic, phosphorous, and nitrogen to be less toxic as one moves up the column; this vertical trend is also observed in the sequence of tellurium, selenium, sulfur, and oxygen. So, he concluded that the best element to work with would be fluorine, the

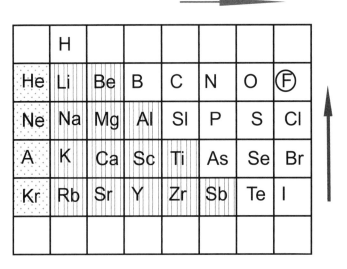

FIGURE 3.6   Midgley exploration with periodic table.

element in the upper right-hand corner, which would make the least flammable as well as the least toxic compound of the lot.

Now Midgley needed to synthesize some compounds, and he decided to start with methane CH_4, with four hydrogen atoms hanging on a single carbon atom. He investigated the effects of replacing one or more hydrogen atoms with fluorine or chlorine. There are 15 such combinations, where one or more H-atoms of methane are replaced by Cl or by F; he arranged them in the form of a triangle, with methane on the top, carbon tetrachloride on the left, and carbon tetrafluoride on the right. This is shown in Figure 3.7. He noticed another trend that the compounds on the

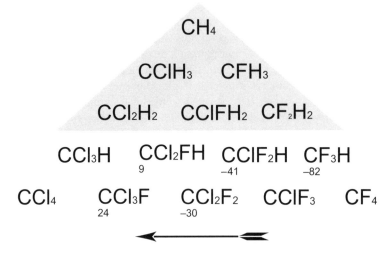

FIGURE 3.7   Midgley exploration with fluorine and chlorine substituted methane.

top three rows of the triangle with two or more hydrogen atoms were quite flammable, so they were dismissed. The band on the left began with carbon tetrachloride at the bottom, chloroform just above it, and dichloromethane above that, all of which were very toxic and were also rejected. That left seven compounds still under consideration. Under each of the remaining compounds, the small number such as 9 is the boiling point in centigrade. Midgley noticed yet another trend: the compounds with more chlorine had higher boiling points. Since he wanted a compound that boiled at around $-10°C$, he chose the compound CCl_2FH.

Then he bought four bottles of antimony trifluoride to react with chloroform, and he finished the synthesis and tested for the desired properties. It was easy to determine that the compound was not flammable. For a toxicity test, he placed a dish of this compound inside a bell jar together with a guinea pig. After 5 min, the guinea pig did not struggle for air or perish, so Midgley decided that the compound was not toxic! That was what passed for a toxicity test in those days, a far cry from the toxicity tests of today, which involve thousands of mice and many other tests.

Midgley had found what he was looking for, and it had taken only 3 days since he had received his assignment from Kettering! He was also a showman and decided to announce his discovery in 1930 at the American Chemical Society meeting in Atlanta. He sat at the podium on stage with a candle and a flask of his compound. He first announced that it was not toxic, and then swallowed a mouthful without suffering any harmful effects. Then he announced that the compound was not flammable, and he blew compound-full breath at the candle and extinguished it, to thundering applause in the room. This was, indeed, the long sought nonflammable and nontoxic refrigerant, which made home refrigerators safe, even if they leaked. This class of compounds was named chlorofluorocarbons or CFCs, also trademarked as Freon.

The refrigerator made it easy to preserve food, even in steamy tropical areas; and CFCs made it safe to have home refrigerators. It has been estimated that before refrigeration became widespread, only two-thirds of all food grown ever reached the dining table, and the rest had to be thrown away as spoilage. Better refrigeration was a tremendous benefit to help feed the world's growing population. It also became possible to eat fresh seafood, vegetables and fruits that were brought in from distant shores and farms.

During the World War II, Midgley became a national hero as the US Army Air Forces were using 100-octane fuel laced with tetraethyl lead for road vehicles and warplanes. The British and American airplanes could derive one-third more power than the German and Japanese planes, which meant they could climb faster, fly higher, and carry more weapons. This advantage often meant winning instead of dying in aerial dogfights. In 1940, Midgley suffered a severe attack of polio and became paralyzed from the waist down. He designed an intricate system of bars, pulleys, and ropes that enabled him to get in and out of bed, a wheelchair, and the swimming pool unaided. In 1944, he was elected president of the American Chemical Society. Tragically, a few months later, his wife found him dead in his bedroom, presumed to have strangled on the pulley that he had designed to get out of bed.

Due to its nonflammable and nontoxic properties, the success of CFCs led to numerous other applications unrelated to preserving food. The first obvious extension was air conditioning to remove heat and humidity, so that it became possible to have environmental comfort in offices, hospitals, restaurants, theaters, and homes even in the tropics. It is said that fire made possible the conquest and human habitation of the cold northern climates, so places like Stockholm, Moscow, and Toronto were livable in the winter; in the same way, air conditioning made possible the conquest of the hot southern climates, so the cities of Houston, Cairo, and Calcutta became more bearable in the summer.

The heat pump uses the same equipment as the refrigerator to pump heat from the colder outdoors to the warmer indoors, and is usually much more efficient than resistance electric heating, especially when the outdoor coils are buried under the ground as in a geothermal heat pump. CFCs were also used in propelling aerosol sprays, in such applications as hair styling, blowing polymers to make cushions and packing foams, dusting computers and electronics, and fighting fire in space capsules. In 1988, world consumption was 1 billion kg of CFCs at a cost of $2 billion. In the United States alone, there were 700,000 jobs that depended on CFCs.

After a period of great success and acclaim, CFCs met the same fate as the other Midgley invention of tetraethyl lead. In 1973, James Lovelock measured atmosphere concentration of CFC-11 and found it to be 0.6 parts per billion. This is not a large concentration, but CFCs are practically indestructible, so this concentration would inevitably increase with time. Ozone O_3 is produced in the stratosphere at 10–30 km above the earth, by the chemical reaction of oxygen molecule O_2 and ultraviolet light. Ozone protects the earth from deadly ultraviolet radiation, so the loss of ozone can have catastrophic consequences on life. In 1974, Rowland and Molina forecast that CFCs would rise to the stratosphere 10–30 km above the earth and then break down by solar ultraviolet radiation to form chlorine radicals, which would combine with and destroy the ozone.

The days of CFCs were coming to an end, and in 1987 a NASA expedition to the South Pole discovered an ozone hole in Antarctica, which confirmed the predictions of Rowland and Molina. The 1987 Montreal Protocol called for substitutes to replace CFCs. Since the problem was with the chlorine atoms in CFCs, the nearest replacement was HFCs or hydrofluorocarbons, without any chlorine. They are the compounds on the right side of the triangle of the 15 compounds, which includes CF_3H and CF_4. But without chlorine, these two compounds have boiling points that are too low to be useful as refrigerants. So the decision was to shift to HFCs based on the larger molecule ethane C_2H_6 instead, and the most suitable compound was HFC 134a, which is CFH_2CF_3 without chlorine. This solution turned out to be equally problematic, as all the HFCs were found to be active greenhouse gases, which contribute to global warming. Now the HFCs are to be banned by the Kyoto Protocol after 2011. It is not clear that we can create any compound that would satisfy all of the requirements; thus, we may have to choose among compounds that are "mildly flammable," "mildly toxic," or "mildly greenhouse." When even the most innocuous objects are used extensively over a long period of time, they could accumulate somewhere and cause problems. Sometimes, today's solutions become tomorrow's problems.

Freeman Dyson posed the question, "How does global warming rank among the worst ills of the worlds?" We are currently dealing with nuclear proliferation;

territorial, ethnic and religious conflicts and atrocities; terrorist attacks on civilians; infant mortality; lack of clean water, nutrition, and medicine in less developed countries; epidemics of AIDS, SARS, avian flu; urban homeless, crime, narcotics; hurricane, earthquake, tsunami; and global financial crises. The people and governments of the world are not well organized, and have limited will and budget to solve all the worst problems on earth. We cannot tackle all the problems successfully at the same time, and may have to assign a ranking to each problem—including global warming—in consideration of the degree of certainty, the immediacy of harm, and the number of injuries and deaths involved. How would you budget national and international expenditures for all these good deeds, and distribute rationally among these problems?

REFERENCES

Alexander, J. W. "Economic Geography", Prentice-Hall, Englewood Cliffs, NJ, 1963.

Baldwin, N. "Edison, Inventing the Century", Hyperion, New York, 2001.

Barker, G. "The Agricultural Revolution in Prehistory: Why Did Foragers Become Farmers?", Oxford University Press, New York, 2006.

Bogucki, P. "The Origins of Human Society", Blackwell Publishing, Malden, MA, 1999.

Brown, S. R. "A Most Damnable Invention", St. Martin's Press, New York, 2005.

Christian, D. and W. H. McNeil. "Maps of Time: An Introduction to Big History", University of California Press, Berkeley, 2004.

Coale, A. J. and S. Watkins, editors, "The Decline of Fertility in Europe", Princeton University Press, Princeton, NJ, 1987.

Constable, G. and B. Somerville. "A Century of Innovation: Twenty Engineering Achievements That Transformed Our Lives", Joseph Henry Press, Washington DC, 2001.

Cowan, R. S. "More Work for Mother: The Ironies of Household Technology from the Open Hearth to the Microwave", Basic Books, New York, 1983.

Daniel, C. "Master Mind: The Rise and Fall of Fritz Haber, the Nobel Laureate Who Launched the Age of Chemical Warfare", Ecco, New York, 2005.

Darwin, C. "The Voyage of the Beagle", Doubleday Anchor, New York, 1962.

Davson, H. and M. G. Eggleton. "Principles of Human Physiology", Little Brown, Boston, 1964.

Dyer, L. D. and T. C. Martin. "Edison, His Life and Inventions", Project Gutenberg, 2006. Available at http://www.gutenberg.org

Gere, J. M. and S. P. Timoshenko. "Mechanics of Materials", Van Nostrand Reinhold, New York, 1972.

Gibbons, A. "First Human: The Race to Discover Our Earliest Ancestors", Doubleday, New York, 2006.

Goran, M. "The Story of Fritz Haber", University of Oklahoma Press, Norman, OK, 1967.

Haber, L. F. "The Poisonous Cloud: Chemical Warfare in the First World War", Oxford University Press, London, 1986.

Hager, T. "The Alchemy of Air: A Jewish Genius, A Doomed Tycoon, and the Scientific Discovery That Fed the World but Fueled the Rise of Hitler", Harmony Books, New York, 2008.

Hughes, T. P. "Networks of Power: Electrification in Western Society 1880–1930", Johns Hopkins University Press, Baltimore, 1983.

Institute of Medicine. "Recommended Dietary Allowance", National Academy Press, Washington DC, 1989.

Jonnes, J. "Empires of Light: Edison, Tesla, Westinghouse, and the Race to Electrify the World", Random House, New York, 2003.

Kobe, K. A. "Inorganic Process Industries", Macmillan, New York, 1948.

Kvavadze, E., O. Bar-Yosef, A. Belfer-Cohen, E. Boaretto, N. Jakeli, Z. Matskevich, and T. Meshveliani. "30,000-Year-Old Wild Flax Fibers". *Science* 325, 1359, 2009.

Linsley, R. K. and J. Franzian. "Elements of Hydraulic Engineering", McGraw-Hill, New York, 1955.

McGrayne, S. B. "Prometheus in the Lab", McGraw-Hill, New York, 2001.

Nobel and Foundation. "Norman Borlaug". Available at http:\\www.nobelprize.org, 2012.

Shreve, R. N. "Chemical Process Industries", McGraw-Hill, New York, 1967.

Smil, V. "Enriching the Earth: Fritz Haber, Carl Bosch, and the Transformation of World Food Production", MIT Press, Cambridge, MA, 2001.

Stanier, R. Y., M. Doudoroff, and E. A. Adelberg. "The Microbial World", Prentice-Hall, Englewood Cliffs, NJ, 1963.

Strahler, A. and A. H. Strahler. "Elements of Physical Geography", John Wiley & Sons, Inc., New York, 1989.

Stross, R. E. "The Wizard of Menlo Park: How Thomas Alva Edison Invented the Modern World", Crown Publisher, New York, 2007.

Ungar, P. S. and M. F. Teaford, editors, "Human Diet: Its Origin and Evolution", Bergin & Garvey, Westport, CT, 2002.

White, A., P. Handler, and E. L. Smith. "Principles of Biochemistry", McGraw-Hill, New York, 1964.

Wrangham, R. "Catching Fire: How Cooking Made Us Human", Basic Books, New York, 2009.

CHAPTER 4

HEALTH, REPRODUCTION

We all desire a long and healthy life that is free from disease and pain. If we do become ill we strive for good diagnosis and treatment as a secondary line of defense. The World Health Organization of the United Nations defines health as "a state of complete physical, mental and social well-being and not merely the absence of disease or infirmity." Most of us also desire to have children to love and to perpetuate our line.

When we suffer physical trauma, like a knife wound or a fracture, the injury clearly requires some intervention, such as surgery, to heal the body. Some form of surgery was practiced by humans as far back as the Neanderthals, as was evident from some skeletons found in their burial grounds; broken bones and ribs had been *splinted* together. Also, skull bones were found that had been drilled open to drain fluid or to remove tumors, showing that trepanation had been practiced. In the *Iliad*, the surgeon Machaon was mentioned as being more valuable than a mere warrior.

When there is internal pain or fever, the cause of illness is not so clear and diagnosis and treatment is more difficult. In the ancient world, a mystical view attributed the origin of illness to random misfortune or to the malicious acts of a god, demon or witch. In the Epic of Gilgamesh, when the hero Enkidu fell ill by the anger of the gods for killing Humbaba and the Bull of Heaven, his best friend King Gilgamesh was unable to do anything for him. When illness struck in everyday lives, people made pilgrimages to healing temples and sought out priests to divine the source of their disease, and learn what prayer, sacrifice, or charm might make them well.

The World Health Organization keeps track of vital statistics, and reported that in 2004, 57 million people died worldwide out of a population of 6.2 billion. The three leading causes of death were noncommunicable diseases (59%), communicable diseases (32%), and injuries (9%). The noncommunicable diseases include hypertension, diabetes, cardiovascular diseases (stroke, heart attack), cancer, asthma, allergy, and mental health problems. In contrast, until recently, the death rate was dominated by communicable or infectious diseases originating from microbes. The greatest glory of modern medicine is the dramatic reduction in the devastation caused by communicable diseases. We have also made improvements in treating noncommunicable diseases such as those caused by genetics, lifestyle, and ageing.

Great Inventions that Changed the World, First Edition. James Wei.
© 2012 John Wiley & Sons, Inc. Published 2012 by John Wiley & Sons, Inc.

Communicable diseases are caused by the invisible microbes: viruses, bacteria, fungi, and protozoa.

- Viruses are curious organisms, not much more than genetic material protected by a protein coat. They cannot even reproduce except by invading host cells. They are responsible for diseases such as flu, chickenpox, smallpox, and AIDS.

- Bacteria are single-cell prokaryotes without a nucleus, including rod-like bacillus, spherical cocci, and coiled spirochetes. They are responsible for diseases such as tetanus, typhoid, pneumonia, syphilis, cholera, and tuberculosis.

- Fungi are larger eukaryotes with a nucleus such as yeasts, molds, and mushrooms; they are responsible for diseases such as athlete's foot and ring worm.

- Protozoa are even larger eukaryotes, many of which are motile; they are responsible for diseases such as dysentery, malaria, and sleeping sickness.

The impact of the revolution in health care and nutrition can be seen when we compare the recent health statistics of the less developed countries (LDC) with those of the more developed countries (MDC), which are shown below. It can be seen that the life expectancy at birth is much higher in Japan and the United States in comparison with sub-Saharan Africa. Even more striking is that the under-5 mortality per 1000, formerly 1/3 to 1/4 of all births, has now fallen to only 1/100 births in the MDC—a tremendous improvement. In contrast to the mortality rate, the crude birth rate and fertility rate per woman in some of the MDC countries is only a quarter of that in the LDC countries. In fact the birth rate in some MDC countries is below 2.1, which is the estimated replacement rate of fertility/woman. This means that many advanced countries cannot maintain their population without immigration (Table 4.1).

TABLE 4.1 Vital Statistics of 12 Nations

Nation	Life expect at birth	Under 5 mortal/1000	Crude birth/1000	Total fertility/woman
Angola	40	260	48	7.1
Mozambique	44	235	40	6.2
Togo	46	177	37	6.4
Niger	47	320	49	7.9
India	59	115	24	3.8
Brazil	67	57	19	2.8
China	60	45	12	2.1
Russia	69	27	10	1.3
United States	78	11	14	2.1
Spain	81	9	11	1.3
Switzerland	82	9	10	1.6
Japan	82	6	9	1.5

4.1 PREVENTION

"An ounce of prevention is worth a pound of cure." Modern healthcare has evolved to placing significant emphasis on preventive medicine for individuals and for entire populations. The simplest methods are directed toward individual habits and lifestyle, such as eating a healthy diet, exercising regularly, avoiding tobacco, drinking moderate amounts of alcohol, hand washing, and other personal hygiene. Prophylaxis is a set of more advanced methods to prevent illnesses, such as vaccination against polio and tetanus, medications to lower cholesterol and blood pressure, and barrier methods to prevent sexually transmitted diseases. Public health organizations endeavor to prevent illnesses at the population level, including the control of communicable diseases and epidemics, environmental health, community health, and occupational health.

When the era of rational medicine arrived, magic and divine anger were banished in favor of more scientific theories. Hippocrates of Cos (460–380 BCE) believed that diseases were not punishments inflicted by the gods, but rather the product of environmental factors, diet, and living habits. He believed in the balance of four humors (Table 4.2).

He taught that illness was caused by the imbalance of the four humors, and prevention consisted of balancing the humors and letting nature take its course. He suggested appropriate diet and exercises, and preached holistic medicine; illness was to be understood in the context of a patient's environment and lifestyle, and riskier interventions such as drugs or surgery were avoided unless all else failed. Since he lived before the age of science, he neither tested his theories with experimental demonstrations, nor suggested ways to quantify the levels of the four humors in a healthy or sick person. In an era when a physician had few effective cures in his medicine bag, the "Do no harm" practice of Hippocrates was often the best course of action.

Modern science or evidence-based medicine finally arrived when the germ theory of disease was established through the work of Louis Pasteur and Robert Koch. This theory postulated that many illnesses are due to the presence of specific germs that could be demonstrated to reside in each diseased person. The prevention and cure of a disease would be based on methods to prevent the entry of germs into a person, and on discovering drugs that would kill the germs without harming normal human tissue.

TABLE 4.2   The Four Humors of Hippocrates and Galen

Blood	Yellow bile	Black bile	Phlegm
Sanguine	Choleric	Melancholic	Phlegmatic
Spring	Summer	Autumn	Winter
Infancy	Youth	Middle age	Old age
Hot, wet	Hot, dry	Cold, dry	Cold, wet
Air	Fire	Earth	Water

4.1.1   Vaccination: Smallpox

Smallpox is a deadly infectious disease caused by the *Variola* virus, which at one time caused more than 10% of all the deaths in the world and killed an estimated 400,000 Europeans each year during the eighteenth century, including over 80% of infected children. It is also a symbol of one of the greatest successes in modern medicine, and in 1979 was declared by the World Health Organization to be eradicated. This amazing turnaround was accomplished by the development of the first vaccine. Early variations on the vaccine were possible because the *Variola* virus occurred in two forms: a major form with an overall mortality rate of 30–35% and a minor form with only a 1% mortality rate.

A form of inoculation against smallpox was performed in Asia and Africa, and spread to Europe later. This practice of inoculation was introduced in England by Lady Mary Montagu who observed its practice by Turkish physicians in Constantinople around 1716. She learned to take liquid from a smallpox blister in a mild case, and carried it in a nutshell to inoculate her family in London. This practice spread among the royal families of Europe. In America, Cotton Mather learned about the African practice of inoculation from a Sudanese slave, and inoculated many people in Boston in 1721. Voltaire in France also advocated inoculation. Inoculation in the East was historically performed by blowing smallpox crusts into the nostril. The European method involved taking material from a smallpox pustule on a mildly infected person and rubbing the material into a scratch on the skin of the person on the receiving end. This method was called *variolation*, and conferred upon the recipient a much milder case of smallpox and thereby immunity from the more deadly major form of smallpox. However, with this method patients still contracted the smallpox infection although in a minor form. This resulted in many weeks of sickness as well as fasting, purging, bleeding, and isolation as was the custom at that time.

Edward Jenner (1749–1823) was a surgeon and general practitioner in England. As a child, he went through the ordeal of *variolation*. He first received recognition and was elected a fellow of the Royal Society in 1788 for his observations on cuckoos and their habit of stealing other birds' nests and pushing their eggs and fledglings from the nest. As an adult physician, he heard that milkmaids who had contracted the much milder sickness of cowpox became immune to smallpox. This piece of folk wisdom had never been tested definitively. In 1796, Jenner treated a milkmaid named Sarah for cowpox. Later, he obtained parental permission to inoculate 8-year-old James Phipps, by taking fluid from a cowpox pustule on Sarah's hand and putting it on two scratches on the boy's arm. Seven days later, James complained of mild discomfort but completely recovered. A month later, Jenner inoculated James on both arms with fresh material taken from a smallpox pustule. James developed blisters within a few days where he had been inoculated, but showed no sign of disease. Jenner was able to demonstrate with a series of 23 subjects that immunity to smallpox had been achieved by this procedure of inoculation with cowpox.

Jenner was convinced that he had discovered a great boon to mankind, and he called this *vaccination*. He drafted a paper for the Royal Society that was rejected, so he published the results privately in 1798. Cowpox vaccination proved to be effective in preventing smallpox and was far safer than smallpox inoculation, and spread throughout the world with great speed. In 1840, the British government banned variolation with smallpox, and provided vaccination with cowpox free of charge. This was one of the first times in human history when a medical doctor had something truly effective to offer the patients, besides high sounding theories and advice on healthy habits. Now there are vaccinations for a wide range of diseases including rabies (Pasteur 1885), tuberculosis, polio (Salk 1953), diphtheria, cholera, typhoid, whooping cough, and influenza. Since 1980, smallpox has been an eradicated disease and remains only in laboratories in the Centers for Disease Control and Prevention (CDC) in Atlanta, Georgia.

4.1.2 Germ Theory: Hygiene

Modern science-based medicine began with the germ theory, which holds that numerous infectious diseases are caused by microbes invisible to the naked eye. Hygienic habits are now part of modern daily life: we wash our hands, boil or chlorinate our water, put antiseptics on cuts, and store food in refrigerators.

The study of microbes began with Antoni van Leeuwenhoek, a cloth merchant (1632–1723) from Delft, Holland. In 1653, he went to Amsterdam and saw a simple three-power microscope being used by textile merchants to examine merchandise. He returned to Delft and started a drapery business, but did not study at a university. At that time, microscope lenses were made from glass disks by laborious grinding and polishing. He began to develop his own microscope lens using a hot flame, instead of grinding and polishing, to create small spheres of glass. The smallest spheres provided the highest magnifications of up to 275 times. Besides textile specimens, he had the curiosity to look at everything else around him, including plants, raindrops, and water from ponds. He described the many "little animals" that he observed under his single-lens microscope: molds, protozoa, bacteria, and sperm cells. He corresponded with the Royal Society of London, but was met with some skepticism. Eventually, a delegation was sent by the Royal Society in 1680 to examine his lens and methods of operation at Delft, and vindicated his observations. He was made a Fellow of the Royal Society the same year.

Most physicians at that time still believed that human diseases were caused by the imbalance of body humors, changing seasons, atmospheric miasmas, and astrological conjunctions. The Hungarian surgeon Ignaz Semmelweis (1818–1865) was head of the obstetrical clinic of the Vienna General Hospital. The clinic was separated into the First Clinic, which was staffed by surgeons and medical students; and the Second Clinic, which was staffed by midwives. In 1847, Semmelweis studied the high incidence of women who died of puerperal or childbirth fever at the clinics, where mortality rates ranged from 10 to 35%. He noticed that women whose babies were delivered by midwives at the Second Clinic were three times more likely to survive than if their babies were delivered by the surgeons and medical students in the First Clinic. The two clinics used identical procedures and appliances, and furthermore, the Second Clinic was more crowded and popular. Women would beg not to be admitted to the First Clinic, and some even preferred to give birth on the streets.

Semmelweis also noticed that after a surgeon from the First Clinic acciden-tally punctured his finger during an autopsy, he died from infection a few days later. He postulated that if the surgeon's contaminated scalpel carried enough contagion to kill him, then his hands and clothing could also be carriers of contagion. Semmel-weis reasoned that when a surgeon rushed from an autopsy to a delivery room with bloody hands and clothing, he would be carrying this contagion that he called "cadaverous particles."

Soon after, he ordered that before entering a ward, all students or doctors must wash their hands thoroughly in a solution of chlorinated lime. The death rate of new mothers had been 18% before this order, and fell to 1% after that. Despite his work, his practice of hand washing was not accepted by the medical world, partly because he did not have a satisfactory theory to explain how the cadaverous particles caused disease or death. After years of ridicule, Semmelweis suffered a nervous break-down, and died in an asylum at age 47. It was not until Louis Pasteur established the germ theory a few years later that the Semmelweis procedure of hand washing became the accepted standard practice around the world.

Another major figure in the pregerm theory days was John Snow (1813–1858), an English surgeon in London famous for administering chloroform to Queen Victoria in 1853 during childbirth. In 1848, cholera struck England and killed many people. The symptoms included severe diarrhea cramps and dehydra-tion. There was no known cure, and the prevailing theory was that "miasma" or bad air was the cause of this illness. In 1854, cholera struck London again, and the death toll increased from 4 to 500 in 10 days. Snow suspected the water supply in the area of Soho was somehow involved. He drew a map of all the fatalities in the area as well as the locations of all the public water supply pumps. He found that most of the deaths clustered around a water pump on Broad Street. That water pump was in turn near a sewage pipe that passed through the area and Snow suspected that this sewage pipe could be carrying the contagion. He brought his findings to the atten-tion of the city authorities, which resulted in an order to remove the pump handle. Within a few days, the cholera outbreak dwindled. To provide further evidence to his findings, Snow noticed that in another part of town there were three sources of water: two of them, Southwark and Vauxhall, drew from the Thames River; and the third one at Lambeth drew from a distant source. He noticed that there were 286 deaths in families drawing water from Southwark and Vauxhall, and only 14 deaths in families drawing water from Lambeth. So even before the germ theory of disease was established, there was evidence that the sources of infectious diseases was some invisible substance carried by unwashed hands and clothing, or by an unclean water supply. Today, Snow is considered to be one of the fathers of epidemiology.

Louis Pasteur (1822–1895) was born near the French Jura Mountains, and trained to be a chemist at the elite École Normale Supérieure in Paris. His photo is shown in Figure 4.1. He became a professor of chemistry at the University of Strasbourg, where he made his first great discovery in 1849 in stereochemistry. He discovered that when tartaric acid is chemically synthesized, it forms into crystals that rotate light to the left or the right, due to their three-dimensional molecular structures. He found that grapes produce only the right rotating form of sugar D-glucose that is the digestible form. The left rotating form of L-glucose cannot be

FIGURE 4.1 Louis Pasteur.

digested by living organisms. This became the foundation of stereochemistry, of molecules produced by living organisms.

In 1849, Pasteur married Marie Laurent, the daughter of the rector of the University of Strasbourg. They had one son and four daughters, but two died of typhoid and one died of a brain tumor, so only two survived to adulthood. These personal tragedies inspired him to study the causes and treatment of diseases. In 1856, he returned to the École Normale Supérieure as director of scientific studies.

Pasteur's next great discovery was in 1857 when he was asked by wine makers to investigate a large and mysterious outbreak of wine spoilage. He made a very important finding by examining the wine under a microscope: in the normally fermented wine, he observed only one type of yeast; but when he examined the spoiled, cloudy, and foul-smelling wine, he found many other types of bacteria. His discovery showed vintners and brewers how the growth of microorganisms was responsible for spoiling beverages. He then invented a process with Claude Bernard in 1864 to reduce the number of microorganisms in a beverage by heating. The pasteurization process applied a high temperature of 72°C to wine for 15–20 s, which proved sufficient to control yeasts, molds, and common bacteria. This technique has been a tremendous lifesaver for millions of people; however in recent years, heat resistant pathogens have been found that can survive pasteurization.

From the time of Aristotle, people believed in spontaneous generation, the theory that living organisms could develop from nonliving sources of decaying organic matter, such as maggots appearing to sprout from rotten meat. Pasteur, however, believed in Biogenesis, the theory that living organisms could only come from other living organisms. Although he was not the first to suggest that infectious diseases were caused by microorganisms, Pasteur produced the first convincing demonstration of this principle. In 1862, he devised two methods to prevent particles in the air from contacting broth that had been boiled to kill any living matter. In one, he exposed the boiled broth to air in a vessel, but covered it with a filter that was impervious to particles. In the second, he put broth in a vessel that only contacted outside air through a long S-shaped tube that would not allow dust particles to pass. Nothing could grow in either of these two vessels until the flasks were broken open. This demonstrated that the source of infection came from the outside, and did not spontaneously appear in the broth. His demonstration gave support to the premise that better sanitary control could reduce infectious diseases in surgery, hospital management, agriculture, and industry. Thus, Pasteur dealt a deathblow to the theory of spontaneous generation and firmly established the germ theory.

There are many more examples of Pasteur's work with infectious microbes. In 1865, a devastating silkworm disease had damaged French silk production, and Pasteur was again called in to save an industry. He found two microbes that had caused the silkworm deaths, developed methods for the farmers to screen silkworm eggs for signs of the disease, and ultimately saved the industry. In the 1870s, anthrax ravaged the cattle population in Europe. Pasteur guessed that a rod-shaped bacillus was the cause of anthrax. He took a drop of fresh blood from an animal that had died of the disease, and placed it in a flask containing a sterile liquid culture medium. When the organism had multiplied, he transferred a drop to a second flask. After a dozen similar transfers, he injected the material into animals. He found that among the animals injected with material from the flasks, 12 died of anthrax within days. In 1881, he developed a vaccine against sheep anthrax, and demonstrated that with vaccination, 24 out of 25 sheep survived; without vaccination, none out of 25 sheep survived. In 1879, Pasteur also invented immunization with weakened or attenuated cultures. Five years later, he developed a rabies vaccine for dog bites, using dried spinal cord from infected rabbits. He died after a series of strokes in 1895. The Institute Pasteur in Paris was established in 1888 in his honor, and is devoted to the battle against infectious diseases.

The English surgeon Joseph Lister (1827–1912) was a professor of surgery at the University of Glasgow. He read Pasteur's publications about fermentation, which discussed three methods of killing bacteria: filtering, heating, and exposing to a chemical solution. Lister determined that the first two methods were not appropriate with human wounds, so he chose to work with carbolic acid (also called phenol), a chemical that had been used to deodorize sewage. He sprayed this chemical on his surgical instruments, surgical incisions, and wound dressings. He found remarkable results in reducing the incidence of gangrene that he published in *Antiseptic Principles of the Practice of Surgery* in 1867. He recommended that surgeons wear gloves and wash their hands with a 5% solution of carbolic acid before and after operations. He also advised against the use of porous natural materials in

FIGURE 4.2   Robert Koch. Reprinted courtesy of National Library of Medicine.

the handles of surgical instruments because they could harbor harmful microbes. The Listerine mouthwash was named after him.

Robert Koch (1843–1920) was a German physician, and with Louis Pasteur is considered to be a cofounder of medical microbiology. His photo is shown in Figure 4.2. He was a country doctor in Prussian Poland with very limited resources, but he became famous for his medical research and received the Nobel Prize in Physiology or Medicine in 1905. Koch was able to isolate the anthrax bacteria from blood and grow pure cultures. In 1877, he found that the anthrax bacteria could not survive outside an animal host for long but could build persistent endospores that would remain dormant in the soil for a long time with the potential to become active again under the right conditions.

For these accomplishments, Koch was promoted to a position in Berlin with much better facilities. While in Berlin he developed *Koch's postulates*, which became the foundation of scientific procedures in modern medicine. *Koch's postulates* state that in order to establish an organism as the cause of a disease, the following conditions must be satisfied:

- The organism is found in abundance in all animals examined with the disease, but is not found in healthy animals.

- The organism can be isolated from diseased animals, and can be prepared and maintained in a laboratory in a pure culture.
- The organism causes disease when introduced into a healthy animal.
- The organism can be inoculated into a healthy animal, then retrieved from the infected animal in an identical form to the original organism.

Another of Koch's important contributions was the development of the Petri dish, which was invented by Julius Petri while he was working as Koch's assistant. Koch went on to discover the tuberculosis bacteria in 1882, for which he received the Nobel Prize in 1905. He also isolated the cholera bacteria when he traveled to Egypt in 1883 during an epidemic. He died in 1910 in Baden-Baden at age 66.

Once the germ theory of infectious diseases had been demonstrated beyond a doubt, it became paramount to develop effective methods to prevent such diseases by avoiding contact with infected material. Medical hygiene practices include isolation or quarantine of infectious persons or materials, sterilization of instruments used in surgical procedures, use of protective clothing and barriers such as masks and gloves, proper bandaging and dressing of injuries, safe disposal of medical waste, disinfection of reusable supplies such as linen and uniforms, and careful hand washing with antibacterial cleansers. Food safety practices include cleaning and sterilization of food areas, washing of hands and utensils, proper storage of food to prevent contamination, and refrigeration of foods. It is said that millions of people are alive today because their grandparents washed their hands.

4.2   DIAGNOSTICS

There are many diseases that are "silent" in the early stages so patients are not aware that they are ill. Healthcare screening is concerned with discovering diseases before they impact a patient's health in an effort to prevent future illness. The annual physical checkup is expensive but can be effective in early detection. The yield and accuracy of early detection is the most cost effective when targeting groups with special conditions that make them more vulnerable to such diseases as cancer, hypertension, and diabetes. The clinical examination by a physician includes checking vital signs such as temperature and blood pressure, and a physical exam of the heart, lungs, and abdomen. The exam is usually supplemented by laboratory tests such as blood glucose and cholesterol levels, and sometimes by other tests such as the electrocardiogram and mammogram.

Physicians in ancient times relied on the physical exam to try to discover what was wrong with patients. In modern medicine, we have many amazing tools in our arsenal to help decipher the causes of illness. When a person becomes sick and visits a primary care physician or nurse, the first duty of the physician is to find out the cause and extent of the illness before deciding on a course of therapy. The techniques of modern diagnosis include the following:

- *Physical Examinations:* The primary care physician would begin by talking to a patient about his symptoms; getting the patient's medical, family, and social history; and consulting the previous records of the patient. This would be followed by examination of the vital signs: temperature, blood pressure, pulse, and respiration rate. The physical examination procedures include inspection, palpation, and tapping of the chest, abdomen, and limbs; listening to the lungs, heart, and abdomen; looking into the eyes, nose, and throat; checking the sensation and strength of the limbs; and assessing the mental status, speech, and state of consciousness.

- *Medical Laboratory:* Samples of blood, sputum, fecal matter, and urine are collected, and diseased tissue are surgically removed for biopsy, where the tissue is taken to the medical laboratories to be analyzed. The microbiology laboratory looks for bacteria, viruses, and other parasites; the chemistry laboratory looks for toxic substances, liver and kidney functions, hormone levels, electrolyte composition, and sugar and cholesterol levels; and the hematology laboratory looks at the blood composition of white cells, red cells, platelets, and rate of coagulation.

- *Medical Imaging and Other Tests:* These measurement techniques subject the body of the patient to light, sound, and radiation to probe for sources of disease. Some of them produce charts such as the electroencephalography (EEG) for the head and brain, and electrocardiography (EKG) for the heart. Endoscopy involves using fiber optic scopes to directly see inside internal organs such as the colon, stomach, or joints. Imaging technologies produce pictures, such as the X-ray (Roentgen 1895), ultrasound (Hertz 1953, Donald 1958), positron emission tomography PET (Kuhl, Edwards 1958), computed tomography CT (Hounsfield, Cormack 1973), and magnetic resonance imaging MRI (Lauterbur, Mansfield 1973).

From these findings, the physician constructs the underlying physiological or biochemical causes of a disease or condition, and considers methods of therapy. Many instruments and procedures were invented to facilitate these examinations. Before the invention of the thermometer and the quantitative measurement of temperature, the physician could only describe the patient qualitatively as either feverish, normal or cold. The principle of the expansion of air or liquid with temperature was known to Hero of Alexandria. Galileo invented a temperature measuring device with many floating or sinking bulbs, which gave semiquantitative information of the ambient temperature according to the number of sinking bulbs, but could not be used to take a patient's temperature. In 1665, Christiaan Huygens suggested using the melting and boiling points of water as a universal standard for all temperature scales and thermometers. In 1724, Daniel Fahrenheit produced the first thermometer that used the expansion of mercury that could be read on a numeric scale.

A person suffering from high blood pressure has an increased risk of developing a stroke or heart attack. These cardiovascular diseases have become some of the most important causes of death in industrialized countries today. In order to regulate blood pressure we need an instrument that is convenient, painless, reproducible, and accessible in a clinic or even at home. The modern sphygmomanometer has reached a satisfactory design after 200 years of inventions and modifications. The modern sphygmomanometer can measure two important components of blood pressure: the systolic, which is the peak pressure near the end of the cardiac cycle; and the diastolic, or steady minimum pressure. A normal resting adult should ideally have a reading of close to 120/80.

The first device for measuring blood pressure was invented by Stephen Hales (1677–1761). A graduate of Cambridge University, Hales studied many natural phenomena such as measuring the effects of the sun's rays on raising the sap in trees. In 1733, he published a paper on his experiments on opening an artery in a live horse, inserting a brass pipe 1/6 of an inch (4.2 mm) in diameter, and attaching a vertical glass tube to this pipe. He then allowed blood from the horse to enter the tube, and found that the blood level rose to a height of 8 ft 3 in. (2.5 m). He also noticed that the height of the blood column rose and fell by 2–4 in. (50–100 mm) with each heartbeat. He was made a Fellow of the Royal Society and his work won the Copley Medal. This was the first quantitative measurement of blood pressure, but was obviously not suitable for use in people.

Jean Poiseuille was a French physician-physicist, famous for his work on the viscosity of fluids. In 1828, he received a gold medal of the Royal Academy of Medicine on the use of a mercury manometer for the measurement of arterial blood pressure. His work enabled the Hales tube of 8 ft 3 in. to become greatly shortened to a more convenient tube of 8 in. (200 mm) of mercury. However, Poiseuille's method was still invasive as it involved puncturing the skin and the artery, subjecting the patients to pain and the associated risks of bleeding and infection.

Samuel Siegfried Karl Ritter von Basch was the first to produce a workable noninvasive sphygmomanometer in the 1870s, by using an inflatable rubber bag filled with water. The bag was pressed over the pulse until pulsation ceased and the height of the mercury column attached to the rubber bag reflected the systolic pressure. Scipione Riva-Rocci in 1896 introduced the familiar air-inflated cuff that surrounds the entire circumference of the arm, rather than resting on top of the artery. In those days the determination of the absence of pulsation was by feel or palpation; this was not very accurate and there was also no way to determine the diastolic pressure.

The sphygmomanometer finally reached its present form in 1905 when the Russian surgeon Nikolai C. Korotkoff reported that by placing a stethoscope over the brachial artery next to the cuff, tapping sounds could be heard as the cuff was deflated, caused by blood flowing back into the artery. When the cuff is initially fully inflated it cuts off all circulation and the examiner hears silence through the stethoscope. The pressure of the cuff is slowly released till blood starts to flow in the artery; this turbulent flow creates a whooshing or pounding, which is called the first Korotkoff sound, and corresponds to the systolic blood pressure. The cuff pressure is further released until again no sound can be heard, and this corresponds to the diastolic arterial pressure. Korotkoff presented a 207-word paper to announce this auscultator or listening method. For more than a century, his method has been used by millions of doctors with essentially no change. Although there are modern digital electronic instruments that are smaller and more convenient to use, the mercury manual manometers are still considered to be the gold standard as they do not need to be calibrated.

4.2.1 Imaging: X-Ray

Wilhelm Röntgen was born in 1845 in Lennep Germany, but grew up in Holland. He loved nature, and was good at making mechanical contrivances. He studied physics in Utrecht, and then studied mechanical engineering at the Swiss Federal Institute of Technology in Zürich. In 1871, he moved to the University of Würzburg in Germany, and then to Strasbourg. He married Anna Bertha of Zürich whom he met in a café run by her father. He was an incredibly versatile researcher, and excelled in many fields including the heat capacity of gases, heat conductivity in crystals, and rotation of polarized light in gases. He developed a high level of experimental skill, and was able to observe and measure phenomena very accurately. For a while, he worked on the relationship between light and electricity at the University of Giessen, but later returned to Würzburg in 1888 as a professor of physics. In 1894, he became the rector of the Physical Institute at Würzburg, and a year after that he made his momentous discovery of X-rays.

In 1895, Röntgen had begun to study cathode rays coming from a Crookes tube, which was the invention of William Crookes around 1870. X-rays had actually been produced by others long before Röntgen; in 1879 Crookes complained of fogged photographic plates that were stored near his cathode-ray tubes. These tubes were partially evacuated glass cylinders with two metal electrodes at each end. When a high voltage was applied, electrons traveled in straight lines from the cathode end to the anode end, and finally hit a metal target creating radiation in the vacuum.

In November 1895, Röntgen discovered that even after the tube was enclosed in a sealed black carton to let no light escape, a nearby paper plate covered with a barium compound became fluorescent and glowed brightly. This was so startling and unanticipated, that he spent the next 6 weeks in absolute concentration, repeating and extending his observations. He found that these invisible rays traveled in straight lines, could not be refracted or reflected, were not deviated by a magnet, and could travel about 2 m through the air. He also discovered their penetrating properties, and was able to produce photographs of various objects of different thicknesses to show their relative transparencies.

In December, Röntgen brought his wife Bertha into the laboratory and made a "photograph" of her hand, which revealed the finger bones inside the flesh and her wedding ring, shown in Figure 4.3. Since the nature of this ray was not known, he called it the X-ray. In January, he was summoned by the German Kaiser to make a demonstration in court, and was given an award. He was showered with numerous honors, including the first Nobel Prize in Physics in 1901 as the father of the X-ray. Röntgen never obtained a patent on his invention, and gave all of his prize money to his university. He remained modest and courteous, did not have an assistant and preferred to work alone, and built much of his experimental apparatus with his own hands. He became professor of physics at Munich in 1900, and he died in 1923 from intestinal carcinoma.

Years later Max von Laue demonstrated that the X-rays are actually a form of electromagnetic wave just like light but much higher in frequency and energy and consequently much shorter in wavelength. The energy of the emitted photons can be

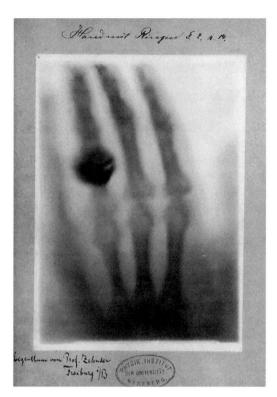

FIGURE 4.3 X-ray of hand of Mrs. Roentgen. Reprinted courtesy of © Deutsches Röntgen Museum.

as high as the energy of the incident electron. When a beam hits matter, part of it is absorbed and part of it continues to move ahead.

SCIENCE OF LIGHT PENETRATION

When a beam of light passes through a material, such as air, flesh, or bone, part of it is absorbed so the remaining light diminishes with distance according to the Beer's Law

$$\frac{I}{I_0} = e^{-\alpha Lc}$$

where I is the intensity of the beam, I_0 is the initial intensity of the beam, L is the thickness of the matter, α is the absorption coefficient of the matter, and c is the concentration of the matter.

The absorption coefficient α is highly variable among matter, and it is ranked in the order gold > lead > iron > calcium > carbon. That is why in the photograph, the X-ray beams do not go through the gold ring, and are partly stopped by the calcium in the finger bones, but easily penetrate the carbon in the flesh.

The X-ray in medical diagnostics has been invaluable, allowing a doctor to see bone fractures below the skin, abscesses in teeth, and tumors in the brains without exploratory surgery. Before taking a surgical knife in hand, the surgeon already knows the medical problem involved, and can formulate a plan on how to remedy matters. Over the years, many advances in medical imaging methods have followed. The ultrasound was introduced in 1959, which utilizes sound waves, and is fondly associated with viewing the fetus inside a mother's womb, even revealing the gender of the baby. Another huge step forward in imaging occurred with the creation of the computerized tomography or CAT scanner by Godfrey Hounsfield and Allan Cormack, for which they won the 1979 Nobel Prize. The CAT scanner works by taking multiple X-ray pictures from different angles, then analyzing the results with computers to determine the two-dimensional cross-section view. Next came the MRI scanner of Paul Lauterbur and Peter Mansfield, which creates images by measuring the interaction of radio waves with the hydrogen atoms in tissues. MRI offers superior imaging of soft tissues, even showing the flow of blood and the functioning of different areas of the brain when certain thoughts or actions are taking place. Lauterbur and Mansfield won the Nobel Prize of 2003 for their work on MRI.

The discovery of X-rays also sparked a new direction in the scientific understanding of nature, especially in crystals and DNA. X-rays have a very short wavelength of 0.01–10 nm, which is in the range of the length of the chemical bonds in molecules. When an X-ray hits a crystal, it is not simply absorbed or reflected back, but also undergoes a process called diffraction to create many waves that may enforce or cancel each other; the result is a pattern of dots that can be used to understand the crystal structure. This was the work of William Lawrence Bragg and his father Sir William Henry Bragg, which won them the Nobel Prize in Physics in 1915. This led to the discovery of the atomic structure of the crystals of salts and inorganic compounds, as well as proteins and DNA. The discovery by Crick and Watson on the double helix of the DNA molecule was made possible by the X-ray diffraction pictures taken by Rosalind Franklin.

4.3 THERAPY

Therapy or treatment comes after diagnosis, where the best objective is a complete cure, and the less desirable palliative objective is to increase comfort and reduce suffering. The most effective treatment methods include surgery, medication, and devices. Surgery has been an effective method to deal with wounds, fractures, and growths for thousands of years. Galen of Pergamum was a follower of the humor theory of Hippocrates, however he recommended aggressive bloodletting as a cure for almost any disease, including hemorrhage. Hippocrates was a big proponent of medicines and suggested a number of medicinal plants, including willow bark and laudanum (opium); however these drugs were chosen for their empirical effectiveness, and do not fit in with the four-humor theory. Prosthetic devices such as the walking cane and reading glasses have also had long histories.

4.3.1 Surgery: Anesthesia and Ether

Surgical operations have been performed since before the Neolithic age, and involved the endurance of excruciating pain in exchange for the chance of recovery.

Many surgical instruments were invented and recorded in history to perform such early tasks as the removal of tumors, defective organs or foreign bodies such as arrows; and the repair of broken parts such as bone fractures. More modern surgical instruments have been developed to unblock blood vessels and to transplant or replace joints, kidneys, and hearts.

Advances in surgery would not have been possible without the development of the field of anesthesia. Early on, the problem of alleviating the pain of surgery was accomplished by the use of plants such as hemlock, mandrake, or the opium poppy. Whisky, hypnotism, and acupuncture were also used for surgery or childbirth. In 1799, Humphry Davy published a book about the properties of the recently discovered gas nitrous oxide, to which he gave the name "laughing gas." He found that he could make himself unconscious by inhaling nitrous oxide vapor, and suggested that it could be used to block pain in surgery. Many of his society friends allowed themselves to be tested as intoxicated human guinea pigs, including such luminary poets as Samuel Taylor Coleridge and Robert Southey. Davy's apprentice, Michael Faraday, published similar observations in 1818 concerning the effects of sulfuric ether. During the first half of the nineteenth century, nitrous oxide and ether were widely used for recreation by stylish British and American partygoers. A flask of ether or a bag of nitrous oxide could provide exhilaration, emotional release, and unconsciousness. Despite the efforts of Davy and Faraday, the use of laughing gas or ether for easing pain in surgery did not come into common use in clinical practice for many more decades.

Crawford Long was a country doctor in Georgia who had access to ether. He and his friends participated in ether-driven frolics in which they banged into things but did not notice any pain, bruising, or scrapes until the ether had worn off. Long realized the potential of ether for surgical pain. He first tried using the ether therapeutically before operating on his friend James Venable to remove a neck tumor in 1842. He poured ether on a towel, had his friend breathe the fumes till he appeared insensible, then excised the tumor. Venable awoke a few minutes later having felt no pain. Incredibly, despite this success, Long did not publish a paper on his ether trial till 1849, and apparently never used it again. So he was the first to use ether for surgical pain, but by not sharing his findings right away, his work did not change the world of surgery.

Horace Wells was a dentist in Hartford, Connecticut who attended a traveling nitrous oxide show and noticed that an injured audience member appeared to have no pain while under its influence. In 1844, Wells himself inhaled nitrous oxide from a leather bag till he nodded off, during which time his student removed his sore wisdom tooth. When he awoke a few minutes later he was elated. He asked his former student William Morton to approach John Warren, the head of surgery at Massachusetts General Hospital in Boston, about performing a demonstration with the nitrous oxide. Warren approved a public lecture and demonstration in January 1845. Wells administered what he thought would be an adequate dose of nitrous oxide and proceeded to extract the tooth. However, as the tooth came free, the young man groaned and cried out. The audience laughed and jeered with shouts of "Humbug!" The demonstration was not deemed a success.

Undaunted after Wells' failed demonstration, Morton continued to experiment. He consulted with Charles Jackson, a Harvard Medical School graduate, who apparently suggested that ether would be easier to administer in liquid form. Warren was willing to try again, and scheduled a second demonstration on October 16, 1846 at the Massachusetts General Hospital. Morton poured ether onto a young man with a tumor in his neck and jaw, and within a few minutes the young man was deeply asleep. Morton stepped aside and Warren operated with speed during which time the patient did not stir. When the operation was over, Warren turned to the audience and announced "Gentlemen, this is no Humbug." The operating room at the Massachusetts General Hospital is named the "Ether Dome" in honor of this historic event. After that the use of ether for anesthesia spread with amazing speed. Oddly, there were initial objections to its use in childbirth, as some felt that God had cursed Eve with labor pains as punishment for eating the forbidden fruit: "In sorrow thou shalt bring forth children." However when Queen Victoria herself used ether for her delivery, it silenced many opponents.

The field of modern anesthesia is divided between central nervous system depressants that induce unconsciousness, such as ether and halothane; and local anesthetics such as lidocaine, which work on a limited area of tissue such as a tooth or a finger, while the patient remains awake. Anesthetics act by blocking the signal of pain generated at nerve endings from reaching the brain. There is frequently a narrow margin between a dosage high enough to stop pain, yet low enough to avoid severe damage or death. Because of this, malpractice insurance for anesthesiologists in the United States has become very expensive. Anesthesiologists must be highly trained physicians, and have to carry very high insurance against malpractice suits.

4.3.2 Medicine: Penicillin

Many plants have been used from ancient times as medicines, including foxglove (digitalis) for heart problems, poppies (opium) and willow bark (aspirin) for pain relief, cinchona tree bark (quinine) for malaria, and chaulmoogra for leprosy. Despite these plant medicines, there was no satisfactory medicine available for a myriad of human ailments. The poet-physician Oliver Wendell Holmes Sr. said in 1842, "If all medicines in the world were thrown into the sea, it would be better for mankind, and all the worse for the fishes."

The era of synthetic drugs began in the twentieth century with Paul Ehrlich, who in 1908 synthesized the compound salvarsan, also called 606, for the treatment of syphilis. Frederick Banting isolated the hormone insulin from dog pancreas in 1923, paving the way for eventual synthetic production of modern insulin using recombinant DNA. Gerhard Domagk discovered the antibacterial effects of coal-tar dyes on infections in mice and developed the sulfa drug Protonsil in 1939, the first antibiotic. All three received the Nobel Prize in Medicine.

Modern drug inventions can be divided into several broad categories:

- *Psychopharmacological Agents Acting on the Central Nervous System:* anesthetics, barbiturates, narcotics, lithium, codeine, antidepressants.

- *Pharmacodynamic Agents Acting on Organs to Block a Normal Function or to Correct a Deranged Function:* Lidocaine, beta blockers, nitroglycerine, antihistamines, digitalis, heparin.
- *Chemotherapeutic Agents Against Infectious Diseases and Cancer:* Quinine, penicillin, cisplatin, Taxol.
- *Metabolic and Endocrine:* Insulin, thyroid hormone, cortisone, diuretics, antiacids.
- *Vitamins:* Vitamin A and D.

Infectious diseases are caused by microbes: bacteria, fungi, protozoa, and viruses. These microscopic and invisible agents were the most important causes of illness and death in the ancient world. The germ theory of Pasteur and Koch did much to clarify the sources of these illnesses, and subsequent antiseptic methods and habits did much to prevent these microbes from invading human bodies. Fortunately, not all exposure to microbes results in infection as most people have a fairly robust immune system; but when antiseptic methods failed and the immune system became overwhelmed, there was not much that a physician could do to get rid of them. Early research looked at chemicals that were poisonous to microbes, but most were also poisonous to human cells and tissues. The search was on to find a medicine with "selective toxicity," which is one that is much more toxic to microbes than to host tissue. The basis of chemotherapy is to exploit the differences in human and microbial cells, such as differences in cell wall structure so that a drug can penetrate the microbial cell wall but not the human cell wall.

Gerhard Domagk was a pathologist inspired by Paul Ehrlich's work on syphilis. Domagk reasoned that some dyes preferentially adhere to bacteria and not to mammalian cells; this suggested to him that these dyes might be modified to become toxic and selective killers of bacteria while leaving mammalian tissue unharmed. He joined the dye firm I.G. Farben in 1927, and screened dyes on mice inoculated with a highly virulent strain of streptococcus. He tested the brilliant red azo dye Protonsil in 1932, and found that it cured mice infected with streptococcus. In 1935, his daughter Hildegard developed severe septicemia after pricking her finger with a needle and he used Prontosil to save her life. It turns out that once this red dye is absorbed into the body, it splits to form a colorless molecule 4-aminobenzene sulfonamide, which is really the active principle. This discovery led to the synthesis of more than 1000 sulfa compounds with variations that could alter its properties. Domagk was unable to secure a patent for sulfonamides as the compound had actually been synthesized in 1908 by chemist Josef Klarer, although Klarer had never discovered its wonderful antibacterial properties. Domagk was, however, awarded the Nobel Prize for Medicine in 1939 for discovering the therapeutic values of the compound. He was initially prevented from receiving the Nobel in person because of an anti-Nobel policy of the German government during World War II, but finally received his prize after the war in 1947.

Penicillin ranks with radar and the atomic bomb as one of the greatest inventions of World War II. It was first discovered by accident, but was subsequently developed into an effective medicine by a large international team after years of

FIGURE 4.4   Alexander Fleming. Reprinted with permission The Image Works.

coordinated and dedicated work. The story of penicillin began with Alexander Fleming, who was born in 1881 in Scotland (Fig. 4.4). He studied at the St. Mary's Medical School of London University, and became a lecturer. During World War I, he was a captain in the Army Medical Corps, and became accustomed to the crude antiseptic methods that were used at the time and observed more soldiers dying in the hospitals than on the battlefields. In 1928, after the war, he returned to do research at St. Mary's Hospital, working on finding antibacterial substances that would not be toxic to animal tissue. His project involved growing colonies of staphylococcal bacteria in Petri dishes in his musty, dusty, and untidy laboratory. After a month-long vacation, he returned to find that one of his Petri dishes had become contaminated with a blue fungus; there was also a circular area where the staphylococcus bacteria colonies were not growing, shown in Figure 4.5. Instead of throwing away the dish, he isolated the effective substance from the circle that turned out to be a penicillium mold. He then grew the mold in pure culture and discovered that it could inhibit not just staphylococcus but other disease-causing bacteria as well. He called his discovery penicillin, and published his results in 1929. Unfortunately at the time his results were largely ignored.

Penicillin works by preventing bacteria from rebuilding their protective cell walls, and is not a threat to human cells since they do not have cell walls. Fleming knew there could be practical applications for penicillin, but he did not have the knowledge and skill to grow or purify enough of this compound, or to test it with live animals or humans. So after his 1929 work was published, it

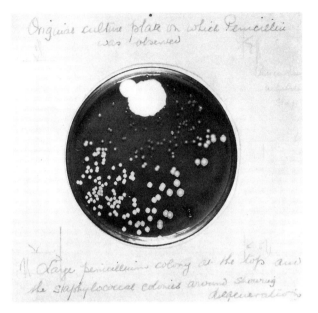

Original culture plate on which Penicillin was observed

Large penicillium colony at the top and the staphylococcal colonies around showing degeneration

FIGURE 4.5 Petri dish of bacillus growth with penicillin circle. Reprinted with permission The Image Works.

languished for many years. At his Nobel Prize ceremony in 1945, Fleming said that, "It arose simply from a fortunate occurrence which happened when I was working on a purely academic bacteriological problem which had nothing to do with antagonism, or molds, or antiseptics, or antibiotics . . . We tried to concentrate penicillin, but it is easily destroyed, and to all intents and purposes we failed. We were bacteriologists—not chemists, and our relatively simple procedures were unavailing."

The development of penicillin from its discovery in a laboratory to saving millions of lives was a group effort involving many investigators with different skills and from different organizations, under the financial support of the US government. When World War II began a team was formed at Oxford University by Howard Florey an Australian pathologist, and Ernest Chain a German biochemist. They were encouraged by the success of the new sulfa drugs of Gerhard Domagk, and were searching for new miracle drugs. In 1939, 10 years after Fleming's paper had been published, they decided to investigate penicillin. First they made and purified sufficient quantities of the substance to demonstrate its therapeutic effectiveness on mice. However it soon became clear to them that there would not be enough resources available to them to solve many manufacturing and application problems much less to develop large-scale industrial plants to produce the penicillin in wartime England.

In the summer of 1941, Florey came to the United States and sought help from the National Research Council, which in turn referred him to the Northern Regional Research Laboratory (NRRL) in Peoria, Illinois. This laboratory was devoted to finding new industrial uses for surplus farm commodities, and had a new fermentation division. In the laboratories of Fleming and Florey back at Oxford, fermentation was carried out on thin solid surfaces in Petri dishes, which could only produce penicillin at a rate of 3 units/mL, with a unit equaling only 0.6 μg of pure sodium penicillin G. The staff at NRRL switched the fermentation process to a broth of corn steep liquor in large flasks, and they were able to increase production to 100 units/mL. They also realized that different strains of mold would produce different quantities of penicillin, and began a worldwide search for high-yield molds from places as far as Chunking and Calcutta. Ironically, after screening thousands of strains of penicillium molds, the best strain found was in a moldy cantaloupe in a Peoria fruit market, which enabled the production rate of penicillin to increase to 1500 units/mL.

Howard Florey needed 1 kg of penicillin for clinical trials. A consortium of pharmaceutical companies including Abbott, Lederle, Merck, Pfizer, and Squibb was summoned by the US Office of Scientific Research and Development under Vannevar Bush in 1941. Initially fermentation was carried out in 1 L bottles, each containing 200 mL of medium; thousands of flasks were mounted on shake tables to stir and to aerate the broth. Merck and Pfizer scheduled experiments by placing 100,000 bottles on shake tables. Ultimately they found that it was most productive to perform deep-tank fermentation in 38,000 L tanks, a process that is still in use today.

The fermentation broth could produce 20–30 mg/L of antibiotics, which was only 30 ppb and required significant extraction and concentration. The most effective method of extraction was developed by Shell Oil and Podbielniack using an oily solvent such as amyl acetate. This separation was based on the principle that when placed in a flask with both water and an oily organic solvent, the penicillin would be more soluble in organic liquids when the broth was acidic, so penicillin would be extracted from water into the organics; but when the oily mixture was placed in a second tank with water and neutralized, the penicillin would return to the water phase. Another major problem was the instability of penicillin, which broke down with a half-life of only 2.5 h even when cooled to 0°C. This made it difficult to manufacture on a large scale and then store for subsequent shipment to battlefields. The winning solution was the freeze-drying method, where penicillin was first frozen in trays at −30°C, then placed in a vacuum chamber at a pressure below one-thousandth of an atmosphere, which would sublimate the ice crystals into water vapor leaving a more stable product.

The production rate of penicillin rose from 1.7 billion units per month in 1943, to 570 billion units per month in 1945. The War Production Board of the United States sponsored the project, and prepared ample stock for the D-Day invasion of Normandy Beach in Europe on June 6, 1944. Penicillin was an effective medicine against throat infections, pneumonia, meningitis, gas gangrene, diphtheria, syphilis, and gonorrhea. In 1945, Fleming, Florey, and Chain jointly received the Nobel Prize for Medicine. Florey said, in 1949, "Too high tribute cannot be paid to the enterprise and energy with which the American manufacturing firms tackled the large-scale production of the drug. Had it not been for their efforts, there would certainly not have been sufficient penicillin by D-Day in Normandy in 1944 to treat all severe casualties, both British and American."

Other antibiotics soon joined the ranks of bacteria fighters, such as Streptomycin that was developed in 1944 by Selman Waksman, Aureomycin by Benjamin Duggar in 1948, and Tetracycline by Lloyd Conover in 1950. By the twenty-first century, sanitation and antibiotics had greatly reduced the death toll from infectious diseases leaving cardiovascular diseases (stroke, heart attack) and cancer as the new leading killers in industrial nations. However, penicillin-resistant bacteria have been found in recent years, which raise serious problems in public health.

4.3.3   Devices

When Oedipus was walking to the city of Thebes, he met the Sphinx who demanded an answer to her riddle: "What walks on four feet in the morning, two in the afternoon, and three at night?" Oedipus answered: "Man—as an infant crawls on all fours, an adult walks on two legs, and an old man uses a walking stick." This is one of the oldest references to prostheses, devices that help us overcome missing or defective body parts. War, mining, and transportation have been major causes of injuries that require artificial organs or limbs. Some devices are used to augment organs that are functioning below expectations, such as hearing aids or eyeglasses. Other devices, such as the kidney dialysis machine and the artificial heart, attempt to replace the failing function of an organ. New research in tissue regeneration is exploring the possibility of growing complete artificial organs, one cell at a time, on a structure called a scaffold. It is important to use materials that are compatible with the human body, and that do not cause negative immune responses.

Eyeglasses or spectacles for vision correction and eye protection have been in use for a long time. The earliest mention was in an Egyptian hieroglyph in the eighth century BCE, which depicted "simple glass meniscus lenses." In the first century, the Roman Emperor Nero is said to have watched gladiatorial games using an emerald as a corrective lens. Seneca the Younger wrote "Letters, however small and indistinct, are seen enlarged and more clearly through a globe or glass filled with water." Corrective lenses were said to be used by Abbas Ibn Firnas of Cordoba in the ninth century, who found a way to produce a very clear glass. This glass was shaped and polished into round shapes for viewing and was called a reading stone. Around 1284, Salvino D'Armate of Florence was credited with inventing the first wearable eyeglasses. The earliest image for the use of eyeglasses is the Tomaso da Modena's 1352 portrait of the cardinal Hugh de Provence reading in a scriptorium. Other clerics in Italy, such as Fra Alessandro da Spina of Pisa, were credited with making glasses. The development of glasses, both of concave lenses for distance vision to help near-sighted people (myopia), as well as of convex lenses for close reading to help farsightedness (hyperopia) and vision in old age (presbyopia), are related to the development of telescopes and microscopes. Benjamin Franklin suffered from both myopia and presbyopia, and is credited with inventing bifocals in 1784 to avoid having to switch between two pairs of glasses.

SCIENCE OF READING GLASSES

Light bends when it enters from air to a denser medium, and the refractive index n is the ratio of the speed of light in the medium divided by the speed in air. The refractive index of air is 1 and water is 1.33. Glass has a refractive index of 1.54–1.61 and plastic polycarbonate is 1.58. Diamond has the much higher refractive index of 2.42. The power of a corrective lens is measured in diopters. The bending of light follows the Snell Law

$$n_1 \sin \theta_1 = n_2 \sin \theta_2$$

where n_1 and n_2 are the refractive indexes of the two media, and θ_1 and θ_2 are the angles between the vertical and the beams. So for a glass with refractive index 1.50, the entering beam bends to a smaller angle of the leaving beam:

θ_1	0	15	30	45	60	75	90
θ_2	0	9.9	19.5	28.1	35.2	40.1	41.8

The work of a lens follows the thin lens equation

$$\frac{1}{S_1} + \frac{1}{S_2} = \frac{1}{f}$$

Here S_1 is the distance from the object to the lens and S_2 is the distance from the focus to the lens, and the power of the lens is measured in diopter $1/f$ when f is measured in meters. When the object is infinitely far away, the focus is sharpest at a distance of f meters from the lens. The relaxed human eye has a power of 60 diopters, and the value of f is about $1000/60 = 16.7$ mm from the lens to the retina. The shape of the human lens can be changed by cilliary muscles in a process called accommodation, by as much as 20 diopters for the very young, but the ability to do so declines with age. Thus, the purpose of spectacles is to supplement the natural lens with the equation

$$\frac{1}{S_1} + \frac{1}{S_2} = \frac{1}{f} + \frac{1}{L}$$

where L is the power of the lens. For a farsighted person and for reading glasses, we need $L = 1.00$–4.00 diopters, which means a lens that is thicker in the center than the rim; and for a nearsighted person, we need $L = -1.00$ to -3.00, which means a lens that is thinner in the center than the rim.

Each kidney has about a million nephrons that filter out the unwanted components of blood, which then drain via the ureters into the bladder as urine. The substances that must be cleared from the bloodstream include products of metabolism, such as urea, creatinine, and uric acid, as well as excess sodium and chloride ions. When kidney function is impaired by injury or by chronic kidney disease, these waste products accumulate in the blood and lead to many serious medical problems, and eventually to death. Kidney transplantation is an excellent option for patients with failed kidneys, however this procedure is dependent on the availability of a donor kidney and can be complicated by rejection of the organ by the patient's immune system. Dialysis is a process that was invented as an alternative method to filter blood outside the body, and has the advantage of being more available than transplants for many patients.

The artificial kidney, now known as hemodialysis, was invented by Willem Kolff during the height of World War II in German-occupied Holland. His story is one of the most amazing tales of achievement under adverse conditions. Willem Kolff was born in the Netherlands in 1911, and by 1938, was a young physician at the University of Groningen. While there, he witnessed a young man's slow and agonizing death from kidney failure. Kolff reasoned that if he could find a way to remove the toxic waste products that had accumulated in a patient's blood, he could keep those patients alive until their kidneys recovered.

For his first experiment, Kolff filled sausage casings with blood, expelled the air, added the kidney waste product urea, then he agitated the casing in a bath of salt water. The casings were semipermeable, so that small molecules of urea could pass through the membrane into the bath, while the larger blood cells and molecules stayed in the membrane. In 5 min, all the urea had moved from the blood through the membrane into the salt water.

In 1940, Germany invaded the Netherlands, and Kolff moved to a small hospital in Kampen on the Zuider Zee to wait out the war. He continued to work on his artificial kidney. One of his early attempts involved wrapping 50 yards of sausage casing around a wooden drum that was immersed in a salt solution. A patient's blood was drawn from a wrist artery and fed into the casings. The drum was then rotated to wash away the impurities that oozed out of the sausage casings. To get the blood safely back into the patient, Kolff used a pump that he had copied from a water-pump coupling used in Ford motor engines. He also used orange juice cans and a clothes washing machine to build his apparatus. The first 15 people tested on the machine eventually died of kidney failure, but Kolff was not discouraged and persisted in his work.

Kolff continued to make improvements, including the use of blood thinners to prevent clotting during the dialysis. In 1945, a 67-year-old woman who had fallen into a coma from kidney failure was put on the machine for a long period and lived. She came out of the coma, and said "I am going to divorce my husband," as apparently he was anti-Nazi while she was a Nazi collaborator. She lived for seven more years. In 1947, Kolff sent one of his artificial kidneys to Mount Sinai Hospital in Manhattan. In 1950, Kolff moved to the Cleveland Clinic and finally to the University of Utah, where he headed the Division of Artificial Organs and helped develop the first artificial heart. He said that "If a man can grow a heart, he can build one." By 2011, tens of thousands of people were undergoing hemodialysis three times a week, often as a bridge to a kidney transplant.

Today we have many successful medical devices not just in hospitals but in homes as well. They are used in the cure, mitigation, treatment or prevention of diseases. Some are removable devices, such as contact lenses and hearing aids, which improve our quality of life. Others are surgically implanted products such as the pacemaker for people with irregular heartbeats or the stent to maintain arterial circulation in blocked arteries. We have dental crowns and implants, artificial eye lenses implanted during cataract surgery to replace cloudy lenses, and cochlear implants for the hard of hearing. Artificial limbs to replace hands and feet have been used from ancient times for people whose extremities were shattered in wars and accidents. Pirates are often depicted in movies with peg legs and hooks. Now prosthetic limbs are so advanced that in the track and field events of the 2008 Olympics, Oscar Pistorius had mechanical ankles that were considered an unfair advantage over runners who had natural ankles.

Knee and hip joint replacements to relieve arthritis pain or fix severe joint damage are becoming more common in recent years. The first recorded hip replacement was by Gluck in 1891, using ivory to replace the femoral head. Austin Moore in 1940 reported the first metallic hip replacement made of a cobalt-chrome alloy. The modern total hip replacement uses a titanium shaft, attached to a ceramic head or ball, designed to rotate within a polyethylene acetabulum cup. This procedure is currently one of the most common orthopedic operations.

4.4 REPRODUCTION

For most people, children and grandchildren confer a form of physical immortality, with a sense of continued existence and influence in the world. For people without children, death could mean the dispersion of possessions and the end of family tradition and memory. The theory of evolution by Charles Darwin emphasized that success and dominance on earth is achieved more by reproductive success than by competitive success. Plant and animal species without reproductive successes eventually disappeared from history. For people, the Hebrew God said, "Be fruitful and multiply, and fill the earth."

In nature there are two strategies for ensuring the survival of species: (1) small seed, or producing small but plentiful seeds such as a fish laying millions of eggs to ensure that a few will survive to adulthood and (2) large baby, or producing large but limited offspring such as a lion that bears one to four cubs per litter, carries the cubs in her body for months, and continues to nurture them after birth. For a species to survive with either strategy, each pair of parents must produce two surviving offspring to replace the parents. For most of human existence on earth, women had to give birth to a dozen or more children in a lifetime in order to have two surviving to adulthood. With modern medicine, nutrition and sanitation, infant and child mortality has plummeted and a dozen babies often survive to become a dozen adults. Thus a new human strategy has emerged: modern woman still built to bear a dozen children, but needing only two babies to ensure survival of the species.

4.4.1 Increase Fertility

There are a multitude of reasons why people want or do not want to have children. Among the discoveries that have changed the world is the technology to help couples have children as well as to prevent couples from having children. More than 10% of all couples are infertile, which is defined as the inability to become pregnant within one year of trying. The resulting involuntary childlessness can be a source of severe anguish and depression for some couples. Human fertility depends on numerous factors including nutrition, sexual behavior, and age. The most fertile period in the life of a woman is between the ages of 19 and 24, and fertility declines after age 30. Advanced education and prestigious employment for women can lead to postponement of pregnancy, thus pushing the limits of age-based fertility. Causes of infertility are not limited to women, however, as sometimes male factors are the cause of inability to conceive. Regardless of the causes, the process of assisting fertility in the United States has become a huge business worth $3 billion a year.

Historically, for couples who were unable to conceive a child, adoption has been the traditional way to form a family. With the explosion of new technology in fertility treatments, some now choose to try to have a child that will carry their genes and can go to a fertility clinic to see what methods are available to help them. Some of the most important methods promote the meeting of sperm and egg under favorable conditions.

The principal causes of infertility include problems with male sperm production (too few motile sperms, genetic disorder), female ovum production (menopause), egg transport from the ovaries to the uterus, endometriosis (growth of endometrial tissue outside of the uterus), coital difficulties, and unknown factors. These problems sometimes can be treated by medication, surgery, or procedures such as intrauterine or *in vitro* fertilization (IVF). Fallopian tube obstructions can be opened by surgery, fertility drugs such as gonadotropin-releasing hormones can induce ovulation, difficulties with coitus can be bypassed by intrauterine insemination (IUI), and counseling can educate couples on reducing stress and on other factors that can affect fertility. There are options even when one or both partners have no viable gametes (egg or sperm). In the Old Testament of the Bible, Abraham's wife Sarah did not become pregnant after years of marriage so she gave Abraham her handmaiden Hagar as a second wife, after which Hagar gave birth to Ishmael. In the famous novel by D.H. Lawrence, the paralyzed and impotent Lord Chatterley urges his wife, Lady Chatterley, to have a child with another man to ensure an heir to his estate. In these two examples no new technologies were involved.

There are now many artificial reproductive technologies (ART) available to help couples who remain sterile and want children. These technologies are also available to postmenopausal women and nontraditional clients who would like to become parents, such as single men or women and same sex couples. In some cases fertility clinics can help clients have a baby that has the genes from at least one member of the couple. The principal methods of treatment include the following.

The method of artificial insemination (AI) involves collecting sperm from the prospective father. The freshly collected sperm are washed and concentrated, then injected into the cervix of the mother, ideally just prior to ovulation. In this method, fertilization still takes place in the uterus and only the method of insemination has changed. When healthy sperm cannot be provided by the prospective father, fresh or frozen sperm from donors can be provided, or obtained from sperm banks where they are stored in small vials in tanks cooled by liquid nitrogen, and could be effective for up to 21 years. The more advanced method of IUI involves injecting sperm with a catheter through the cervical into the uterus. The client father may or may not be the genetic father, but the client mother retains the role of the genetic mother and also carries the baby to term as usual. This method was first developed for dairy cattle to allow a bull with superior traits to impregnate many cows. If the client mother is unable or unwilling to carry the pregnancy, a variant method is for another woman to be a surrogate. The surrogate mother would be artificially inseminated with the client father's sperm that would fertilize her own egg and she would then carry the baby to term, surrendering the baby to the client parents at birth.

The method of IVF requires obtaining both sperm and eggs and fertilizing them in a dish. If fertilization takes place the resulting embryo or embryos are inserted into the uterus of the prospective mother. This technology was developed after more than 20 years of research. In 1935, Gregory Pincus described experimental conditions required for rabbit eggs to mature *in vitro*. Min-Chueh Chang showed in 1959 that matured rabbit eggs could be fertilized *in vitro* giving rise to viable 2-cell stage embryos.

The next challenge was to see if these results could be extended to humans to produce babies. Robert G. Edwards, a biologist who began his research at Edinburgh on the fertility of mice, moved on to do research at Cambridge with immature human eggs released from ovarian tissue. He found that eggs in the immature stage are dormant and even with incubation *in vitro* under a suitable environment, they do not develop beyond the 2-cell stage. Between 1969 and 1978, the gynecologist Patrick Steptoe worked with a laparoscope that could be inserted through the abdominal wall to directly visualize the uterus and ovaries with the additional ability to extract and remove eggs from the ovaries.

Edwards and Steptoe teamed up to try to remove mature eggs at the appropriate time in the menstrual cycle, then exposed the eggs to sperm in a dish. With this method they were able to achieve fertilization and embryos with eight cells. Steptoe then returned the 8-cell embryo to the prospective mother's uterus. He attempted this procedure more than one hundred times, all leading to discouragingly short-lived pregnancies. They persevered, trying many combinations and approaches, and finally one embryo developed into the first "test tube baby"— Louise Brown, who was born on July 25, 1978. In 1980, Steptoe and Edwards established Bourn Hall Clinic in Cambridgeshire to continue to develop the IVF method, achieving 139 births by 1983 and 1000 births by 1986. Although the use and development of *in vitro* fertilization has never been without controversy for religious and other reasons, it undoubtedly changed the landscape of human reproductive science. Steptoe continued as the Medical Director at Bourn Hall Clinic till his death in 1988. In 2010, Edwards received the Nobel Prize in Physiology or Medicine, "for the development of *in vitro* fertilization"; by the time of his award more than 4 million babies had been born through IVF.

Since the first *in vitro* baby, there have been many variations of the process including donor sperm, donor eggs, or a donor uterus. Many ethical and legal issues have been debated over IVF, and the more parties that are involved, the more complicated these debates have become. Laws passed before these new artificial production technologies were not designed to cope with all of these new possibilities, and additional complications come from the possibility of different state laws for the various parties. A sampling of these issues include the following:

> *Who is Entitled to Custody of the Child:* The default legal assumption in many states is that the woman giving birth to a child is that child's legal mother. Even with a contract between the surrogate and the clients, it is difficult to take custody away if a surrogate decides to keep the baby. In

some states, the legal rights of a gestational surrogate are recognized and regulated; in other states their rights are denied; and yet other states may allow only altruistic surrogacy where no money changes hands and commercial surrogacy is banned. What are the responsibilities of the donors and the clinic if the client parents turn out to be seriously ill or irresponsible? If the client parents prove incompetent, who has the responsibility to provide for the child? Can the surrogate mother retain custody if she discovers that the client parents are mentally ill, and which state court has jurisdiction if they reside in different states? Should the donors of eggs, sperm, or uterus have residual rights of contact and visits with the child?

Anonymity of the Parties: Should the identity of the sperm donors be totally concealed from the resulting children forever, or told only to the clients and the child upon adulthood? What are the clients entitled to know about the characteristics of the sperm or egg donors: race, age, education, IQ? Must the child be told of the facts of the artificial birth, and does the child have the right to contact and visit the donors? When a sperm donor fathers children with different mothers, such as through a sperm bank, should the children be informed so they can avoid choosing mates among half-siblings?

4.4.2 Decrease Fertility

Confucius said "The great needs are drink, eat, men-women." People have sought effective means of preventing pregnancy perhaps since before history was written. Documents from ancient Egypt describe the use of acacia gum that worked as a spermicide. Plants with various properties were used in ancient Greece as contraceptives. The possibility of reliably controlling when to have children is another technological advance that has changed the world. Arguably the sexual revolution was made possible because of birth control. Even couples who do want children could use birth control to regulate the time interval between children or to have only as many children as they felt they could reasonably provide for.

Contraception technologies involve preventing sperm from fertilizing an egg. The methods range from low cost and minimally effective methods that are reversible, to high cost and highly effective methods that are irreversible:

- *Behavioral Methods:* Sexual abstinence, coitus interruptus or withdrawal, the rhythm method based on the menstrual cycle, and prolonged lactation as breastfeeding retards ovulation.
- *Physical Barriers:* Spermicides, condoms, the diaphragm, and the cervical cap.
- *Hormonal:* Oral, intramuscular, or intravaginal estrogen and progesterone.
- *Devices:* Intrauterine device (IUD), inserted by gynecologist.
- *Sterilization:* Female tubal ligation or hysterectomy, male vasectomy.
- *Abortion:* Not preventive, occurs after conception.

There is a wide range of political attitudes and support for birth control. Not all parties are happy with these new technologies. Some groups voice objections on the grounds that the sole justification of coitus is for procreation and that birth control would encourage promiscuity and casual extramarital sex. They also object to abortion as a form of birth control, equating it with murder.

The Chinese government, on the other hand, has legislated mandatory birth control with its "one family, one child" policy. Families that have more than one child, especially in urban areas, can be punished and fined. Chinese authorities estimate that this policy has prevented up to 300 million births, reducing the huge demand of overpopulation on resources and the environment in China.

Currently in the United States, about two-thirds of the women use contraceptives, and about one-fourth of all pregnancies end in abortion. The current methods used in world contraception are shown below in the table.

	MDC (%)	LDC (%)
Male sterilization	7	7
Female sterilization	12	38
Intrauterine devices	8	26
Pill	24	11
Condom	19	4
Injection	0	4
Others	30	10

The most groundbreaking invention in birth control is the oral contraceptive pill, better known simply as the Pill. It is a combination of the female hormones estrogen and progestin, taken by mouth daily, and is currently used by more than 100 million women worldwide. Many people were involved in developing the pill and, ironically, several of them actually started off in fertility research.

Katharine Dexter McCormick (1875–1967) was the second woman to graduate from MIT, and the first to earn a bachelor's degree in biology there. She married Stanley McCormick in 1904, who was an heir to the International Harvester fortune, and a star athlete in tennis who graduated cum laude from Princeton. In 1908, Stanley was diagnosed with schizophrenia, and for this reason Katharine decided not to have children. Katharine became a suffragette, holding the positions of vice president and treasurer of the National American Woman Suffrage Association, and later of vice president of the League of Women Voters.

Another prominent activist and the founder of Planned Parenthood, Margaret Sanger (1879–1966) was born into a Roman Catholic family. Her mother had 18 pregnancies that resulted in 11 live births, but died in her forties from tuberculosis and cervical cancer. Margaret was the sixth child and spent much of her youth doing household chores and taking care of her siblings. She encountered a number of women who suffered from or died of self-induced abortions. She formed the opinion that for women to have more social and healthy lives, they would need the freedom to decide when to become pregnant. In 1916, Sanger opened a birth control clinic in Brooklyn, which was raided by the police and closed in 9 days; this resulted in her being sentenced to 30 days in a workhouse.

In 1917, Katherine McCormick met Margaret Sanger for the first time, and together they forged an alliance to develop reliable birth control. In 1937, Katharine's mother died and left her an estate of more than 10 million dollars. In 1947, her husband Stanley also died leaving her an estate of 35 million dollars. Thus, she had the means to support birth control research that had no government funding. Eventually she spent 2 million dollars funding Gregory Pincus' ultimately successful research program to develop and test the first oral contraceptive pill. She also donated the Stanley McCormick Hall to MIT, as an all female dormitory for 200 female students. When she died in 1967, she willed 5 million dollars to the Planned Parenthood Federation of America and 1 million dollars to the Worcester Foundation for Experimental Biology.

Gregory Pincus (1903–1967) studied at Cornell and Harvard, where he became an assistant professor in reproductive physiology in 1931. His photo is shown in Figure 4.6. In 1934, he successfully performed *in vitro* fertilization in rabbits and later also reported the first asexual production of a rabbit. His work was

FIGURE 4.6 Gregory Pincus. Reprinted courtesy of University of Massachusetts-Worcester Archives, Lamar Soutter Library, University of Massachusetts Medical School, Worcester, Massachusetts.

controversial but he persevered, founding the Worcester Foundation for Experimental Biology in Massachusetts in 1944 to further his research. Around that time the pharmaceutical firm GD Searle was interested in investigating the use of steroid-based hormones and supported the research at the Worcester Foundation. In 1951, Margaret Sanger met with Pincus and requested his help in developing practical contraceptives. He had already demonstrated that injections of progesterone suppressed ovulation in rabbits, but he needed to find an oral contraceptive that would be far more acceptable in the marketplace.

Previously in the 1920s, it was discovered that when an ovary from a pregnant rat was transplanted into another sexually mature but not pregnant rat, the second rat would not ovulate. The substance suppressing ovulation was found to be coming from the empty sac or corpus luteum that had held the recently released egg. Once isolated, the substance was found to be a steroid hormone and was named progesterone. A year later, a second female hormone secreted by the ovary was discovered, which stimulated sexual maturation and estrus in rats and was named estrogen. When Pincus decided to investigate using progesterone for use as a contraceptive, he knew that he would require a cheap and plentiful supply of progesterone. Hundreds of rat ovaries were required to even produce enough hormone for his experiments; he knew he would have to look elsewhere to find enough estrogen or progesterone for large-scale clinical applications.

There is a tradition in Aztec folk medicine that led some Mexican women to eat wild yam for contraception. In 1940, Russell Marker was an organic chemist at the Pennsylvania State University, working on the synthesis of progesterone from plants, and on the extraction of steroids from the urine of pregnant mares. He went on to invent a method to make synthetic progesterone from the giant Mexican yam in Veracruz. In 1944, he cofounded the company Syntex, but left the company after one year following a dispute. Natural progesterone taken by mouth has a half-life of only 5 min in blood, and required modification to last longer. Carl Djerassi (1923), an organic chemist from Vienna, was recruited to work for Syntex in Mexico. In 1951, he led a team that synthesized a form of progestin that could be taken orally. Syntex began producing oral progestin for the market, and in 1964 moved to Palo Alto in California. The work at Syntex created a cheap and plentiful source of progesterone that supplied the research needs of Pincus. Djerassi later became a professor of chemistry at Stanford University, collecting art and writing novels such as *Cantor's Dilemma* and *The Bourbaki Gambit* in his spare time.

By 1952, Pincus and his assistant, Min-Chueh Chang, had done enough experimentation with animals to show that high doses of progestin could shut down the ovaries, preventing ovulation. The effectiveness of combining progestin with estrogen was tested on beagles, but there were concerns about possible side effects on breast tissue. Fortunately, this concern was later identified as being caused by impurities rather than the drug itself, so they were able to proceed to human trials.

John Rock (1890–1984) was a professor of clinical gynecology at Harvard and a practicing Roman Catholic. He believed that couples should have as many children as they had means to support, but should also have the power to limit the size of their families. He worked for a fertility clinic and helped countless couples to have children through *in vitro* fertilization and other methods. High doses of progesterone reduced fertility, but Rock discovered that low doses of progesterone are effective in *increasing* fertility. In 1954, Rock was recruited to work with Pincus to test the use of progesterone for *decreasing* fertility, under the umbrella of the fertility clinic. He provided 50 volunteer women in Brookline, Massachusetts with 5–50 mg doses of three forms of progestin for 3 months; the volunteers took a daily pill on days 5–25 for a total of 21 pills, and nothing for the other 7 days of the cycle. The test was a complete success. Next they needed to test the progesterone on many more women for longer periods of time in order to have a chance of winning approval by the US Food and Drug Administration. They formulated a combination of sex hormones to create the first birth control pill, and gave it the name Enovid.

Due to increased opposition in Massachusetts, they had to go to Puerto Rico in 1956 to conduct larger trials, worked out some side effect issues and ultimately came up with an effective birth control pill. Despite condemnation by the Catholic Church, women flocked to the clinics in the hope of limiting the sizes of their families. They found that the pill was 30 times more effective in preventing pregnancies than were condoms or diaphragms. Enovid was approved in 1957 by the Food and Drug Administration initially only for "menstrual disorders," but it carried a warning that the drug would prevent ovulation! The FDA approval occurred only 6 years after the first meeting between Margaret Sanger and Gregory Pincus. In the next decade, three of the six principle players in this invention passed away: McCormick, Sanger, and Pincus. The FDA approval of Enovid was followed by a dramatic rise in women who claimed to have menstrual disorders, and were thus in need of prescriptions for Enovid! When President Eisenhower declared in 1959 that "birth control is not a proper political or government responsibility," this paved the way for even broader acceptance of the pill. In 1960, the FDA finally approved Enovid for birth control as well, so it was no longer necessary for a woman to pretend to have a menstrual disorder to obtain a prescription.

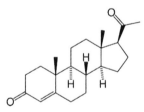

All of the sex hormones involve a compound with four rings, and are related in structure to cholesterol. There is only a small difference between the female progesterone and the male testosterone. The menstrual cycle lasts 28 days, very much like the phases of the moon. For the average woman, menstruation takes place on days 1–5 and ovulation on day 14. Ovulation is triggered by the presence of a combination of two hormones LH and FSH, which are influenced by the levels of progestin and estrogen in the blood. When the pill is taken, the levels of estrogen and progesterone are similar to the levels during pregnancy; thus the ovary is tricked into behaving as if the woman is already pregnant: FSH is suppressed so new follicles do not develop and there is no ovulation.

After the FDA approval of Enovid, birth control methods continued to develop with new variations. There are now birth control skin patches, intravaginal rings, and emergency postcoital pills such as Plan B that can be effective for up to 72 h after intercourse. Another drug, RU-486 or mifepristone, can induce abortion as much as 2 months after conception.

In 1999, the Economist magazine named the pill the most important scientific advance of the twentieth century. The pill has been credited with many changes in the structure of society. Couples can now control the timing and number of children, and women can choose between having a large family or a fulfilling career. In 1960 before the pill, the typical American woman had 3.6 children; by 1980, that number had dropped below 2.0. Detractors blame the pill for contributing to moral decline and promiscuity. The future development of technology to enhance or block fertility will undoubtedly continue to improve available choices while changing lifestyle and creating controversy.

The total fertility rate (TFR) is the average number of children that would be born to a woman in her lifetime. There is a strong negative correlation between the TFR and GDP/capita or personal wealth—in other words, the rich have fewer children. The highest values of TFR are in tropical Africa. The replacement fertility rate is the value of TFR that would keep the population stable and is held to be 2.1. Total fertility rates are much lower than 2.1 in some countries, especially in Europe, Japan, China, and Singapore. The decline in the birth rates of advanced nations, combined with falling death rates and longer life expectancies, is making a major impact on the age distribution of the population; this trend increases the burden on the working population (~age 15–64) to support the retired older population. The United Nations has determined that since the year 2000, more than half of the world's population lives in countries or regions where the fertility rate has fallen below the replacement value. On the other hand, there are zones where the fertility rate remains greater than 4: northern India and Pakistan, and the Arabian Peninsula through sub-Saharan Africa. The following table of TFR and age distribution is compiled from CIA data, together with a forecast for Taiwan in the year 2060.

Nation	TFR	Age 0–14 (%)	Age 15–64 (%)	Age 65+ (%)	Median age
Niger	7.2	50	48	2	19.2
Egypt	3.0	33	63	5	24.3
United States	2.1	20	67	13	36.9
Japan	1.2	13	64	23	44.8
Taiwan 2060	1.3	9	49	42	

This unsettling trend in industrialized nations will lead to the closing of many schools that will be matched by the opening of many old age facilities. The declining populations of young workers will lead to a shrinking national economy. Young workers could be replaced with large-scale immigration of young workers from countries with excess population, but this might in turn raise the potential for conflicts due to ethnic differences in culture and values. Even without immigration, an aging population in itself changes the character of a country—is an aging population as creative or optimistic as a youthful population? There are many different views on the desirable age distribution of human populations as well as the more controversial question of the ethics of artificially manipulating the size of the population.

REFERENCES

Adler, R. E. "Medical Firsts", Wiley, New York, 2004.

Albert, A. "Selective Toxicity: The Physico-Chemical Basis of Therapy", Chapman and Hall, London, 1973.

Bogue, D. J. and J. Nelson. "The Fertility Components and Contraceptive History Techniques for Measuring Contraceptive Use-Effectiveness", University of Chicago Press, Chicago, 1971.

Booth, J. "A Short History of Blood Pressure Measurement". Proceedings of the Royal Society of Medicine 70, 793–799, 1977.

Cape, J. and A. Maurois. "Life of Sir Alexander Fleming", Dutton, New York, 1959.

Claxton, K. T. "Wilhelm Roentgen", Heron Books, London, 1970.

Davis, M. L. and D. A. Cornwell. "Introduction to Environmental Engineering", McGraw-Hill, New York, 1991.

Davson, H. and M. G. Eggleton. "Starling's Human Physiology", Lea & Febiger, Philadelphia, 1962.

de Kruif, P. "The Microbe Hunters", Washington Square Press, New York, 1926.

Diczfalusy, E. "The Contraceptive Revolution: An Era of Scientific and Social Development", Parthenon, New York, 1997.

Dubos, R. "Pasteur and Modern Science", Anchor Books, Garden City, New York, 1960.

Food and Drug Administration. "Medical Devices ". 2010. From http://www.fda.gov/MedicalDevices.

Goldberg, D. R. "Aspirin: Turn-of-the-Century Miracle Drug", Chemical Heritage, pp. 26–30, Summer 2009.

Goldstein, A., L. Aronow, and S. M. Kalman. "Principles of Drug Action: The Basis of Pharmacology", Wiley, New York, 1974.

Goodman, J. and V. Walsh. "The Story of Taxol", Cambridge University Press, Cambridge, 2001.

Hayward, J. A. "The Romance of Medicine", George Routledge & Sons, London, 1937.

Hyde, J. S. and J. D. DeLamater. "Understanding Human Sexuality", McGraw-Hill, New York, 2006.

Jeffreys, D. "Aspirin: The Remarkable Story of a Wonder Drug", Bloomsbury, New York, 2004.

Katchadourian, H. A. "Fundamentals of Human Sexuality", Holt, Rinehart and Winston, New York, 1972.

Korolkovas, A. "Essentials of Medicinal Chemistry", Wiley, New York, 1988.

Landau, R., B. Achilladelis, and A. Scriabine. "Pharmaceutical Innovators", Chemical Heritage Foundation, Philadelphia, 1999.

Macfarlane, G. "Alexander Flemings: The Man and the Myth", Harvard University Press, Cambridge, Massachusetts, 1984.

Magner, L. N. "A History of Medicine", Taylor & Francis, Boca Raton, Florida, 2005.

Margotta, R. "The Hamlyn History of Medicine", Reed International Books, London, 1996.

Maurois, A. "The Life of Sir Alexander Fleming, Discoverer of Penicillin", Dutton, New York, 1959.

McPhee, S. J., M. A. Papadakis. et al. "Current Medical Diagnosis & Treatment", McGraw-Hill Medical, New York, 2008.

Ostdiek, V. J. and D. J. Bord. "Inquiry into Physics", West Publishing, Minneapolis/St. Paul, 1995.

Porter, R. editor, "Medicine: A History of Healing", Michael O'Mara Books, London, 1997.

Richards-Kortum, R. "Biomedical Engineering for Global Health", Cambridge University Press, Cambridge, 2010.

Snowden, R., G. D. Mitchell, et al. "Artificial Reproduction: A Social Investigation", Allen & Unwin, London, 1983.

Stanifer, R.Y., M. Doudoroff, and E. A. Adelberg. "The Microbial World", Prentice-Hall, New York, 1963.

Suffness, M. "Taxol: Science and Applications", CRC Press, Boca Raton, Florida, 1995.

Wain, H. "A History of Preventive Medicine", Thomas, Springfield, IL, 1970.

Waller, J. "Discovery of the Germ", Totem Books, New York, 2002.

World Health Organization. "WHO Statistics". Available on http://www.who.org, 2012.

CHAPTER 5

SECURITY

Mankind faces many threats to their lives, as well as to their health, freedom, family, employment, income, possessions, and standard of living. In Paleolithic times, the big threats came from hostile animals and locusts, wind and rainstorms, floods and droughts, and volcanic eruptions and earthquakes. In modern times, one of the new and more important threats is to economic well-being, from loss of income and savings. But the most dangerous of all is human violence, from robbers and gangs, and invading troops. Such threats can easily overwhelm the defensive resources of individuals and communities, and require the collective organizations of many people to establish a police force and army. Many kings and rulers have styled themselves as defenders and protectors of the people, and have used this protection from misfortunes to give them the legitimacy to require services and taxation.

The forces of nature are generally too strong for our efforts to prevent. We are dimly aware that government fiscal and monetary policies may influence business cycles, but we cannot predict financial meltdowns ahead of time. Most of our inventions and methods of dealing with threats to security are defensive in nature, such as predicting and giving advanced warnings, building dams and dikes against floods, and investing in private insurance and public welfare against economic threats. Human violence often occurs when the victims give an impression of weakness and vulnerability. The best deterrent seems to be a good shield against attacks and the ability to retaliate and cause severe damage to the attackers, which makes the act of aggression less profitable.

5.1 NATURAL THREATS

Human history is full of stories about the wrath of nature that has given mankind overwhelming hazards and disasters. Many of these disasters are geological, such as earthquakes, volcanoes, and landslides; some are hydrological, such as the flooding of rivers and lakes and tsunamis from the oceans; some are biological, such as plagues of locusts and epidemics; and some are climatic events, such as hurricanes and storms, as well as droughts and heat waves. For defense against fierce animals in the Epic of Gilgamesh, the heroes had axes, knives, quivers, and bows to battle Humbaba and the Bull of Heaven. Against threats from geology and hydrology, our ancestors had little defense except to heed early warnings and flee to safety.

Great Inventions that Changed the World, First Edition. James Wei.
© 2012 John Wiley & Sons, Inc. Published 2012 by John Wiley & Sons, Inc.

Utnapishtim and Noah both had early warning from the gods that a great flood was coming, which gave them time to build their boats, load them with supplies, and float to safety. The last century alone brought many devastating natural disasters: in 1953, the Zuider Zee in Holland had an unprecedented storm; in 2004, the Banda Aceh earthquake created a tsunami in the Indian Ocean; and in 2005, the Hurricane Katrina slammed into New Orleans.

Despite our progress in anticipating these threats of nature and organizing recovery efforts, we have very few triumphant solutions and need more effective inventions. For the threat of earthquakes, for example, we have the following defenses:

- *Identification of Natural Risks and Human Vulnerabilities:* Such as keeping historic records of the California earthquakes on San Andreas Fault and noticing the types of damage to collapsed buildings, highways, and bridges.

- *Preparations:* Reducing vulnerabilities such as the design of earthquake-resistant structures in earthquake zones and the construction of evacuation facilities and shelters.

- *Forecasts:* Prediction methods based on past records and leading indicators, such as earth tremors and animal behaviors.

- *Crisis Management:* Execution of evacuation, search and rescue, and shelter.

We may understand the origin of earthquakes in the movement of tectonic plates, but we can neither predict the time nor the place of the next big quake, such as the Haiti earthquake of 2010 and the Tohoku earthquake of 2011, nor can we stop or lessen the power of destruction.

We can also identify major areas vulnerable to flooding, such as the lowlands near river valleys that are fed from mountain snowmelt in the spring and the low-lying seacoasts such as Venice, Amsterdam, New Orleans, and Bangladesh. We have weather models to predict the coming of storms and we have radar and earth satellites to check the formation and status of hurricanes. Nevertheless, we underestimated the severity of Hurricane Katrina in 2005 that flooded New Orleans. We have invented and built dikes and dams that are partially successful. Nature continues to give us punishing blows, and societies usually do not invest enough resources in such preventive measures, as we cannot predict with accuracy when and where disaster could strike and how severe the strike could be. There is also a human tendency to spend money on immediate enjoyments that are certain and neglect distant calamities that may never happen. The construction of dams and levees also creates displacement, such as moving the residents in low valleys and installing obstacles to the spawning of returning salmon from the sea. Outrage over human or environmental displacement can generate strong political opposition.

5.1.1 Dikes and Dams

A dike is an embankment or raised platform parallel to the shores of a sea or to the banks of a river to prevent flooding; a levee is a natural or artificial embankment next to a river. A dike is built to protect the houses and farms along the river from flooding. They are usually built with compacted earth and are seldom taller than 15 m, but can be hundreds of miles long. A dam is a perpendicular barrier across a river to control the water flow, to create a lake or reservoir upstream for irrigation, and to generate electricity. Dams are built with stronger material such as steel-reinforced concrete to hold back the water. They are sometimes as high as 300 m, but they are no wider than a river. A photograph of a tall and narrow dam can be much more dramatic than that of a low and long dike.

The first levees were constructed over 3000 years ago in ancient Egypt, where a system was built along the west bank of the River Nile for more than 1000 km, stretching from modern Aswan to the Nile Delta on the shores of the Mediterranean. The Mesopotamian civilizations also built large levee systems. Because a levee is only as strong as its weakest point, the heights and standards of construction have to be consistent along its length. Some historians have proposed that such large-scale hydraulic efforts would have required a strong governing authority to compel each person in the area to contribute time and effort and may have been the most important reason for the development of governments in early civilizations. In modern times, prominent levee systems exist along the Mississippi River in the United States, as well as the Po, Rhine, Meuse, Loire, Vistula, and Danube in Europe.

The Yellow River was a source of great sorrow in Chinese history, as it often flooded and destroyed many homes, farms, and lives. Emperor Shun needed a minister to tame the river, so he appointed Kun for the job, who tried to control the flood with barriers. After 9 years with no success, Shun executed Kun but appointed Kun's son Yu to continue the job of taming the waters. Yu inspected the rivers to discover the failure of Kun, and decided to change the control method from barriers to diversion of water. He traveled extensively to establish the elevation of the land and dug canals to divert the water to the sea. He did not rest and took time out even when he was newly married. He even passed his home three times in subsequent travels, but did not go in to visit his family. For 13 years, he traveled for his work, wore the coarsest clothes, ate the poorest food, and slept in the lowest hovels. After 13 years, all the rivers flowed smoothly to the sea, proving that he had solved the flooding problem for a while. When Emperor Shun completed 33 years on the throne, he gave the throne to Yu, who became the founder of the Xia Dynasty in the year 2070 BCE, which lasted for over 400 years. Becoming emperor is probably the highest award ever received by any inventor or engineer in history.

Consider a river with a V-shaped channel, which flows at a normal height and width under normal flow conditions. When the flow is increased by a factor of 4, the height and width of the river will double. This doubling leads to inundation of the adjacent land, which are highly valuable due to the fertility of the soil and the convenience in transportation. Levees are usually built by piling earth on a cleared, level surface. They are broad at the base and taper to a flat and level top, where temporary embankments or sandbags can be placed. The levee surfaces must be protected from erosion, so they are planted with vegetation such as Bermuda grass in order to bind the earth together. On the landside of high levees, a low terrace of earth is usually added as another antierosion measure. On the riverside, erosion from strong waves or currents presents an even greater threat to the integrity of the levee. The effects of erosion are countered by planting with willows, weighted matting, or concrete revetments. Levees turn the river into a U-shaped channel and prevent the river from becoming wider, which protects the valuable land near the river; but the constant deposition of silt causes the river level to become higher and increases the flow velocity. After many years of silt deposits at the bottom of the channel, the levees have to be raised higher than before; eventually, the river and its levees can become higher than the surrounding land, such that a break would cause much greater damage. Sometimes a levee is deliberately breached to flood over less inhabited farmlands in order to save the levees in crowded cities.

The Mississippi River levee system for the main river and principal tributaries represents one of the largest systems found anywhere in the world. It comprises over 5800 km of levees extending some 1600 km along the Mississippi, stretching from Cape Girardeau, Missouri, to the Mississippi Delta. In the eighteenth century, the levee system was started by French settlers in Louisiana to protect the city of New Orleans. The first Louisiana levees were about 1 m high and covered a distance of about 80 km along the riverside. After many decades of confinement behind levees, the deposition of silt raised the bottom of the river that then required the building of higher levees, so that the river eventually became higher than the surrounding land. By the mid-1980s, the Mississippi levees had reached their present extent and averaged 8 m in height, while some levees are as high as 15 m.

Before we design a dike to keep out high water, we must ask how high a dike is needed to stop any foreseeable disaster. The best design depends not only on the available material and structure and on the reliable forecasts of likely hurricanes but also on what we can afford and are willing to spend. The design of a dike or levee is geared to the highest historic level of water, but the future expected level in the river or lake cannot be predicted with precision. A future storm can be greater than any in previous recorded history. Since record-keeping began in 1920, the greatest Category 5 Atlantic hurricanes have been Camille in 1969 and Allen in 1980, both with a top wind of 190 miles/h (317 kph). Consider the following list of Atlantic hurricanes that reached Category 5:

Decade	Number	Max speed (mph)	Max speed (kph)
1920	1	160	267
1930	3	160	267
1940	1	160	267
1950	4	185	308
1960	6	190	317
1970	3	175	292
1980	3	190	317
1990	2	180	300
2000	8	185	308

Should we design the dikes in New Orleans to keep out 200 mph hurricanes, or even higher, for greater safety? What if the appropriated funds were only sufficient for 180 mph? Who would be blamed if the next hurricane turned out to have top winds of 220 mph? In the next 50 or 100 years, what is the likely maximum flood level, and how much safety margin should we build into the design? What does "likely" mean, and what about "Black Swan" events that happen less than once a century? The Tohoku earthquake of Japan in 2011 was a magnitude 9.0, which is higher than any previously recorded Japanese earthquake. The penalty for building a dike wall too low is flooding and the loss of properties and lives; the penalty for building the wall too high is wasting resources that may be better spent for other urgent matters, such as food and medicine for the poor. Reliable forecast methods are comparable in importance with the best material and structure for the dikes.

5.2 ECONOMIC THREATS

A person in modern society needs a steady stream of cash for daily necessities, as well as for major expenditures from time to time—such as buying a house or car, providing for severe illness, paying college expenses, or starting and expanding a business. The steady stream and major expenditures can come from a source of income from employment, withdrawing from savings, or borrowing from others. In the Paleolithic world, people foraged every day and stored what they could spare for winters and for the years of drought and famine. In the modern money economy, income is generated mostly by employment and the surplus of income not spent can be saved and invested in assets such as bank deposits and properties. Tangible assets include land, houses, cars, furniture, and other durable goods; financial assets include bank accounts, stocks and bonds, pension plans, and equity in small businesses.

Income is threatened when an individual is unable to work because of illness, accident, or old age; the individual can also be laid off by the employer due to downsizing, outsourcing, business failure, and general economic recession. A temporary loss of income can be offset by withdrawing from personal savings and from unemployment insurance and social security. The inventions of banking, insurance, and pension plans provide networks for partial protection, by sharing risks against unpredictable events. A severe international recession and depression may overwhelm the local networks, and the national government or international organizations, such as the World Bank and the International Monetary Fund are called upon to protect those affected.

An individual asset such as a house or shares of stock can lose value due to external economic conditions or internal problems; an investor can reduce the risks by a diversified portfolio with different assets that are not correlated. For instance, when the interest rate on fixed bonds goes up, there is often a drop in the stock market, as many investors may choose to sell their stocks to buy the higher yield bonds. So if the investor has both stocks and bonds in his portfolio, he would not have "all his eggs in one basket."

5.2.1   Banks

When there are surpluses that are not required for the moment, we look for places of safekeeping. Squirrels store nuts in the hollows of trees, bees store honey in their hives, and some people store surplus cash under the mattress at home. Thieves and predators may rob the owners of these stored treasures. The very first banks may have been the religious temples of the ancient world, for the storage of gold and jewels. Their owners may have felt that the temples were the safest places available in which to store gold since the temples were constantly attended, well built, and sufficiently sacred to deter some would-be thieves. Modern workers receive pay-checks periodically, such as weekly or monthly, and deposit them in a local bank to be withdrawn later when needed. It is safer to put the money in the bank that is better guarded than a mattress at home, with the added advantage of being able to pay bills by checks rather than carrying a lot of cash through the streets.

The concept that cash saving deposits could be used to generate loan income was recorded from the eighteenth century BCE in Babylon, in the form of loans from temple priests to merchants. One justification for charging interest is to compensate for the lender's risk that the merchant may not be able to pay back the loan; another justification is the time value of money, as people usually feel that a nugget of gold today is worth more than a nugget next year. A bank combines these two functions: accepting deposits that are liquid and can be withdrawn at any time, and making loans with longer fixed periods at high interest rates. When a farmer needs funds in the springtime to buy seeds and fertilizers, which can be repaid in the autumn when the crop is sold, he can borrow from a bank for 6 months and pay interest. When a store wants to buy inventory in time for Christmas shopping, or to open a new branch, it can finance these activities from savings, investors, or a bank loan. A homebuyer finances the purchase with a long-term mortgage where the house is the collateral in case the borrower is unable to pay back the loan and interest on time.

The ascent of Christianity in Rome exerted an influence and led to restrictions to banking, as the charging of interest and usury were seen as immoral and sinful. Jewish entrepreneurs were not restricted from lending money to Christian customers, and established themselves in the provision of financial services, which were in increasing demand by the expansion of European trade and commerce. The pilgrims and Crusaders needed money to finance their journeys from Western Europe to Jerusalem. The Templars became the bankers of kings in the Holy Land and the financiers of pilgrims and crusaders. After the loss of Jerusalem and the Holy Land to the Moslems, the Templars concentrated on banking in Europe and established the first multinational corporations. King Philip IV of France was deeply in debt to the Templars from his war with the English, so he destroyed the Templars to acquire their wealth in 1307 and had the Grand Master burned at the stake.

The first Medici Bank was established in 1397 in Florence with branches in Rome and Venice, became a patron of art, and also made intellectual contributions to the accounting system of double entry of credits and debits. The Medici family created a dynasty of wealth and influence and produced Dukes, Popes, and Queens of France. Modern Western economic and financial history traced back to the coffee houses of London, and the London Royal Exchange was established in 1565. At that time, moneychangers were already called bankers with a hierarchical order: at the top were the bankers who did business with heads of state, next came the city exchanges, and at the bottom were the pawnshops with the three spheres symbol attributed to Lombard banking and adopted by the Medici family. Many European cities today have a Lombard Street where the pawnshops were located at one time. Banking offices were usually located near centers of trade, and in the late seventeenth century, the largest centers for commerce were the seaports of Amsterdam, London, and Hamburg.

There are two principal risks to a banker: (i) The "liquidity risk" is when the bank has tied up too much of its assets in long-term loans that cannot be readily recalled, and runs out of cash when too many depositors want to withdraw funds at one time. (ii) The "credit risk" is when the borrowers are unable to pay back the loans and interests on time. As long as the bank is only a depository of money, the depositors have complete liquidity since the bank can pay back all the deposits at once. However, an observant banker with an almost constant 2 million dollars in deposits might note that all the new depositions of another million tend to arrive on the 30th day of each month, and that the withdrawals are distributed at an average rate of $30,000 or 3% each day. Thus, it would seem safe to lend out up to 1 million dollars on loans of 6 months to 20 year contracts and run very little risk of a "bank run" when too many depositors demand withdrawals on the same day. A conservative banker may think that it is safe to lend out 50% of the stable deposits, to earn 5% interest or $25,000 a year; as he reasons that the daily withdrawal of 3% would not make a big disturbance on the 50% cash in the bank. An aggressive banker may think that it is more profitable to lend out 90% of the stable deposits, and earn $45,000 a year. The aggressive banker runs a bigger risk, as there is only 10% cash left in the bank to service the 3% withdrawals from depositors, which become critical if the withdrawal rate is tripled to 9% per day or even higher. The rate of withdrawal usually goes up many times near Christmas when there is a large need to buy presents for the holiday; the retail stores also have to borrow money from the banks to buy inventory and get ready for the biggest sales month of the year. If there is a rumor that a bank is running out of money, many depositors will converge on the bank and demand their money back, creating a panic. A conservative banker is safer but makes less money, while an aggressive banker runs a greater risk of bankruptcy but makes more money. Such risky behaviors are encouraged when the bank board gives a bonus to the managers that bring in unusually high profits; so the risk-taking bankers are paid more than the conservative bankers, but a failure can bring the entire bank down. The bank managers that can receive a bonus for higher profits may value their bonuses more than they value the welfare of the owners of bank shares. There were numerous times in history when bankers became too aggressive and risk-taking, which led to bank failures as well as a general economic depression on the entire community and world, such as in the global financial crisis of 2008.

A method to reduce credit risks is the system of credit rating by an independent agency, to evaluate whether the borrower is likely to pay back on time, to determine the appropriate interest rate to charge the borrower, and to identify adequate collateral in case of a failure. This is usually based on the financial history of the borrower, as well as the nature of the assets and liabilities. The best-known credit rating agencies include Moody and Standard & Poor's; these agencies issue credit ratings to individuals, corporations, and even sovereign nations. On August 11, 2009, the consumer rate for a 15 year home mortgage was 4.93%, a 30 year mortgage was 5.68%, and a 48 month auto loan was 7.24%. The corporate investment grade AAA bonds, of the highest quality with the smallest degree of risk, had the lowest yield at around 4.20%. The Baa bonds are medium grade and subject to moderate credit risk and had an intermediate yield of 6.25%. The speculative grade are known as high yield or "junk" obligations subject to high risk, and the lowest rated CCC bonds that may be in default have yields of up to 17.06%.

The system of credit rating cannot protect the banks or the savers when there is a general economic recession or depression, and when the largest banks run out of money so that many consumers and businesses have no money to spend or become too worried to spend. At such times, many businesses contract or close down, leading to unemployment and loss of income. In the Depression of 1929, the conservative US government did nothing, and the situation became worse. John Maynard Keynes came out with the theory that the government needed to spend money to stimulate the economy, even by borrowing money. This policy was adopted by Franklin D. Roosevelt from 1932 to 1941; then World War II took over and became an economic stimulant. In the economic downturn of 2008, the US government once again became the spender of the last resort to stimulate the economy and loaned massive amounts of money to shore up stricken banks who were considered to be "too big to fail"—which is interpreted to mean "not permitted to fail."

5.2.2 Insurance

It is very risky to make long distance trips to transport valuable goods, as there is danger to caravans of desert storms, ships capsizing in the sea, illness and accidents, and robbers and pirates. The concept of insurance is to share risks among a large enough pool of people, so that the few unfortunate individuals are compensated by the many fortunate individuals. Early methods of transferring or distributing risk were practiced by Chinese and Babylonian traders as long ago as the second millennia BCE. Chinese merchants traveling treacherous river rapids would distribute their wares across many vessels, to limit the loss due to any single vessel capsizing. There are other insurance methods that do not involve money, such as a group of people forming a "mutual aid society" so that if one person loses his or her house, the others will come to help rebuild the house.

When money was invented, the more familiar forms of money-based insurance became possible. The Babylonians developed a system that was recorded in the famous Code of Hammurabi in 1750 BCE, which was practiced by early Mediterranean sailing merchants. When a merchant received a loan to fund his shipment, he would pay the lender an additional premium in exchange for the lender's guarantee to cancel the loan should the shipment be stolen. A thousand years later, the inhabitants of Rhodes invented the concept of the "general average." Merchants whose goods were being shipped together would pay a proportionally divided premium, which would be used to reimburse any merchant whose goods sank during a storm. It is a prudent thing to pay an insurance fee of 10% to avoid the possibility of a 100% loss of your goods.

The Greeks and Romans introduced health and life insurance around 600 when they organized guilds called "benevolent societies" who took care of the families and paid the funeral expenses of members who died. Guilds in the Middle Ages served a similar purpose. Before insurance was established in the late seventeenth century, "friendly societies" existed in England, in which people donated money to a general account that could be used in case of emergency. Independent insurance contracts, which were not part of loans or other contracts, were invented in Genoa in the fourteenth century. These new insurance contracts allowed insurance to be separated from investments. Insurance became far more sophisticated in post-Renaissance Europe and many specialized varieties were developed.

Toward the end of the seventeenth century, the growing importance of London as a center for trade led to rising demands for marine insurance. In the late 1680s, Edward Lloyd opened a coffee house that became a popular haunt of shipowners, merchants, and ship captains, and thereby a reliable source of the latest shipping news. It became the meeting place for persons wishing to insure cargoes and ships, and those willing to underwrite such ventures. Today, Lloyd's of London remains the leading market for marine and other specialized types of insurance.

The Great Fire of London took place in 1666, which devoured 13,200 houses. In the aftermath of this disaster, Nicholas Barbon opened an office in 1680 to insure brick and frame homes. The first insurance company in the United States provided fire insurance and was formed in Charleston, South Carolina in 1732. Benjamin Franklin helped to popularize and made standard the practice of insurance, particularly against fire in the form of perpetual insurance. In 1752, he founded the Philadelphia Contributions for the Insurance of Houses from Loss by Fire. Franklin's company was the first to make contributions toward fire prevention; his company warned people against fire hazards, and refused to insure certain buildings where the risk of fire was too great, such as all-wood houses.

People at the age of 50 have an expected death rate of 5 per 1000 in the United States. A person may buy life insurance for $100,000 in case of death, to help the survivors to pay funeral expenses and to cushion the transition. The insurance company may have 1000 insured customers, each paying a premium of $700 per year, so that the company collects $1000 \times \$700 = \$700,000$ per year and expects to pay out $5 \times \$100,000 = \$500,000$ per year. This leaves $200,000 for expenses and profit.

PROBABILITY AND INSURANCE

There are many reasons for death, which can be random and not connected, such as heart attack, stroke, cancer, accidents, pulmonary, diabetes, homicide, liver, and kidney problems. The causes for death can also be coordinated, such as earthquake and war when many people die at the same time for the same reason. If the cause is random and not connected, the frequency of occurrence may follow the Poisson distribution where $P(\lambda, k) = (\lambda^k/k!)e^{-\lambda}$. This shows that there are a number of possibilities for death next year:

k Number of death	Probability	Cumulative probability	Company profit, 1000 dollars
0	0.0067	0.0067	700
2	0.0842	0.1247	500
4	0.1755	0.4405	300
6	0.1462	0.7622	100
7	0.1044	0.8667	0
8	0.0653	0.9319	−100
10	0.0181	0.9863	−300
12	0.0034	0.9979	−500
14	0.0005	0.9998	−700
16	0.0001	1	−900
17	0	1	−1000

There is an 86.67% probability that the number of deaths is no more than seven and the company will not lose money; thus, there is a 13.33% probability that there will be eight or more deaths and the company will lose money. There is a 6.53% probability of eight deaths where the company would suffer a loss of $100,000 and a 0.34% probability of 12 deaths where the company would lose $500,000.

Of course, the number of deaths is not always the expected five, but can be much higher or lower. The highest probability is that there would be four–six deaths, and the company would be able to make expenses and profits of $100,000–$300,000. When the actual number of deaths the following year is greater than seven, then the insurance company would suffer a loss, to the great displeasure of the stockholders who invested in the company. To decrease the probability of a loss, the company could raise the premium to $800 per year, then the probability of a loss would be lowered to 6.81%, or 1 loss year out of 15 years. Raising the premium to $900 per year would further lower the probability of a loss to 3.2% or 1 loss year out of 33 years. Since there are other insurance companies in the city, how many customers would be lost by increasing the premium?

The other method to increase the chance of profitability is to enlarge the size of the pool of customers, and let the Law of Large Numbers take care of the company. In fact, the size of customers has a critical bearing on the probability of loss.

Customer size	Probability of loss (%)
1,000	13.33
3,000	5.31
10,000	0.30
30,000	0.0002

An insurance company also has the option of buying reinsurance by paying a premium to a reinsurance company. The reinsurance companies are huge organizations built to insure a large number of insurance companies, which reduces the risk of a single insurance company becoming insolvent. The reinsurers with the largest assets include the Swiss Re and the Berkshire Hathaway/General Re. If 10 small insurance companies bought reinsurance from the American Re, then the risk of 0.30%, or 1 bad year out of 333 years, would be quite small for the American Re.

However, not all deaths are random and uncoordinated events, such as the wave of fatalities in an earthquake or in a war. At such times, even with millions of insurers, the American Re may still get into trouble, and the US government may have to step in to save the situation. In a major economic crisis, such as in the 2008 failure of Lehmann Brothers and the bankruptcy of AIG, the US government had to step in as the banker and insurer of "last resort." There is greater safety in larger numbers, so the financial crisis of 2008 was much harder on small nations such as Iceland and Ireland and less devastating on a large economy such as the United States and the European Union. We hope that the day will never come when even the US government collapses, such as in a worldwide nuclear war.

5.3 HUMAN VIOLENCE: WAR

War is of vital importance to the State, the matter for life and death, the road to safety or ruin, and must not be neglected.

Sunzi, *Art of War*

No doubt, the greatest threat to security is from other men, whether they are local robbers, terrorist groups, or foreign nations. National victory on the battlefield is the most important determinant of future independence and power, and history shows that superior inventions and technology have frequently played critical roles. No wonder rulers and governments have been more willing to pay for military inventions than for any other types of inventions. In 2009, the US Federal R&D obligations were 157 billion dollars, with 86 billion dollars or 55% allocated to war and only 40 billion dollars to health.

A good defense is essential to shield against aggressors, but most military inventions are designed instead for attacks to overwhelm the enemy. There is a tacit assumption that the victim invites aggression by being weak and vulnerable, and the best deterrent is a strong attack capability that would retaliate and inflict great pain on aggressors. There is a Chinese legend about a weapon maker who advertised that his new spear was strong enough to break all existing shields; after he sold all his new spears, he began to advertise his new shields that would stop all existing spears. History is full of examples of innovations in walls and castles designed to be impregnable and immune to sieges, followed by innovations in siege machinery to overcome these walls, followed by yet further defensive innovations to stop these attacks, and the cycle continues.

5.3.1 Ancient Weapons and Defense

The *Iliad* is about warfare, and here Homer gave detailed information on the war equipment of that time. The wall around Troy was a critical factor for 10 years, which stopped the Greeks and their superior forces from storming the city and carrying back Helen. Apparently, the Greeks had no siege equipment that could threaten or overcome the wall. The wall was eventually penetrated only when they employed deception in the form of hiding warriors inside a wooden horse, which was unwittingly taken inside the city walls by the unsuspecting Trojans. In Book 3, Paris prepared for war with equipment that included a pair of greaves or shin guards, corselet to protect his chest, a sword, shield, helmet with horsehair, and a spear. In Book 7, Aias of Telamon carried a shield that looked like a wall, made of an outer layer of bronze and seven inner layers of oxhide. When Hector threw a spear against this shield, it tore through the bronze layer and six layers of leather, but was stopped by the seventh layer of leather.

Before Achilles set out to fight Hector in the climactic duel, his mother Thetis went to see the god Hephaestus to make him the most wonderful set of armor that anyone could hope to wear. She asked for a shield, a helmet, two greaves with clasps for the ankles, and a corselet. So for this masterpiece, Hephaestus went to his 20 bellows to blast the crucibles and blow the flames high. He cast bronze and tin on the fire and the valuable gold and silver. He set up his great anvil, gripped in one hand the ponderous hammer, while in the other he grasped the pincer. First he forged a shield that was huge and heavy and threw around it a shining triple rim that glittered and then cast the shield strap of silver. There were five layers on the shield, and upon it he elaborated many things with his skill and craftsmanship. From lines 483 to 607, Homer described the scene on the shield in such astonishing detail that you wonder if this was meant to stop a spear thrust from Hector or to cause Hector to stop and admire its beauty and forget about the fight. It described the sky and numerous stars; one city was in peace with a scene of a wedding and festival and a dispute in front of judges; another city was at war; there were even vineyards and dancing floors. The outer rim of the shield's strong structure was the Ocean River. To fit all this detail, the scenes must have been microscopic in size, or the shield itself was enormous.

As an afterthought, Hephaestus wrought for Achilles a corselet brighter than fire, a helmet massive and fitting close to his temples, and out of pliable tin wrought his leg armor. Did Hephaestus think that the main purpose of the armor was to protect the wearer from harm or to make a fashion statement and proclaim him as the most stylish warrior in Greece? What made him think that a silver handle was right for the shield, and tin was right for the greaves? So Hephaestus, or Vulcan, was more a sculptor than an engineer, as a competent engineer would have designed the armor to be strong and light. Let us look at the table of material properties:

Metal	Density (g/mL)	Young's modulus (GPa)	Mohr hardness	Melting point (°C)
Leather	1.0			
Tin	7.3	50	1.5	232
Iron	7.9	211	4.0	1535

Copper	8.9	130	3.0	1083
Silver	10.5	83	2.5	961
Lead	11.3	16	1.5	327
Gold	19.3	78	2.5	1063

Gold is much heavier than iron, which is in turn much heavier than leather. Leather is by far the lightest material here, but not very strong for stopping blades and spears. Tin is the lightest metal here with a modest Young's modulus to resist compression, but it is too soft. Iron has the highest Young's modulus and the highest hardness, and it should have been used for everything, but because its melting point is also the highest, it is more difficult to work with. If meteoric iron was not readily available, the next best idea was bronze from copper and tin. Under no circumstances would a good engineer use silver or gold, as they are both heavy and soft.

How well did this suit of armor perform in combat, which was supposed to be the purpose of the request of Thetis? You would be disappointed to hear that no one stopped in their tracks to admire the marvelous and beautiful scenery on the shield, as most warriors ran away when Achilles showed up, and the few warriors who stayed to fight were not connoisseurs of fine art. Later in Book 20, the armor of Achilles finally received a real test in his encounter with Aeneas who threw his ponderous pike against the shield. The shield had two folds of bronze on the outside and two folds of tin on the inside, and between them a single gold fold. In fact, the spear of Aeneas was stopped by the gold, but it must have made a mess of the pretty pictures on the outside, which Homer did not describe.

The earliest bow was a simple device from a single piece of wood, such as yew bent in the arc of a circle, so that the archer could store up energy slowly and discharge the arrow suddenly. The string was made from the gut of animals. The big advantage of the bow and arrow is the ability to throw projectiles from a safe distance, which is why Apollo was such a cool and distant fighter, in comparison to Ares, who did hand-to-hand combat with the risk of a blow to the head. Strong archers can manage a pull of 45 kg and even more with special diet and exercise. The simple wood bow was gradually replaced by the composite bow made of several layers of lamination. When the bow is bent, the belly or the inside closer to the shooter is compressed and shortened and the back closer to the target is stretched and lengthened. It became apparent that different materials excelled in different functions, such as animal sinew being better in tension but animal horn being better in compression. A classical composite bow has a layer of horn at the belly, a layer of wood in the middle, and a layer of sinew in the back. These three layers are glued together with a gelatin, bound together with fibers, and lacquered to make the bow waterproof.

Historians differ on whether the crossbow was invented independently in the West or in China, which dates back to at least 450 BCE when Sun Zi wrote about the crossbow. The crossbow can be pulled to a higher load of 91 kg, by putting both feet against the bow and pulling with both hands or using a crank for the string. When the string is held by a cocking mechanism, one can aim at leisure without straining or expending any force. The crossbow has a trigger mechanism that is made of bronze or iron and can have three parts. The first part is the tumbler or stock, with an upright tooth that can hold the string and can also rotate forward to release the string. To arm the crossbow, the archer pulls the bowstring back toward the trigger mechanism and rests the string on the tooth of the stock. The stock is prevented from rotating forward by the second part, the sear, which is in turn controlled by the third part, the trigger. When the trigger is pulled, the sear escapes and allows the stock to release the string and the arrow. After the arrow has left, the archer can recock the trigger by pulling the stock up and then reposition the sear into a notch on the trigger. Then the archer is ready to shoot again.

The Welsh longbows were adopted by the English against the French in the Hundred Years' War, with great success at Crecy (1346). The English army of 12,000 relied on the longbow and managed to slaughter the elite French army of 36,000. The English longbow archers were deployed on foot in a V-formation and could shoot five or six arrows every minute. The French cavalry repeatedly charged into this V-formation and the storm of arrows struck and killed their horses, so they were unhorsed in the mud. The French hired Genoese crossbowmen who could shoot only one or two arrows every minute since the reload mechanism was much slower. It is astonishing that in the two subsequent encounters at Poitiers (1356) and at Agincourt (1415), the English longbow continued to slaughter the French cavalry, who had learned nothing by these defeats and did not change their fighting methods.

To prevent the entry of intruders, you may enclose your family and possessions inside a barrier, such as a fence, a ditch, or a wall. Uruk in Sumer is one of the oldest walled cities in the world, which was built of stone and brick and was lovingly described in the Epic of Gilgamesh. Jericho in Palestine had a wall that dates from 7000 years ago. Nebuchadnezzar built a wall around Babylon, which has the beautiful Ishtar Gate as the entry point. On the other hand, the palace of Knossos on the island of Crete had no wall around it, presumably trusting its navy to keep away intruders.

The Great Wall of China was started in 750 BCE as an extremely ambitious project to enclose and protect not only a city but also the entire country of farmers from the marauding nomad tribes of the steppes and deserts. To construct this Great Wall far from the capitol, only local materials were used that included some stone, but was more often rammed earth—made by mixing earth with chalk and lime, as well as straw, dung, and gravel as binders. This mixture was loaded between vertical frames, and rammed from above to compact it and to reduce its porosity. The walls were often 12 m high and 12 m wide at the bottom, but narrowed to 7 m wide at the top. The earlier wall of the Emperor Qin that dated from 220 BCE had mostly eroded away in the succeeding centuries, even though the climate is relatively arid. The current wall dates back to the Ming dynasty in 1368, when much more brick and stone were used, especially at the watchtowers.

A fortress should be placed on high hills and have tall walls, so that the defenders can see the invaders from a long distance, shoot the arrows far, and drop stones down to demolish approaching enemies. The attackers hope to climb the wall with ladders, smash the gate with battering rams, and dig trenches underground to enter the walls. The Trojan wall withstood the Greek army for 10 years, who had no siege machinery and had to depend on the ruse of the Trojan Horse to enter the gates of Troy. The Romans built Hadrian's Wall in England in 122, to keep the wild Scots from raiding. It was 120 km long, and cut Scotland off from England. On the east side, it was made of stone and was 6 m high by 3 m wide; but on the west, it was made of turf and was 4 m high by 7 m wide. Small milecastles were placed approximately 1 Roman mile apart and guarded by garrison. Vitruvius was one of the most famous architect–engineers who wrote about the craft of fortification for defense as well as siege for attack—with weapons such as the catapult, battering ram, and siege towers.

The great age of European castle building began when the Crusaders returned from the Orient and brought many ideas. One of the best preserved castles is the beautiful Krak des Chevaliers in southern Syria, situated at a critical crossroad and recognized by UNESCO as a World Heritage Site. T.E. Lawrence described it as "Perhaps the best preserved and most wholly admirable castle in the world." It is situated on an isolated hill commanding a vast plain, so that invaders are exposed and required to climb upward. It was designed as a concentric castle with an outer wall, separated by the ground in the outer ward and a ditch from the inner wall, which enclosed several imposing towers as a last defense. The outer wall is steep and up to 3 m thick, and enforced with 13 towers. There is a barbican or fortified gateway on the outer wall, which can open to allow visitors into the outer ward; thereafter, there is still a ditch to cross and then there are the inside walls girded with seven even stronger towers. There is a separate outside entrance to the inner ward, leading to a very long and narrow ramp that climbs gradually up and makes a hairpin turn to another narrow ramp, which is exposed to missiles from above. The final defense is in the towers that also serve as the loge of the master and the knights.

In 1482, Leonardo da Vinci wrote an often-quoted letter to Ludovico Sforza, the Duke of Milan, offering his service. He listed his numerous outstanding skills as a military engineer: building and destroying bridges for war, removing water from trenches, designing catapults and engines to fling stones, planting mines to pass under trenches or rivers, and designing military chariots. Finally, at the end of the letter, he said incidentally that in times of peace, he could give perfect satisfaction equal to that of any other in the architecture of buildings public and private, in sculpture with marble and bronze, and in painting, as well as any other. The letter was successful, and Ludovico employed Leonardo for 17 years including the commission for the Last Supper.

Walls and towers can be attacked by siege machinery, which includes the catapult to throw rocks at the walls, battering ram to smash the gates, siege towers to allow archers and scalers to be at the same height as the defenders on the walls, and sappers to dig tunnels under the walls. A strong wall can stand a siege for a long time, as long as the defenders have stores of food and a source of water, although they will eventually be starved out by a long siege. When explosives arrived around 1200, the tall and thin towers and walls could easily be pounded by cannons. The land and sea walls of Constantinople were 20 km long, had stood for more than a 1000 years, and were taken only once during the Fourth Crusade. Sultan Mehmed II approached with his huge army in the siege of Constantinople in 1453. Orban the Hungarian tried to sell his services to the Byzantines, who were unable to secure the funds to hire him. Then, Orban turned to join Mehmed II and designed cannons up to 9 m long that could hurl a 600 kg projectile over 1.6 km. Mehmed's massive cannons fired on the walls of Constantinople for weeks, and his army also engaged to break through the walls by digging and mining underground tunnels. The fall of Constantinople also meant the end of the Eastern Roman Empire.

The cannons forced a new age of castle building, when defensive walls were redesigned to offer much lower and thicker targets, which are not as heroic looking as taller targets. When attackers concentrated their efforts on a section of a circular wall, the defenders stationed at other parts of the wall could not help to fire on the attackers. This gave rise to the famous star forts, which were employed by Michelangelo in the defense of Florence. The points of the star that jut out from the core were heavily fortified bastions that were protected by ditches and palisades and manned by defenders with arrows and cannons. When an invading army attacked the weaker walls connecting the bastions, they would be raked by overlapping cross fire at different angles from two bastions. Perhaps the most famous innovator was Sebastien Le Prestre de Vauban (1633–1707) who was a French military engineer, a marshal of France, and the greatest expert in both building and attacking castles. He made numerous innovations in the design of castles. Between 1667 and 1707, he upgraded the fortifications of 300 cities and directed the building of 37 new fortresses.

5.3.2 Horses, Cavalry, and Stirrup

The introduction of horses revolutionized warfare, as the speed and strength of horses could not be matched by the infantry. Lynn White explained that the historic use of horses in battle was divided into three periods: (i) the chariot, (ii) the mounted warrior who clung to his steed by pressure of the knees, and (iii) the mounted warrior with his feet in stirrups. Each introduction of new technology led to far-reaching social and cultural changes and elevated the status of people with horses. There are numerous wall paintings in Mesopotamia, Egypt, and China with images of triumphant kings on their chariots, sharing with drivers and archers and trampling on their prostrate enemies. This technology required horses and two-wheeled chariots and gave the owner a high platform from which to observe the battle and rapid movement to arrive at the critical places. The chariot was also an excellent platform for shooting arrows, throwing javelins, and cutting down enemy infantry with spears. The Greeks and Macedonians developed the infantry phalanx in formation, where each soldier carried a spear as long as 7 m in the right hand and a shield in the left hand. As long as the phalanx was locked in order, it was very difficult for chariots and mounted cavalry to break through this formation of spears. The Greeks and Macedonians repeatedly proved the superiority of the phalanx over the chariot and less organized infantry.

During the second period of Lynn White, the mounted cavalry sat on a horse with nothing more than a piece of cloth between them. Before the invention of the stirrup for the rider's feet, a horseman was primarily a mobile bowman and hurler of javelins. He could do limited swordplay on his horse, but a swipe that missed might throw him to the ground, and his spear could only be thrusted with the force of his shoulder and biceps. He could not charge and ram his spear at the infantry without

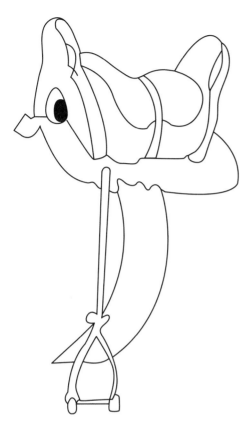

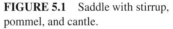

FIGURE 5.1   Saddle with stirrup, pommel, and cantle.

being thrown out the back of the horse at the same time. So the purpose of riding a horse was speed and a high vantage point, but heavy hand-to-hand combat often meant dismounting first and fighting as infantry.

When the horse stirrup was invented and adopted, and the saddle was strapped to the horse, the knight with his feet in the stirrup could keep his seat safely despite violent actions. In Figure 5.1, the saddle is strapped to the horse and has stirrups for the feet, so that the rider can resist forces that push him to the left or right. The saddle also has a high pommel in the front and a high cantle at the back, which prevent him from sliding forward or backward if the horse suddenly changes speed, such as during the impact in a joust. In fact, his most lethal attack would be to rest his lance in the forward position and allow the combined weight of his body and the charging stallion to thrust the lance with much greater force than swinging his lance with his biceps.

White attributed the invention of the more primitive big-toe-stirrup to India in the second century BCE, and the invention of the foot stirrup to China that was first mentioned in 477. The use of the stirrup spread to Byzantium by 582 and to Western Europe at the time of Charles Martel, which led to the institution of knighthood. The early Franks mainly fought on foot. In the year 732, Charles Martel met the Saracens near Poitiers and the infantry line stood like a wall against the Saracen horses, but Martel's victorious infantry was unable to pursue the fleeing Saracen horses. The support of horses and heavy armory required more investment by a dedicated and professional army of a small elite group, which led to feudalism. With the acquiescence of Pope Zacharias on the grounds that it was necessary to fight the Saracens, Saxons, and Frisians, Martel confiscated church lands to distribute among his fiefs as an endowment to support horses and armor. Feudalism meant that these fiefs who accepted the land would swear an oath of allegiance that they would come with horses and armor when called by their lords. This was a small professional army that the more numerous yeoman with light armor on foot could not hope to compete with. The army of Charlemagne was dominated by cavalry.

The Norman conquest of England and the Battle of Hastings of 1066 could also be attributed to this change in technology. The Anglo-Saxons knew about stirrups, as Harold and his housecarls rode horses with stirrups, but they did not develop feudalism and did not have an elite corps of mounted cavalry. At Hastings, the Saxon lords dismounted to do battle on foot, in the old Germanic style of shield wall. The Saxons also had the advantage of a higher position on the hill of Senlac and they probably outnumbered the Normans. But Harold fought without cavalry and had few archers, and could only stand in line to resist this mobile striking power, without the ability to rapidly maneuver and counterattack. When William won his victory and the crown of England, he feudalized his new kingdom and created more mounted knights.

The period of the dominance of the cavalry lasted for several centuries till the arrival of the long bow. The Battle of Crecy took place in 1346, when 12,000 English foot soldiers decimated the French cavalry. The Age of Chivalry was over when common soldiers could stand up to their furious charges with long bows, and later with the much more lethal explosives and guns.

5.3.3 Explosives

There is a long history of war by fire and incendiary material, which may be considered the forerunners of gunpowder. The siege of Tyre in 332 BCE by Alexander featured blazing pitch. The famous Greek Fire was used by the Byzantines against the Arabs as early as 671. A secret formula was involved, but it probably utilized a jet of petroleum oil, forced out of a bronze tube by a pump, and mixed with substances such as quicklime that can ignite when mixed with water.

Warfare in China used smoke screens and toxic smoke, which was mentioned in *Mo Zi* in the fourth century BCE, and was created with mustard, lime, arsenic, and other irritants. The burning of incense was found in the *Shi Jing*, earlier than the seventh century BCE. Daoist alchemists searched for immortality and had many chemical substances at their command, including cinnabar, nitrates, and many other chemicals or mixtures. In their mixing of various substances, they produced many elixirs as early as 200 BCE for the First Emperor of Qin, but these elixirs actually may have accelerated his death. Perhaps another objective of this experimentation was the religious and public health tradition of "smoking out" undesirable things such as worms with smoke and fumes. Fumigations were also carried out for medicinal purposes, as well as for reducing bookworms that damage books. Sulfur can be used by itself for such fumigation, but the addition of saltpeter will make the smoke far more impressive. Gunpowder was invented in China in around 850 during the Tang dynasty, not by artisans or soldiers but by a number of Taoist alchemists. That year, a Taoist book mentioned a mixture of saltpeter (potassium nitrate), sulfur, and carbonaceous material, and cautioned readers that the mixture might catch fire and burn down the house.

Early gunpowder had a very low proportion of saltpeter and could be used only as an incendiary or as a flamethrower, not as an explosive bomb or to propel bullets from a gun. By 1000, gunpowder with higher saltpeter content was being placed in spherical bombs and grenades, and lobbed from trebuchets or catapults at enemies. The first mention of a formula in print was from 1044, and contained a very low percentage of saltpeter. The ideal mixture for gunpowder is, according to the chemical reaction formula, made with two moles of saltpeter, one mole of sulfur, and three moles of carbon

$$2KNO_3 + S + 3C \rightarrow K_2S + N_2 + 3CO_2$$

So the ideal weight ratio of saltpeter/sulfur/carbon is 75/12/13. The purpose of the saltpeter or potassium nitrate KNO_3 is to provide oxygen for the rapid combustion of carbon C, which burns at a high temperature; the purpose of the sulfur S is to lower the ignition temperature to something like 250°C. Sodium nitrate can serve the same purpose as potassium nitrate, but it readily absorbs water from the moist air and becomes less effective. We can divide the speed of combustion or explosions into various grades:

(i) *Slow Burn:* Such as a lamp with oil, pitch, or sulfur.

(ii) *Quick Burn:* Such as fire with distilled petroleum, naphtha, or the Greek Fire.

(iii) *Deflagration:* Rapid burning with sparkles, suitable for launching rockets and fire lances, and requires at least 33% saltpeter.

(iv) *Explosion:* Very rapid burning with a bang, suitable for guns and cannons to shoot a projectile out of a metal tube, and requires 50% saltpeter.

(v) *Detonation:* Extremely rapid explosion with sufficient forces to burst metal containers, suitable for a bomb, and requires 75% saltpeter.

There is a silk banner from the Dunhuang caves in China from around 950, which is now at the Guimet Museum in Paris, and shows a devil menacing the Buddha with a fire lance, while another devil is about to lob a spherical bomb. The fire lance requires a cylindrical container, and the most easily available cylinder in China was a bamboo pole with the interior separators removed. In a chronicle of 1130, a fire lance with low nitrate powder, presumably stuffed in a bronze or cast iron cylinder, was attached to the end of spear; it had an effective burning time of 5 min and was used to discourage troops from storming a city. These flamethrowers are comparable to the Greek Fire, which used liquid naphtha as the combustion mixture. Since the range was very small, the fire lance was more of a threat than a highly effective weapon.

The next invention in explosives was the rocket, which turns the fire lance upside down, lets the jet of exhaust gases shoot downward, and then propels the cylinder upward into the air by recoil. The rocket can also be used to carry smoke, incendiary substances, and explosives. The true handgun or bombard appeared in 1280, and had three ingredients:

 (i) A metal cylindrical barrel.

 (ii) A gunpowder with high nitrate content.

 (iii) A projectile that seals the muzzle, so that the full force of the powder cannot blow through gaps between the projectile and the muzzle, and will propel the projectile at great force.

There was a slow rate of invention from gunpowder to guns, taking some 400 years. By 1350, the large cannons began to appear. The Arabs first learned about these advances through contacts and warfare with the Mongols. The first mention of gunpowder in Arabic was in the 1280 work of Hassan al-Rammah of Syria, which appeared about the same time as *Liber Ignium ad Comburendos Hostes* in Byzantium. In contrast, the development of gunpowder, muskets, and cannons progressed rapidly in Western Europe. Roger Bacon (1219–1292) talked about the wonders of gunpowder. A true cannon requires a very strong bronze cylinder that is closed on one end, to form a chamber for the gunpowder, and a projectile that is jammed into the cylinder for a close fit. The bronze casting technology had already been developed in the casting of bells. On the land, the cannon led rapidly to the downfall of castles and aristocratic feudalism; on the sea, the multioared war galley with battering rams was displaced by the full rigged sailing ships with cannons. By the time of the visit of Matteo Ricci to China in 1582, the West had far superior guns and cannons to the slowly evolving Chinese weapons.

> *It was a great pity, so it was,*
> > *This villainous saltpetre should be digg'd*
> > *Out of the bowels of the harmless earth,*
> > *Which many a good tall fellow had destroyed*
> > *so cowardly*

So lamented William Shakespeare, in *Henry IV, Part I.*

Of the three ingredients of gunpowder, carbon or charcoal was the easiest to obtain by simply burning wood with insufficient air. Sulfur was more difficult to find, but obtainable from volcanoes, hot springs, and from metal sulfide ores. The most difficult ingredient was saltpeter, which takes up 3/4 of the weight of gunpowder and was not readily available in the West. It is formed in nature by the decomposition of organic matter by bacteria, such as excreta in an environment of high temperature and humidity. It appears as a "ground frost" of small white crystals on the soil and in mountains and marshes. The white crystals were swept and collected, dissolved in hot water, and then filtered to remove impure particles. It was then boiled to remove the water and the saltpeter thus precipitated into crystals in the shape of pine needles, which the Arabs called "Chinese snow."

To supplement this slow and uncertain supply of nature, nitre beds were prepared by mixing manure with wood ashes, common earth, and organic materials such as straw to give porosity to a compost pile, typically more than 1 m high. The heap was usually covered from the rain, kept moist with urine, turned often to accelerate the decomposition, leached with water, and harvested after 2 years. The liquid containing various nitrates was then converted with wood ashes to potassium nitrates, crystallized, and refined for use in gunpowder. By 1626, saltpeter began to reach Europe from India, mined in the more arid regions as a monopoly of the British East India Company. The General Court of Massachusetts passed an order in 1642 requiring every plantation within the Colony to erect nitre sheds. It was reported that only 5 g of saltpeter were recovered per liter of the nitre bed every 2 years! It is hard to see how they could prepare for a war with such an inefficient method of production. Later, potassium nitrate came from the vast deposits of sodium nitrate in the Chilean deserts, formed by centuries of guano from birds that caught fish in the rich Humboldt ocean currents. Today, ever since Haber invented the method of nitrogen fixation to make ammonia, nitrates are made by oxidation of ammonia over platinum gauzes into nitrogen oxide gases NO and NO_2, which is reacted with water to make nitric acid.

The high explosives were inventions of the mid-1800s, with an explosive power at least five times higher than that of black gunpowder. The first high explosive was nitroglycerine, invented by Ascanio Sobrero in 1847 at the University of Turin. Nitroglycerine is produced by adding sulfuric and nitric acids to glycerin and keeping it below 10°C to prevent explosion. It is an unstable product that can explode suddenly, causing great destruction of lives and properties. The explosion, as shown below, produces carbon dioxide, water, nitrogen, and oxygen gases.

$$C_3H_5(NO_3)_3 \rightarrow 3CO_2 + 2.5H_2O + 1.5N_2 + 0.25O_2$$

Dynamite was invented by Alfred Nobel in 1866, and is made by absorbing three parts nitroglycerine in one part diatomaceous earth or kieselguhr. It was much safer to handle than nitroglycerine alone, and was in high demand for peacetime construction and tunneling. Guncotton was invented in 1845 by Christian Friedrich Schönbein, a Swiss chemist from Basel. He accidentally spilled concentrated nitric acid on his kitchen table, reached for a cotton apron to wipe it, and hung the apron on a stove door to dry, which exploded when it was dry.

TNT is 2,4,6-trinitrotoluene, invented in 1863 by the German chemist Joseph Wilbrand, who used it as a yellow dye. Its explosive powers were discovered later, after which TNT was used by the German armed forces to fill artillery shells and armor-piercing shells. It is insensitive to shock and friction, which reduces the risk of accidental detonation. It melts at 80°C and can be used in a wet environment. Its explosion produces the gases nitrogen, water, carbon monoxide, and carbon.

$$2C_7H_5N_3O_6 \rightarrow 3N_2 + 5H_2O + 7CO + 7C$$

Alfred Nobel was the inventor of dynamite, which revolutionized both war and peace, as explosives are also used for excavation, tunneling, and mining. He is shown in Figure 5.2. Besides being an inventor, he was also an entrepreneur and businessman who worked round the clock and also filled the roles of general manager and manager of advertising, sales, and public relations. He tirelessly traveled to many cities of Europe and America, hauling padded suitcases filled with cans of

FIGURE 5.2 Alfred Nobel. Reprinted courtesy of © The Nobel Foundation.

blasting oil, to make demonstrations in mines. He achieved great fame and wealth, but had no family and no fixed residence, and was called "the loneliest millionaire." Nobel lived through numerous accidents and explosions where many people were killed, and earned another name: "the merchant of death." He also founded the Nobel Prizes, which have been awarded since 1901 to celebrate the highest achievements in science, literature, and peace.

Alfred Nobel was born in 1833 as the third son to Immanuel and Andriette Nobel in Sweden. He was a sickly child, and had poor health throughout his life. Later, he wrote a poem in English about his childhood:

> My cradle looked a death-bed, and for years
> a mother watched with ever anxious care,
> so little chance, to save the flickering light.

Immanuel was an inventor and entrepreneur who constructed bridges and buildings in Stockholm. He experienced wide fluctuations in his fortune, and was bankrupt during the year of the birth of Alfred. In 1837, he fled to Russia without his family, where he was successful in providing equipment for the Russian army; so he sent for his family to join him in 1842. Alfred and his brothers were given first-class education by private teachers. By the age of 17, Alfred Nobel was fluent in Swedish, Russian, French, English, and German. His primary interests were in English literature and poetry, as well as in chemistry and physics. He was withdrawn and seldom played with other children, but he was an excellent student.

In 1850, Nobel was sent on a study tour abroad and visited Germany, France, and the United States. In Paris, he worked in the private laboratory of Professor T.J. Pelouze, where he met the young Italian chemist Ascanio Sobrero, the inventor of nitroglycerine. Although its explosive power greatly exceeded that of gunpowder, the liquid would explode in a very unpredictable manner if subjected to heat and pressure. Sobrero considered it to be too dangerous to use in industry and felt that it should be used only in medicine. Nobel became very interested in nitroglycerine for practical use in construction work, but he needed to solve the safety problems.

The Crimean War began in 1853, and Alfred returned to his family in St. Petersburg to work on ammunition for the Russian army. Immanuel convinced the Tsar and his generals that naval mines could be used to block enemy naval ships from threatening the city. The naval mines were simple devices consisting of submerged wooden casks filled with gunpowder. Anchored below the surface, they effectively deterred the British Royal Navy from moving to St. Petersburg during the Crimean war. Alfred experimented with nitroglycerine as a possible explosive for his father's sea mines. He tried for years to control the erratic and wild substance, and he began experimenting with a revolutionary idea of using black powder to ignite the more powerful nitroglycerine.

The war ended in 1856 and the army war contracts ended, so Immanuel Nobel was again forced into bankruptcy. Alfred returned to Sweden in 1863, and concentrated on developing a detonator for nitroglycerine as an explosive. He did many experiments, including one where he placed a small metal tube of black powder inside a large tube of nitroglycerine, and one involving fulminated mercury. After successfully developing a detonator, he filed for a patent and obtained financing from Paris for the construction projects of Napoleon III. He began to manufacture and ship his high explosives around the world. But in 1864, his Heleneborg plant had a horrendous explosion, in which his brother Oscar-Emil and several other persons were killed, and the building and plant were scattered in shreds. Alfred Nobel narrowly escaped being charged with homicide, and public opinion in Sweden turned against him. The authorities were convinced that nitroglycerine production was exceedingly dangerous and forbade further experimentation with nitroglycerine within the Stockholm city limits, so Alfred Nobel had to move his production factory to a covered barge anchored in a lake outside city limits and quickly fulfilled contracts to the state railway for blasting tunnels. He was extremely unpopular with locals who called his mobile factory a "death ship," so he was constantly on the move, floating his laboratory to new anchorages and fleeing mobs of angry peasants brandishing pitchforks. Alfred reacted to these setbacks with remorse and depression. He wrote 10 years later, "I have no memories to cheer me, no pleasant illusions about the future to comfort me, or about myself - to satisfy my vanity. I have no family to provide the only kind of survival that concerns one; no friends for my affections, nor even enemies for my malice . . . "

In 1865, Alfred Nobel bought 42 hectares of land at Krümmel, an isolated hilly area about 30 km from the center of Hamburg, on the shore of the Elbe River.

The following year, the Krümmel works were destroyed by explosions. He tried to improve the safety of nitroglycerine by mixing it with stabilizing inert materials such as charcoal, cement, sawdust, and so on. Finally, he tried the sand from the Krümmel dunes, the kieselguhr that is a porous rock powder of fossilized remains of diatoms, and here he found the perfect solution. The nitroglycerine and kieselguhr mixture was a soft pliable material like dough, and could be shaped into any form. It was patented under the name "dynamite" in 1867 and was shaped into rods suitable for insertion into drilling holes. The best combination is three parts nitroglycerine to one part kieselguhr, which has less power but much more safety than pure nitroglycerine.

The inventions of detonator and dynamite drastically reduced the cost of blasting rocks, drilling tunnels, building canals, and many other forms of construction work. The market for dynamite and detonating caps grew very rapidly and Alfred Nobel also proved to be a very skilful entrepreneur, promoter, and businessman.

Nobel became very rich and moved to Paris in 1873, but much of his life he was still constantly traveling. Victor Hugo at one time described him as "Europe's richest vagabond." Many of the companies founded by Nobel have developed into industrial enterprises that still play prominent roles in the world economy today, including Imperial Chemical Industries (ICI) of Great Britain, Société Centrale de Dynamite of France, and Dyno Industries of Norway. He continued to do experiments, in addition to traveling and founding plants. In 1876, he invented and patented the blasting gelatin in Paris, which is much more powerful than dynamite. He once wrote that he had accidentally cut his finger and had applied collodion (low nitration nitrocellulose) to form a rubber protective coating over the cut. Although he was in pain from the cut, he rushed to his laboratory at 4:00 a.m. in his nightclothes and began working. By morning he had a dish of stiff jelly, which could easily be molded into any shape for blasting. In 1891, he moved to San Remo in Italy, after a dispute with the French government. After suffering a stroke and partial paralysis, Alfred Nobel died in San Remo, Italy, on December 10, 1896. By the time of his death, he held 355 patents.

During most of his life, Nobel had complained of indigestion, headaches, and occasional spells of depression. Toward the end of his life, he suffered from a heart condition marked by paroxysms of intense pain (angina pectoris). Nitroglycerine was being used in the treatment of several diseases, and had been found to be effective in relieving pain in angina pectoris. In 1890, Nobel's physicians recommended nitroglycerine as a remedy for his heart disease, but he declined to take it. In a letter to Ragnar Sohlman dated October 25, 1896, Nobel wrote, "My heart trouble will keep me here in Paris for another few days at least, until my doctors are in complete agreement about my immediate treatment. Isn't it the irony of fate that I have been prescribed nitroglycerine, to be taken internally!"

Nitroglycerine had been used in the treatment of angina pectoris without anyone knowing its physiological mechanism of action. Thanks to the work of Robert Furchgott, Louis Ignarro, and Ferid Murad, who share the 1998 Nobel Prize in Physiology or Medicine, we now know that nitroglycerine acts by releasing nitric oxide, NO, a common gas and environmental pollutant. The gas is released and diffuses into the smooth muscle cell layer, triggering a relaxation of its myofila-ments. As a result, the blood vessel widens, which allows more blood to pass and more oxygen to flow to the heart muscle and reduce the pain.

On November 1895, Nobel signed his final will and testament in Paris, which was only one page long. He declared that his entire remaining estate should be used to endow "prizes to those who, during the preceding year, shall have con-ferred the greatest benefit to mankind . . . . It is my express wish that in award-ing the prizes no consideration whatever shall be given to the nationality of the candidates, but that the most worthy shall receive the prize, whether he be a Scandinavian or not." The Nobel Prizes have been given since 1901 for the cate-gories of Physics, Chemistry, Medicine, Literature, and Peace. In 1969, the Swedish Bank introduced a prize for Economics, which is also administered by the Nobel Foundation. The Nobel Prizes are now announced in early October to the press, and the potential candidates dream about receiving that phone call from Stockholm. There can be up to three winners for each prize, and the value of the prize is about 1 million dollars. It is always presented on December 10, on the anniversary of Nobel's death, by the king of Sweden, except for the peace prize that is given in Norway.

Explosives changed the nature of both war and peace in the world. Wars became much more destructive of lives and property. On the positive side, a series of civil engineering projects that were considered to be too difficult or too expensive became feasible and within reach; tunnels across the Alps and the construction of the Panama Canal all benefited from powerful and inexpensive high explosives.

Cordite is a medium deflagration explosive that burns at subsonic speed and is used in rifles and cannons to propel bullets. It is less powerful than the high detona-tion explosives at supersonic shock waves, used to blow things up. Cordite is a smokeless powder, which makes it less easy for enemies to locate the shooters, and is made of nitroglycerine and nitrocellulose with petroleum jelly. This mixture is dissolved in the solvent acetone and extruded as spaghetti-like rods. Cordite is now obsolete and no longer being produced, but during World War I, it was greatly needed by the British war efforts against Germany.

Chaim Weizmann (1874–1952) was a chemist who developed a new process of producing acetone through bacterial fermentation of starch. Weizmann's process was highly appreciated by the British government since it gave them a source of acetone to manufacture cordite. He also became a Zionist leader, and his invention gave him the opportunity to meet with political leaders. He was given credit for persuading the British Prime Minister Arthur Balfour to move the proposed Jewish homeland from Uganda to Palestine. The Balfour Declaration of 1917 said that "the British government views with favour the establishment in Palestine of a national home for the Jewish people." He also continued to do chemistry research, and created a research institute that was later renamed the Weizmann Institute of Science in his honor in 1949. Weizmann met with Harry Truman, and worked to obtain the support of the United States for the establish-ment of the State of Israel. He became the first President of the State of Israel in 1949–1952. Cordite was soon replaced by more effective explosives, and the fer-mentation method of making acetone was also replaced by a better oxidation method. So, Weizmann's invention had a short run in the history of technology, but turned out to have a much longer run in the history of politics.

5.3.4 Atomic Bomb

In the history of the world, there is no invention that is more feared and loathed, nor one that has created more remorse among its inventors. After 2 million years of inventions and human progress, there is now a genuine possibility of being "bombed all the way back to the Stone Age" and an even more apocalyptic vision of wiping out humans on earth. But some believe that the doomsday vision of the end of earth is so frightening that it might have contributed to the current stalemate and absence of major wars since the end of the World War II in 1945, a peace of about 70 years. This uneasy peace should be compared with the only 20 years of peace from the end of the World War I to the start of the World War II.

A great number of advances in physics took place at the end of 1890 when only 92 elements were known, from the lightest hydrogen to the heaviest uranium. For thousands of years, alchemists have sought ways to change one element into another, such as lead into gold, which is called transmutation. By 1890, the atom was thought to be impossible to split, to have no internal structure, and was considered a permanent building block of the universe. The ideas of transmutation and alchemy were also discarded.

Wilhelm Röntgen discovered X-ray in 1895 by bombardment of metal targets with cathode ray, which created the mysterious radiation that could penetrate into the flesh but was stopped by the bone. Uranium was found in pitchblende from a mine in Joachimsthal in Czechoslovakia. Antoine Becquerel discovered in 1896 that the uranium salts also emitted a mysterious ray that caused photographic plates in a drawer to fog and demonstrated great penetrating power comparable to X-ray. Then came the discovery of the electron by J.J. Thompson in 1897. Pierre and Marie Curie made further studies with uranium, and in 1898 found another element in pitchblende that is even more radioactive than uranium, and called it radium.

There is a tremendous amount of energy in the nucleus of some atoms, waiting to fly apart. In 1906, Ernest Rutherford bombarded thin gold foil with positively charged alpha particles and found that most particles went through but a few were bounced straight back to the source. This suggested that the gold atom has a very small nucleus that contains most of the mass and is positively charged, surrounded by mostly void space occupied by the negatively charged electrons. The structure of an atom was proposed by Rutherford in 1911 to have a small and heavy nucleus with positive charges, surrounded by a large cloud of negatively charged electrons.

It used to be thought that matter and energy are different concepts that are eternal and unchangeable. In 1905, Albert Einstein proposed that matter can be changed into energy, and described the conversion by the famous equation $E = MC^2$. In this equation C is the speed of light, which is rated at 186,000 miles/s or 300 million m/s so that a tiny amount of mass is equivalent to a great deal of energy. In fact, if 1 kg (2.2 pounds) of matter were converted entirely into energy, it would produce 25 billion kWh of energy that is equivalent to 2.7 million tons of oil. However, at that time, there were no practical methods to change mass into energy.

Nevertheless, in 1914, H.G. Wells wrote a science fiction story called, "The World Set Free," which discussed an air-dropped atomic bomb!

In 1932, James Chadwick bombarded beryllium with alpha particles and discovered the neutron, which has no electrical charge and can be used to penetrate atomic nuclei as it is electrically neutral and would not be repelled by the positive atomic nucleus. In 1933, Leo Szilard in London conceived the idea that if you irradiate a nucleus with a neutron and cause it to create more than one neutron, this could set off a "chain reaction" with rapid increases in neutrons, which is similar to the chemical reactions in flames. Szilard filed for a patent on the concept in the following year, but in subsequent experiments with beryllium, he failed to create more neutrons. In 1934, Enrico Fermi bombarded uranium with neutrons and appeared to have created new elements that were heavier than uranium. For this discovery, Fermi was rewarded with the Nobel Prize in Physics in 1938. He was a physicist, and without the help of a chemist to analyze the properties of the reaction product, he did not realize that he had actually split the atom! The German chemist, Otto Hahn, and his assistant, Lisa Meitner, had worked together for 30 years; but Meitner was Jewish, which was becoming dangerous in 1938 Nazi Germany, so Hahn helped her to leave for Sweden. Hahn and his student, Fritz Strassmann, continued the project to bombard uranium with neutrons. Using chemical methods for analysis, they found barium among the products. Hahn passed the news to Lisa Meitner, who discussed the results with her nephew, Otto Frisch. Since barium has a lower atomic weight of 141 than uranium at 235, this had to mean that the uranium atom had been split in two, which was interpreted as *nuclear fission*, that is, the nucleus of uranium had split into two smaller atoms, such as barium and krypton.

$$U_{92}^{235} + n_0^1 \rightarrow Ba_{56}^{141} + Kr_{36}^{92} + 3n_0^1 + 180MeV$$

Two numbers are attached to the symbol U that stands for uranium: the upper number 235 is the atomic mass, which is the sum of the number of neutrons and protons in the nucleus, and the bottom number 92 is the atomic number, which is the number of just protons. Thus, this atom has $235 - 92 = 143$ neutrons in the nucleus, as well as 92 electrons whirling around the nucleus. Neutron has the symbol n and MeV is the million electron volts, which is a measure of the energy release and an enormous value compared to burning a fuel such as carbon, which makes only about 10 eV. Uranium has two principal forms: U-235 can undergo fission but has a natural abundance of only 0.7%, and U-238 that does not undergo fission and has a much greater natural abundance of 99.3%.

This news about the splitting of the atom was brought by Niels Bohr from Denmark, who went to Princeton University in 1939 to give a seminar and to discuss this development with Albert Einstein. There was great fear that Germany would build an atomic bomb soon, led by Werner Heisenberg, a famous founder of quantum mechanics with a Nobel Prize in Physics. In July 1939, Leo Szilard and Eugene Wigner conferred with Albert Einstein, followed by discussions with the financier, Alexander Sachs. In the fall, Sachs handed a letter from this group to President Roosevelt, urging the United States to work on the atomic bomb.

After Roosevelt approved the project to develop an atomic bomb, the project was entrusted to the Uranium Committee under the direction of the National Bureau of Standards, which moved so slowly that very little progress was made after one year. Adding to the urgency was a report that the Kaiser Wilhelm Institute in Berlin had begun major research on uranium. Two senior government scientists, Vannevar Bush of MIT and James Conant of Harvard, worked to rescue the bomb project. President Roosevelt soon announced the formation of the National Defense Research Committee in June 1940, to be chaired by Vannevar Bush, who would also manage the Uranium Committee. At that time, the research works were undertaken by several widely separated and independent groups:

- Columbia University, under Harold C. Urey, concentrated on the separation of uranium by diffusion and centrifuge.
- University of California at Berkeley, under Ernest O. Lawrence, concentrated on the production of fissionable elements and on electromagnetic separation.
- University of Chicago, under Arthur H. Compton, concentrated on the chain reaction to explosion and was later joined by Fermi from Columbia.

The bomb material U-235 is fissionable and has to be separated from U-238, which is not fissionable. Another bomb material plutonium was discovered in 1940 by Glenn Seaborg and Edwin McMillan at Berkeley by bombarding U-238 with neutrons.

$$U^{238}_{92} + n^1_0 \rightarrow 2\beta + Pu^{239}_{94}$$

The most important isotope of plutonium is Pu-239, which can undergo fission when bombarded by neutrons and release energy and more neutrons. It is also very toxic and is much more dangerous when inhaled, carrying the risk of lung cancer.

A controlled nuclear chain reaction was theoretically possible, but had not yet been demonstrated. Enrico Fermi began his work on nuclear chain reaction at Columbia University, and later shifted to the University of Chicago "Metallurgical Laboratory," which was directed by Arthur Compton. Fermi had a stellar staff including Leo Szilard and Eugene Wigner. The minimum size of an explosive bomb or a stable reactor is called the "critical mass." From the inside of a sphere of uranium, the neutrons created in a nuclear fission reaction will fly off in all directions and may (i) collide with a neighboring U-235 atom leading to more fission, (ii) be captured by a neighboring U-238 or impurities, or (iii) leak out the surface so as to be unavailable to split other uranium atoms. So, the critical reactor should have less surface per volume, which means a larger sphere with more interior volume than exterior surface.

The chain reaction is self-sustained if each primary neutron created in fission leads to more than one secondary fission. This ratio is called the "multiplication factor" k, so that $k > 1$ is the condition of sustained nuclear chain reaction, and $k < 1$ means eventual extinction. When $k = 2$, the number of reactions and neutrons increases with time by the sequence:

---

1   2   4   8   16   32   64

However, if out of each primary neutron released only half of them lead to a secondary fission or $k = 1/2$, the intensity of the chain reaction will die down in time according to the following:

1	1/2	1/4	1/8	1/16	1/32	1/64

A stable reactor should have $k = 1$ or a reaction rate that is stable with time according to the following:

1	1	1	1	1	1	1

So, accelerating the reactor involves increasing the probability of fission, while suppressing capture and leakage. This can be achieved by overdesigning the rate of the reactor and giving it brakes or "control rods"—which are materials that absorb neutrons such as cadmium and boron. The operators start the reactor with the brakes on and get the reaction going by injecting a dose of neutrons by an "initiator," such as a pack of radium and beryllium. When the operators gradually pull out the control rods, the value of k should gradually rise till it reaches and even surpasses 1; conversely, the operators can slow down the reaction by pushing the control rod gradually.

Nuclear fission produces neutrons that are very fast, moving at a speed of 1/10th the speed of light. After a number of collisions, the fast neutron loses energy and slows down to a slower speed, called "thermal neutron." Bohr and Fermi discovered that slow neutrons are more efficient for fission than fast neutrons, so they needed "moderators" to slow down the speed of the neutrons. The moderator should be a light, solid element to deflect the neutrons and should not be much heavier than a neutron such as hydrogen, deuterium, beryllium, and carbon; however, lithium and boron would not work, as they absorb the neutrons. Thus, light water containing hydrogen, heavy water containing deuterium, and graphite containing carbon became favorite moderators to mix with uranium in a reactor pile.

Bohr and Fermi had no experience or data on the size of the smallest sphere or critical mass and used several theories to estimate whether a bomb would require a few kg or perhaps thousands of kg! Enrico Fermi and his team built a nuclear reactor at the University of Chicago in a squash court below the stands of the football stadium Stagg Field. They put together 400 tons of graphite, 58 tons of uranium oxide, and 6 tons of natural uranium metal (0.7% U-235). The pile was a round ball that was 8 m in diameter and 7 m tall, constructed of 57 layers of graphite blocks with uranium in the center. It was equipped with cadmium control rods to slow down the nuclear reaction if it got too high. On December 2, 1942, Fermi ordered the control rods to be withdrawn in gradual stages and the radiation level indicated in Geiger counters rose. After a while, this pile produced more neutrons than was added by the starter package, the value of k rose to 1.006 as the neutron concentration was doubling every 2 min, which was heading for a nuclear runaway that would kill everyone at the site. Fermi ordered the control rods to be pushed in again and the reaction was stopped. Eugene Wigner produced a bottle of Chianti wine and everyone drank a small amount in paper cups. Arthur Compton phoned James Conant at Harvard and told him in a coded message, "The Italian navigator has just landed in the New World." Conant asked, "How were the natives?" Compton replied, "Very friendly." They had demonstrated a nuclear chain reaction with tons of uranium, but they had not produced a bomb that could be easily carried and dropped by an airplane.

Before a bomb could be designed, manufactured, and delivered, there was a long and hard road ahead of solving development problems and the massive task of large-scale industrial production of the bomb material that would tax the resources of any nation, especially during the times of war. The US resolve to move ahead was spurred by the Pearl Harbor attack by Japan on December 7, 1941. In September 1942, the US Army was given jurisdiction over the bomb effort, which was renamed the Manhattan Engineering District. General Leslie R. Groves of the Corps of Engineers was appointed project director. The assignment to Groves was to build bombs that could be dropped on Germany to shorten the war, certainly before the Germans could build a bomb to drop on the United States.

General Groves was 46 years old, the son of a Presbyterian army chaplain, and had studied engineering at the University of Washington and MIT. He had graduated fourth in his class at West Point and had become a good construction engineer who could get things done. He was in charge of the construction of camps, airfields, chemical plants, and so on. He was not thrilled by this assignment to a desk job in Washington as he preferred the glory of commanding troops in the field. He also knew nothing about nuclear physics.

First, Groves took a trip to visit the various parts of his command. He went to Chicago and asked Compton how much uranium or plutonium would be needed to make a bomb, and was given some unproven theoretical estimates. He then asked how accurate was the estimate, and was told that it was within a factor of 10. He wrote, "I am in the position of a caterer who was asked to prepare a dinner, to serve 10 guests or 1000 guests!" This was just one of the many very difficult problems that had to be solved in time and on budget.

The first problem was procurement of sufficient material for the bomb and other essentials, the most important being uranium and plutonium. At that time, the production of uranium in the United States was not more than a few grams and was of doubtful purity. The United States had access to uranium mining in Colorado, as well as at the Great Bear Lake of Canada; but it was impossible to acquire the more desirable ores from Joachimsthal in Czechoslovakia or to develop a remote mine at Katanga in Belgian Congo.

As described earlier, there are two main types or isotopes of uranium: uranium ore is 99.3% in U-238 and is not immediately useful as it absorbs neutron without fission and 0.7% in U-235, which would fission and produce more neutrons. Clearly, it would be better to make a bomb with pure U-235, which has to be separated from U-238. But these two isotopes have identical chemical properties, and their physical properties are different by only a small amount, so the separation is very difficult. A number of separation methods were considered, and the most promising ones were centrifuge, gaseous diffusion, and electromagnetic. It is ironic that the centrifuge method was discarded early in the development effort, as it is today the method of choice. After a tremendous investment in laboratory-scale experiments and plant construction, gaseous diffusion was determined to be the most successful method.

Gaseous diffusion uses the compound uranium hexafluoride UF_6, which is a gas above $60°C$. When the gas is placed in front of a porous membrane, the slightly lighter $U^{235}F_6$ with a molecular weight of 349 can diffuse through slightly faster than the heavier $U^{238}F_6$ with a molecular weight of 352. The difference in speed is minuscule, which is proportional to the square root of their molecular weights, at a ratio of 1.0043. This is known as Graham's law of diffusion, resulting in a tiny difference of less than half a percent. So in stage one of separation, the feed is 0.700% U-235, and the maximum theoretical output is 0.703%; this separation process is repeated in stage two, and the concentration is upgraded to 0.706%, and so forth. It takes 140 theoretical stages to upgrade to 3% for reactor-grade uranium, and 1150 stages to reach 97% for weapons-grade uranium. Since each stage of separation was not perfectly efficient, it actually took 1270 and 4080 stages. This required the construction of an industrial plant with a cascade of several thousand stages.

This compound UF_6 is extremely corrosive, so the barrier material was a difficult problem. After many trials, the best material turned out to be fused nickel powder, coated with the extremely inert polymer of Teflon. Each stage of separation required maintaining a higher pressure upstream and lower pressure downstream, which required inexpensive electric power obtained from cheap water power, such as from the dams of the Tennessee Valley Authority. The production K-25 plant was constructed from 1943 to 1945 in Oak Ridge, Tennessee It is a four-story U-shaped building occupying $200,000 \, m^2$, which provided U-235 for the Hiroshima weapon. The production rate was very slow, and most likely the plant was able to make only enough fuel for one bomb before the war ended.

Plutonium metal does not exist in nature, but can be manufactured from a nuclear reactor pile by radiating U-238 with neutrons. After a period of radiation, the plutonium created can be separated from the unconverted uranium by chemical methods. Glenn Seaborg of Berkeley developed a method involving extraction with the compound bismuth phosphate. The Purex method involves extraction of plutonium by sulfuric acid, followed by solvent extraction with tributyl phosphate (TBP), which is an oily substance that does not dissolve in water. When a reduction agent was added to the extract, the plutonium compounds would dissolve in water, but the uranium compounds would stay in the oily phase.

A plant was constructed from 1943 to 1945 at Hanford, Washington, on the Columbia River, where plutonium was separated from the reactor fuel rods and provided the fuel for the Trinity test and for the Nagasaki bomb. The plant was constructed by the chemical giant E.I. duPont de Nemours, and was undertaken without profit and without patent rights, at a fixed fee of $1.00. It required river water for cooling and electric power from the Grand Coulee Dam, situated on the Columbia River between Oregon and Washington.

An ideal bomb would be small and stable to be safely transported by an airplane, could be detonated at a specific instant, and would produce as much energy as possible. The critical size for natural uranium is something like 8 m, but the critical size for highly enriched uranium or plutonium was not known, and the bomb developers hoped for less than 0.3 m and a weight of less than 45 kg. In a bomb, there would be no need for moderators to slow down the neutrons, and fission worked with fast neutrons. After detonation, there would be so much heat and pressure that the fuel sphere would expand and burst apart, so that the many fuel fragments would become smaller than the critical size and stop the nuclear fission. Thus, the "yield" in energy produced would be much smaller than the theoretical yield with complete consumption of all fuel. The question was how to store and transport the bomb without danger of detonation in advance, as well as how to achieve instant detonation to obtain the highest yield. The best idea was to keep several subcritical fragments of the bomb far apart before the explosion and to bring the fragments together very quickly for detonation.

Bomb design was the most critical component of the Manhattan Project, so the leadership and location of the laboratory were critical decisions. After considering several choices, Groves selected J. Robert Oppenheimer to head the bomb design group and decided to put all the designers in one isolated place, so that they could work together in great secrecy. Groves and Oppenheimer found a site in late 1942 on a flat mesa near Santa Fe around the Los Alamos Boys School in New Mexico. It was 2100 m up in the mountains, isolated by surrounding valleys, and could be approached only by a gravel road. This location was good for secret research, but only 100 km to a railroad junction in Albuquerque for transportation. After a massive construction effort for housing and the lab, one hundred scientists, engineers, and support staff arrived in March 1943. By the summer of 1945, there were 4000 civilians and 2000 soldiers, living in 300 apartment buildings, 52 dormitories, and 200 trailers. The research staff consisted of many current and future Nobel Laureates, and an "extraordinary galaxy of stars": regulars such as Hans Bethe, Edward Teller, and George Kistiakowsky and consultants Enrico Fermi, John von Neumann, James Chadwick, I.I. Rabi, and Niels Bohr. They would all live behind barbed wire fences to keep out intruders.

Besides uranium and plutonium, the bomb mechanism had three important components. The "initiator" was a mixture of radium and beryllium, as radium produces alpha particles that react with beryllium to make neutrons

$$Ra_{88}^{226} \rightarrow Rn_{86}^{222} + \alpha_2^4$$

$$Be_4^9 + \alpha_2^4 \rightarrow C_6^{12} + n_0^1$$

A heavy and dense material called "tamper" was placed all around the bomb that would reflect some of the escaping neutrons back into the bomb, and its inertia would delay the flying apart. Since the tamper needed to be heavy, the best materials were gold, tungsten, rhenium, and uranium—the last one being available as depleted from uranium enrichment with around 0.3% U-235. The third component for the bomb mechanism was high explosives to drive the subcritical pieces together to form a critical mass and thus detonation.

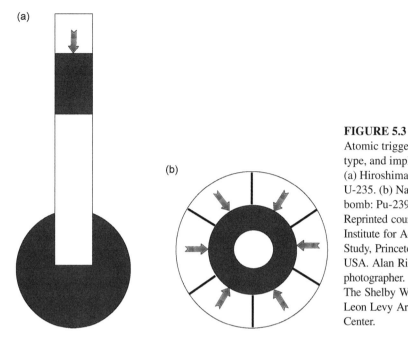

FIGURE 5.3
Atomic triggers, gun type, and implosion. (a) Hiroshima bomb: U-235. (b) Nagasaki bomb: Pu-239. Reprinted courtesy of Institute for Advanced Study, Princeton, NJ, USA. Alan Richards photographer. From The Shelby White and Leon Levy Archives Center.

Two different methods of detonation were developed. For the explosion of the uranium bomb, the gun-type trigger shot a slug of subcritical uranium by a gun assembly toward another subcritical piece (Fig. 5.3). This simple method works with uranium but does not work with plutonium, as plutonium has much higher radioactivity and greater tendency to predetonate; so, a plutonium bomb required the more difficult solution of "implosion." The plutonium bomb involved distributing detonators around the spherical shell to create many shock waves, to compress the plutonium shell uniformly into a small sphere about the size of an orange. These detonators had to be very symmetrical and precisely timed so that the plutonium core would stay spherical as long as possible. The group leader for this project was Seth Neddermeyer. This was a very difficult problem, as the detonators created many divergent semispherical shock waves, but the goal was to make all of them converge into a spherical compression wave for the spherical core of plutonium.

In the autumn of 1943, John von Neumann was brought in to estimate whether implosion was possible; he developed a method of computation and showed that it was possible only if the lack of symmetry was kept below 5%. He received help from Stanislaw Ulam and an IBM computer with punched cards. The progress under Neddermeyer was too slow, so Oppenheimer brought in George Kistiakowsky, the only chemist in the group, during 1944. Kistiakowsky suggested the concept of "shape charge" or lenses, which derived from experiences in armor-piercing explosive warheads. For each segment of the sphere, there would be a region of fast explosive surrounding a region of slow explosive, in analogy with an optical lens, so that the shock waves could be focused from divergence into convergence. Oppenheimer backed Kistiakowsky and had to remove Neddermeyer. Kistiakowsky recalled that he made a bet with Oppenheimer: "One month of my salary for $10 that the bomb will work."

In the end, the enriched uranium core would be the size of a cantaloupe and weighed 15 kg. The plutonium bomb core would be 5 kg and of the size of an orange. But the entire bomb, with all the explosives and tampers, would weigh about a ton and could be delivered by the biggest bomber B-29 that had just been introduced by Boeing.

The device had to be tested before it could be used in the battlefield. In the meantime, President Roosevelt died on April 12, 1944; Harry S. Truman took over and was told, for the first time, about the atomic bomb. Germany surrendered on May 7, 1944, so the only target left for the atomic bomb was Japan.

The first test of an atomic bomb took place on July 16, 1945 at a site called "Trinity," which is near Alamogordo, New Mexico The test took place on top of a 20 m steel tower, using an implosion plutonium bomb, the same kind as the bomb later dropped on Nagasaki that was called Fat Man. Rain and lightning all night hampered the start of the demonstration, but finally came to a stop at around 4:00 a. m. and the explosion occurred at 5:30 a.m. Numerous observers were posted 17–34 km away. The mushroom cloud reached a height of 12.5 km, and the fireball was 200 m wide. Before the test, there was a wide range of predictions about the power of the blast, from a total dud to the destruction of New Mexico or perhaps the entire world. Oppenheimer entered a very pessimistic estimate of 0.3 kilotons of TNT; Kistiakowsky entered 1.4, Bethe 5, Rabi 18, and the optimistic Teller 45. The actual yield was 18.6 kilotons.

At the explosion, Oppenheimer quoted, "Now I am become death, the destroyer of worlds," from the Hindu epic *Bhagavad Gita*. Kistiakowsky was thrown to the ground by the blast, and then slapped Oppenheimer on the back and said, "Oppie, I won the bet." Oppenheimer pulled out his bill folder and turned around to say, "George, I don't have it, and you will have to wait." Later, back in Los Alamos, Oppenheimer made a ceremony of presenting Kistiakowsky with an autographed 10 dollar bill.

American bombers were systematically firebombing Japan by fleets of B-29s, dropping thousands of 3 kg incendiary bombs on flimsy wood and rice-paper housing that started deadly firestorms. Hiroshima, Kyoto, and Nagasaki were the only cities that had survived intact because the Army Air Force had been ordered to set them aside. The United States was preparing for an invasion of Japan, which might result in the deaths of tens of thousands of Americans, as well as hundreds of thousands of Japanese. Was there a chance that the atomic bomb could cause the war to end before the invasion and save thousands of lives?

The first atomic bomb was loaded on the cruiser, Indianapolis, in San Francisco bay, and carried to the island of Tinian in the Pacific. It was then carried to Hiroshima by a B-29 bomber named "Enola Gay," captained by Colonel Paul Tibbetts, and was dropped on Hiroshima on August 6, 1945. It was nicknamed the "Little Boy," with U-235 gun-type explosion. It was a cylinder 3.3 m long, 0.74 m in

FIGURE 5.4 Atomic bomb explosion at Nagasaki. Reprinted courtesy of U.S. National Archives.

diameter, and weighed 5 tons. The bomb contained some 55 kg of U-235, but only 1% was ignited and the yield was 15,000 tons of TNT. In comparison, the entire war consumed only 3 million tons of high explosives over 5 years! When the news circled the globe, the Soviet Union declared war on Japan.

The second atomic bomb was dropped on Nagasaki 3 days later on August 9 (see Figure 5.4), and was a plutonium bomb nicknamed the "Fat Man," 1.5 m in diameter and 3 m long. It was an implosion bomb with 6.2 kg of plutonium, of which 20% fissioned at a yield of 20 kilotons of TNT. The total death toll from the two bombs was around 340,000. This should be compared with the total deaths of 20 million soldiers and 40 million civilians in the world during this war, including 6 million in Japan.

Japan surrendered 6 days later, on August 15. The Emperor, in a formal broadcast to his people, specifically cited "a new and most cruel bomb" as the reason for surrender. These atomic bombs had been accomplished at the staggering cost of 2 billion dollars, according to Harry Truman. Only a dozen years had elapsed from the time that Szilard had thought about the possibility of nuclear chain reaction and the explosion of the two atomic bombs. *Life* magazine wrote, "Modern Prometheans have raided Mount Olympus again and have brought back for man the very thunderbolts of Zeus." Oppenheimer said on October 16, 1945, when he was accepting a Certificate of Appreciation from General Groves, "If atomic bombs are to be added as new weapons to the arsenals of a warring world, or to the arsenals of nations preparing for war, then the time will come when mankind will curse the names of Los Alamos and Hiroshima."

J. Robert Oppenheimer (1904–1967) has often been called "the father of the atomic bomb," although he made none of the key discoveries nor solved the major problems in the production of key material and the creative design of the bomb. His picture is shown in Figure 5.5. It was General Leslie Groves who was the manager of numerous aspects of the largest and most complex development project in history, while Oppenheimer had the narrower role of scientific director of the Los Alamos Laboratory charged with bomb design, and his contribution was to manage the effort successfully.

Oppenheimer was born in 1904, in New York City. His father Julius had immigrated from Germany in 1888 and had become a wealthy textile importer and his mother Ella was a painter. He had one brother Frank who was 8 years younger. The family had an upscale home in New York City with many servants and a summer home on Long Island. Oppenheimer was a star pupil in school, who also studied many subjects on his own, including poetry, languages, religion, and philosophy. In

FIGURE 5.5   J. Robert Oppenheimer.

his youth, he had no interest in any sports except sailing, and on his 18th birthday, his father gave him an 8.5 m sloop that he sailed on Long Island. He grew up to be tall and thin, not well adjusted socially, polite, bookish, arrogant, and snobbish.

In 1922, he enrolled Harvard College, but had severe dysentery during the summer and took time off to recuperate in New Mexico, where he did horseback riding and traveled over mountains and plateaus. Eventually he returned to Harvard to study chemistry and finished in 3 years, *magna cum laude*, in 1925. He did post-graduate work at Cambridge University and changed his field to theoretical physics, and then went to Gottingen to study with Max Born and obtained a PhD in 1928. He returned to the United States to teach at Caltech and at the University of California at Berkeley as an assistant professor of physics. He was diagnosed with TB, and spent time again at a ranch in New Mexico. He was described during that period as a chain-smoker with self-destructive tendencies and a melancholy and insecure dis-position. While in California, he created an important group of theoretical physics students and did important physics research in many areas, but not enough to win a Nobel Prize. Murray Gell-Mann once said that Oppenheimer had a brilliant mind, but did not have sufficient staying power or perseverance to stay in one field long enough to make a really big contribution.

Oppenheimer was not interested in politics, but had a growing hatred of German treatment of Jews, which included many of his relatives; he also sympa-thized with the difficulties of his students in finding jobs during the Depression. He became involved with left-wing politics and the Communist Party when he had a romance with Jean Tatlock, a student in psychology in 1936. Tatlock was involved in politics and the Communist Party and supported the Spanish Civil War on the side of the Republicans against Francisco Franco. When Oppenheimer's father died, he inherited 300,000 dollars and made contributions to the Republican side of the Spanish Civil War. Then he met Kitty Puening, who had been married three times already: first to a musician, then to a Communist party member who had fought and died in Spain, and most recently to a British doctor named Richard Harrison. After meeting Oppenheimer, she divorced Harrison to marry Oppen-heimer in 1940. Their first child Peter was born in 1941 and the second child Katherine in 1944. Oppenheimer became associated with many members of the Communist Party, including Jean Tatlock, his brother Frank and Frank's wife Jackie, and his own wife Kitty.

He was drawn into the atomic bomb project when Ernest Lawrence asked him for help in using cyclotron as a means for separating uranium isotopes for the bomb. In October 1941, he went to a meeting at the General Electric laboratories in Schenectady, where he provided the calculations for the amount of U-235 that would be needed for an effective weapon. This is the "critical mass," and his esti-mate was a 100 kg of uranium. By December 1941, he had stopped making financial contributions to the Communist Party and no longer attended regular meetings. Then the Pearl Harbor attack took place and Oppenheimer decided to concentrate on America's own war. By January 1942, Arthur Holly Compton, then director of the University of Chicago's Metallurgical Lab, asked him to join in their efforts.

Oppenheimer received the greatest challenge of his life when he was chosen by General Groves to be the scientific director of the Los Alamos Lab, where he served from 1942 to 1945. Robert Oppenheimer really impressed Groves with his overall command of the problems ahead. However, he had several serious problems: (i) He did not have a Nobel, but he had to assume command over many Nobel Prize winners. (ii) He had no administrative experience in directing a large group of people. (iii) He was considered a possible security risk, due to his leanings toward the Communist Party and Russia. Nevertheless, Groves selected Oppenheimer as scientific director of the lab by October 1942. There, he was director of the greatest scientific and engineering development project in history, and he delivered the bomb as hoped for. His main efforts were concerned with the bomb's design and fabrication. Many books have since been written about his role in the Manhattan Project. The following are his leadership qualities:

- *Personal Quality:* He was very intelligent, had high scientific achievements, had charm and integrity, maintained good relations with colleagues and subordinates, and inspired trust. On the other hand, he had family and friends in the Communist Party and was considered arrogant and snobbish by many. His friend Hans Bethe said, "Robert can make people feel they were fools."

- *Customer Relations:* He was very skilled at communicating with Groves, his military boss, to establish objectives, obtain the finance needed, set realistic timetables, and satisfy the political and military agendas. His work involved the US Army, which had a different set of rules and culture than his scientists, and he often had to resolve conflicts between the two groups.

- *Recruitment:* He was fabulous at selecting a stellar team and at persuading the top scientists to join his team. When his scientists became discouraged, he was also good at convincing them not to leave.

- *Administration:* He was decisive on acquiring land in Los Alamos and Alamogordo, NM, and in procuring plants, equipment, and raw material. He moved resolutely in the Teller and Bethe battle over whether to pursue the hydrogen bomb over the atomic bomb and also over the Kistiakowsky and Neddermeyer dispute about the development of the implosion device.

After the Trinity Test and the Hiroshima and Nagasaki bombs that had been dropped on Japan, Oppenheimer's work was complete in 3 years. He resigned from the Manhattan Project in 1945 and returned to teaching at Caltech. Now he was world famous, and much sought after as a scientific advisor and speaker. However, he suffered from remorse and felt guilty about the deaths in Japan, and said to Harry Truman that he had blood on his hands. He devoted much time to promoting international control of atomic weapons. He was often restless and smoked a great deal, and his wife Kitty became a heavy drinker. From 1947 to 1966, he was the Director of the Institute for Advanced Studies (IAS) at Princeton. The position was offered to him by Lewis Strauss, a businessman who was a trustee at the IAS. There, Oppenheimer sought to balance the traditional emphasis of the IAS on mathematics with more physics, a balance that was not always accepted by the mathematicians. He recruited many outstanding temporary visitors, as well as excellent permanent members, including Freeman Dyson, C.N. Yang, T.D. Lee, and George Kennan.

Oppenheimer was opposed to Edward Teller's efforts to make the hydrogen bomb, arguing that it would further increase the danger to life on earth. Nevertheless, President Truman decided to make the hydrogen bomb during 1949–1950, and Oppenheimer became more distant from the US policy establishment. The Cold War with the Soviet Union brought a much higher degree of concern about nuclear secrecy, especially after the sensational disclosure about the spy Klaus Fuchs. Fuchs had been at Los Alamos and had passed along atomic secrets to the Soviet Union, and thus shortened the time for the Soviets to develop atomic bombs.

President Dwight D. Eisenhower appointed Lewis Strauss to be the chairman of the Atomic Energy Commission (AEC), who began working with J. Edgar Hoover and the FBI to investigate Oppenheimer and to find ways to diminish his stature and influence. Strauss and Hoover believed that Oppenheimer should be denied access to secret information. A hearing took place where many people testified, some for and some against Oppenheimer. On December 23, 1953, the AEC voted 4 to 1 to revoke his security clearance, which was publicly humiliating and caused him to lose his government connections and standing. His friends compared his public humiliation with the 1633 trial of Galileo by the Catholic Church. His friend I.I. Rabi said, "In addition to being very wise, he was very foolish." Oppenheimer's administration of the Institute for Advanced Studies also involved struggles with Lewis Strauss, who was by now chairman of the Trustees.

After the election of John F. Kennedy in 1960, Oppenheimer was gradually rehabilitated and brought back to the mainstream, and was even awarded the Enrico Fermi Award in 1963; but Kennedy was assassinated before the ceremony, so the award was given by President Lyndon Johnson instead. In February 1966, Oppenheimer was diagnosed with throat cancer, and underwent radiation therapy at the Sloan-Kettering Institute in New York, which improved his health for a while. But that October, his cancer returned, and he died in his sleep in February 1967 at the age of 62. He was cremated and his ashes were taken to his summer home in the Virgin Islands and thrown into the sea. At his funeral, Hans Bethe said, "Without Oppenheimer, the bomb would never have been finished in time for its use in the war." George Kennan, the veteran diplomat and father of America's containment policy against the Soviet Union said of him, "The dilemma was evoked by the recent conquest by human beings of a power over nature out of all proportion to their moral strength. No one ever saw more clearly the dangers arising for humanity from this mounting disparity."

The atomic bomb is both praised and loathed. It led to the peaceful use of nuclear power in energy, which currently provides for 15% of electricity in the world, without the emission of carbon dioxide and contribution to global warming. Furthermore, nuclear power can be the Peaceful Atom and the greatest hope for energy resources and global warming. Controlled plasma fusion is being pursued in Europe, America, and Japan by fusing deuterium and tritium together to produce peaceful energy.

$$D_1^2 + T_1^3 \rightarrow He_2^4 + n_0^1 + 17.6\,\text{MeV}$$

Nuclear materials have also found use in medicine, as tracers in diagnostic tests, and to destroy cancer.

In August 1949, during the Cold War and arms race with the Soviet Union, the American monopoly on the atomic bomb was broken when the Russians exploded their first atomic bomb at Semipalatinsk, Kazakhstan. Then, President Truman approved the development of the hydrogen bomb, which had the potential of creating bombs that would have a thousand times more power than the atomic bombs. This was carried out under the leadership of Edward Teller, with the help of Hans Bethe and John von Neumann. In 1951, Teller and Stanislaw Ulam designed the first thermonuclear bomb and exploded the first test on the Eniwetok atoll in 1952, with a yield of 10.4 megatons, which was a thousand times stronger than the Little Boy. The Russian scientist, Andrei Sakharov, led the effort for the first Soviet hydrogen bomb, which exploded 3 years later in 1955.

This arms race raised the possibility of world annihilation. If 35 kilotons at Hiroshima and Nagasaki could kill 340,000 people, would 500 megatons of hydrogen bombs kill most of the US population of 180 million in 1960? Henry Kissinger calculated that in 1957 a 10 megaton bomb in New York City at 42nd street and Fifth Avenue would cause 3 million deaths in a 5 km radius, which would have been all of Manhattan south of 96th Street plus part of Queens. Would the survivors envy the dead, as they would have to struggle to find food and shelter and care for the wounded and dead; live in a world without cities, manufacturing plants or transportation hubs; be doused with radioactive fallout of Sr^{90} leading to bone cancer, leukemia, and genetic mutations; and endure dust blocking the sun and causing nuclear winter and crop failures.

As a deterrent against Soviet attack on American cities and military installations, the American Strategic Air Command bombers kept some of their bombers in the air at all times, to rain bomb on Soviet cities, if necessary, as punishment and a deterrent. According to the "Fail Safe" procedure, they would fly halfway to Moscow and return to base if they either received a coded message to return or received no message; however, they would proceed to bomb only if they received a coded signal to go ahead. Then came Atlas, Titan, and Minuteman as intercontinental ballistic missiles (ICBM), which could be mounted on railroad cars and constantly moved around the country or launched from silos deep underground or inside mountains. The Polaris, Poseidon, and Trident were submarine launched ballistic missiles (SLBM) that roamed the oceans of the world and were difficult to detect under the waves and icebergs. Soviet bombers were stationed on the Kola Peninsula and at the Chukchi Sea. It would seem that "Mutual Assured Destruction" or MAD was like two scorpions in a bottle, and neither would strike first. This hopefully meant that no sane national leader would ever start a nuclear war, so the world was safe.

In 1960, Herman Kahn published *On Thermonuclear War* and pioneered a quantitative analysis of strategy and consequences in future nuclear wars. He asked what would happen if the enemy made an attack anyway, and what insurance would we have of short-term survival and eventual recovery? Such an attack by the enemy may be due to the following:

  (i) Accident due to unauthorized behavior or true mechanical error.
 (ii) Miscalculation due to escalation, irrationality, and overconfidence.

(iii) Calculated move due to domestic or international crisis or world domination.

(iv) Ambition or desperation of a third nation or a terrorist group.

Thus, evacuation of Soviet civilians from a city could suggest an imminent intent to strike first and be considered in the United States as a very aggressive move! In April 1961, the CIA was involved in the unsuccessful Bay of Pigs invasion intended to topple the Fidel Castro government of Cuba. The event that was closest to triggering a worldwide catastrophe of nuclear war was the October 1962 Cuban Missile Crisis, when Soviets started to build nuclear missile sites in Cuba pointing toward the United States to counter the US Jupiter missiles in Turkey pointing toward Russia. President Kennedy demanded that the Russians remove their missiles and he ordered a naval blockade on Cuba to prevent the shipment of missiles. To the relief of the world, Nikita Khrushchev negotiated and agreed to remove the Cuban missiles in exchange for the secret US dismantling of missiles in Turkey. This also stopped all US plans to invade Cuba for an indefinite period of time.

The Bulletin of the Atomic Scientists has been published since 1945 to engage in nuclear policy debate and to inform the public. The cover of the magazine features a Doomsday Clock, which uses midnight to symbolize the end of the world. The original setting was 7 min to midnight, which was shortened to 2 min in 1953 when there was more Soviet bomb testing. In 1991, the clock relaxed to 17 min when the United States and Soviet Union signed a Strategic Arms Reduction Treaty. But the clock tightened again when India and Pakistan tested nuclear weapons in 1998, and North Korea's testing plus the Iran allegations moved the clock to 5 min in 2007.

The current Nuclear Club includes some nations from the "Axis of Evil" of George W. Bush (North Korea, Iran, and Iraq) and John Bolton (Cuba, Syria, and Libya). In 2009, the negotiations with North Korea to remove their nuclear bomb capabilities in return for aid and to end economic sanctions continued without resolution. The current stockpiles of nuclear warheads are approximately as follows:

Warheads	Nations
More than 1000	United States and Russia
More than 100	United Kingdom, France, China, India, and Israel
More than 10	Pakistan and North Korea

The world is concerned that the decaying Russian stockpile of nuclear material may be smuggled into the hands of terrorists.

Other disadvantages of the nuclear age include the safety of nuclear reactors, the disposal of radioactive material, and the potential for use by terrorists. Even the low-level wastes are hazards to skin, ingestion, inhalation, absorption, or injection. The short-term wastes are iodine-131 with a short half-life of 8 days and strontium-90 with a half-life of 28.9 years. The long-lived Pu-239 will remain hazardous to humans and other living beings for hundreds of thousands of years. Other nuclear wastes are hazardous for millions of years, so they must be shielded and isolated.

The most famous nuclear plant accidents include the 1979 incident of Three Mile Island in Pennsylvania, and the 1986 incident at Chernobyl in the Ukraine. The 2011 Tohuku earthquake reached a scale of 9.0, gave rise to a destructive tsunami and to explosions in the Fukushima Daiichi nuclear power plant, compelling Japan to ask for international help for the first time.

Popular culture has had a strong fascination with the threat of nuclear war and accidents. Tom Lehrer wrote these lyrics in 1952:

> Along the trail you'll find me lopin'
> Where the spaces are wide open,
> In the land of the old A.E.C. (yea-hah!)
> Where the scenery's attractive,
> And the air is radioactive,
> Oh, the wild west is where I wanna be.

Many movies and even one opera have the theme of nuclear war, such as 1957's "On the Beach," about a group of people waiting to die after a nuclear war; 1964's "Dr. Strangelove" by director Stanley Kubrick, about nuclear war and an intellectual with a German accent; 1964's "Fail-Safe" by director Sidney Lumet, about an accidental nuclear war; 1979's "China Syndrome," with Jane Fonda and Jack Lemmon about a nuclear plant accident, which was released 12 days before the Three Mile Island accident; 1983's "Silkwood," directed by Mike Nichols with Meryl Streep, about worker health in a uranium factory; and 2005's "Doctor Atomic," which is an opera by John Adams about Robert Oppenheimer and the events shortly before the Trinity test.

REFERENCES

Benedict, M., T. Pigford, and H. Levi. "Nuclear Chemical Engineering", McGraw-Hill, New York, 1981.

Bernanke, B. "Essays on the Great Depression", Princeton University Press, Princeton, NJ, 2000.

Bird, K. and M. J. Sherwin. "American Prometheus: The Triumph and Tragedy of J. Robert Oppenheimer", Alfred A. Knopf, New York, 2005.

Bown, S. R. "A Most Damnable Invention: Dynamite, Nitrates, and the Making of the Modern World", St. Martin's Press, New York, 2005.

Evlanoff, M. and M. Fluor. "Alfred Nobel: The Loneliest Millionaire", Ward Ritchie Press, Los Angeles, 1969.

Ferguson, N. "Ascent of Money: A Financial History of the World", Penguin Press, New York, 2008.

Frangsmyr, T. "Life and Philosophy of Alfred Nobel", Available at http:\\nobelprize.org, 1996.

Glasstone, S. "Principles of Nuclear Reactor Engineering", D. van Nostrand, New York, 1955.

Goodchild, P. "J. Robert Oppenheimer: Shatterer of Worlds", Houghton Mifflin, Boston, 1981.

Groves, L. M. "Now It Can Be Told: The Story of the Manhattan Project", Da Capo Paperback, New York, 1962.

Halasz, N. "Nobel", The Orion Press, New York, 1959.

Hecht, G. "The Radiance of France: Nuclear Power and National Identity After World War II", MIT Press, Cambridge, MA, 2009.

Kahn, H. "On Thermonuclear War", Princeton University Press, Princeton, NJ, 1961.

Kissinger, H. A. "Nuclear Weapon and Foreign Policy", Harpers, New York, 1957.

Knief, R. A. "Nuclear Energy Technology", McGraw-Hill, New York, 1981.

Mehr, R. I. and E. Cammack. "Principles of Insurance", R. D. Irwin, Homewood, IL, 1976.

Muller, J. Z. "Capitalism and the Jews", Princeton University Press, Princeton, NJ, 2010.

National Academy of Engineering. "The New Orleans Hurricane Protection System: Assessing Pre-Katrina Vulnerability and Improving Mitigation and Preparedness", National Academy Press, Washington DC, 2009.

Partington, J. R. "A History of Greek Fire and Gunpowder", Johns Hopkins, Baltimore, 1999.

Pauli, H. E. "Alfred Nobel: Dynamite King-Architect of Peace", L. B. Fischer, New York, 1942.

Rhodes, R. "The Making of the Atomic Bomb", Simon & Schuster, New York, 1986.

Ringert, N. "Alfred Nobel: His Life and Work", Available at http:\\nobelprize.org, 2006.

Serber, R. "The Los Alamos Primer", University of California Press, Berkeley, CA, 1992.

Smyth, H. "Atomic Energy for Military Purposes", Princeton University Press, Princeton, NJ, 1945.

Sohlman, R. and H. Schuck. "Nobel: Dynamite and Peace", Cosmopolitan Book, New York, 1929.

Teller, E. "The Legacy of Hiroshima", Doubleday, New York, 1962.

Vaughn, E. E. "Risk Management", John Wiley & Sons, Inc., New York, 1997.

White, L. T. "Medieval Technology and Social Change", Oxford University Press, London, 1963.

CHAPTER 6

TRANSPORTATION

For all the plants and animals, the ability to travel and migrate is fundamental in finding food and water, seeking mates, dispersing from crowded family grounds, escaping from predators and environmental threats, and colonizing better habitats.

Plants send out pollen, seed or fruit; the methods of dispersal can be passive attachment to bees and butterflies, to the feathers of birds and the fur of mammals, or in their stomachs; others are borne by the wind and currents of water. Random dispersal does not give control over the path or destiny, and a huge number of travelers are required for a few to survive and succeed. Animals can actively travel on their own, thus gaining greater control on the path and destination. Some animals migrate between two habitats on a regular or annual basis, such as monarch butterflies, storks, and arctic terns that fly round trip each year from the North Pole to the South Pole. Some animals colonize new territories, such as Darwin's finches from South America to the islands of Galapagos.

In the primitive world of Gilgamesh from Uruk in Iraq, the people seldom traveled very far and lived mainly on local resources. Gilgamesh took a memorable journey with his companion Enkidu, and walked to the cedar forest of mountainous Lebanon that is some 1200 km away. They killed the monster Humbaba to gain eternal fame, but they also cut down cedar for lumber to build a temple in Mesopotamia, which had no local stones or timber. Later, Gilgamesh traveled alone to visit his ancestor Utnapishtim in a far away land that no one ever visited, to seek wisdom about the origin of mankind and the knowledge of immortality.

An isolated people have limited resources and exposure to ideas, and can eat only foods that are found or grown locally, use only local materials for tools, hear and learn only local ideas, and mate only with locals and risk inbreeding. When transportation was difficult, only the most valuable and imperishable goods were transported over long distances, such as obsidian and flint in the Neolithic Age, copper and tin in the Bronze Age, and turquoise and lapis lazuli in historic times. Rome had glass and China had silk, so trade along the Silk Route from the first century BCE made them both live better. Besides the better-known land route that Marco Polo took to China, there was also a maritime Silk Route that sailed along the southern coast of Asia from China to India, and then to the Persian Gulf or the Red Sea, which Marco Polo took on the way home. Just as important as goods are the spreading and sharing of ideas, inventions, and technologies. New ideas and religions, such as Buddhism, Christianity, and Islam were carried and spread by travelers. Great travelers in history gave us much of the knowledge of the world, such as Marco Polo, Ibn Batuta, Christopher Columbus, Vasco da Gama, and

Great Inventions that Changed the World, First Edition. James Wei.
© 2012 John Wiley & Sons, Inc. Published 2012 by John Wiley & Sons, Inc.

Ferdinand Magellan. Western ideas such as democracy and capitalism have also played principal roles in changing the world.

Transportation has a principal role in tying people from far-flung lands together socially and politically. The Nile River played a critical role in unifying the Upper Egypt of Luxor and Aswan together with the Lower Egypt of the delta. Roman roads were responsible for the unity of the empire for a thousand years, by collecting taxes and resources to Rome, and by sending legions to pacify and conquer frontier provinces. The Portuguese, Spanish, Dutch, and British Empires were all created and maintained by large ocean-going vessels. Travel also opened opportunities for new habitats, conquests, and colonial empires. The ancestors of men came from East Africa, and the *Homo erectus* ventured into Europe and Asia. The Polynesians journeyed to populate the islands in the vast Pacific Ocean, and the Native Americans walked 17,000 km from the Bering Strait in the north to Tierra del Fuego in the south.

Transportation also creates larger markets, making available a greater diversity of goods and services for consumers; this larger market is also responsible for a greater diversity of employments and specializations. In an isolated community, a person has to be a jack of all trades, including growing food, making clothes, building a shelter, delivering babies, healing the sick, and educating children. A market of larger size can afford a few specialists, such as a toolmaker or a healer, with enough demand to justify an investment in special training and tools. A metropolitan market can afford brain surgeons and fertility clinics with superior knowledge and experience, as well as special tools. Inside a community, good transportation makes possible the separation of workshops away from homes—the workshop can be noisy and stressful, as well as dirty and dangerous, but the homes should be quiet and clean. When a city has a good transportation system, it can create separate business and industrial zones, away from restful residential areas.

A traveling vehicle needs to carry its own fuel and engine, where the measure of efficiency is the power/weight ratio of the engine-fuel system. A floating ocean ship can afford to carry heavier engines and fuels; land vehicles such as automobiles, trucks, and buses need engines with higher power/weight ratios; and airplanes need the most power to stay in the air and require the highest power/weight ratios.

There is also a negative side to good transportation and globalization of the world. The introduction of foreign species of plants and animals often lead to the decline or obliteration of native species. In the human world, globalization often means the decline or disappearance of regional differences in languages and cultures, as well as that of indigenous ethnic groups that are unable to compete. The Chinese philosopher, Laozi of 500 BCE, rejected commerce and technology. He described his ideal as a small state in Book 80 of his canon, the *Dao De Jing*:

> *A small state with few inhabitants*
> *No special employment for exceptional individuals*
> *People do not flee to avoid danger and death*
> *They do not use boats and carts, armors and weapons*
> *They reject writing, and return to tying knots*
> *They enjoy simple food, clothing, dwelling and ways*
> *A neighboring village is within sight*
> *They can hear cocks crow and dogs bark*
> *People grow old and die, but do not visit each other*

6.1 LAND TRANSPORTATION

Walking and running are the oldest methods of transportation on land, with free hands to carry infants and objects. Human power is low in speed and carrying capacity, and animal power was added later with the domestication of oxen, donkeys, camels, and horses. The backs of animals were used to carry people and packed goods. Another method involved the invention of sleds and poles to drag goods, especially over reasonably smooth surfaces, such as sand and ice. The ancestors of men were able to walk from Kenya to western Asia, and then disperse west into Europe and east into East Asia. The most astonishing accomplishment was that they walked from East Asia north to the Arctic, across the Bering Strait to North America, and then continued to walk south to Patagonia at the tip of South America. It would have been very difficult to carry anything beyond the essentials on such journeys.

The wheel was a great invention, which permitted more rapid transportation of heavier goods over longer distances. Bridges are needed to cross rivers and other obstacles, and require long-term investments by more mature and organized societies. The first great modern invention on transportation was thermal power to propel steamships and steam locomotives. The coast-to-coast American railroad network was critical to creating a unified nation and a society with shared values; the Siberian railroad made possible the Russian Empire from St. Petersburg to Vladivostok on the Pacific. The main shortcoming of steam locomotives is their dependence on railroad tracks, which require expensive investments and inflexible routes, so they are only suitable for large volume travel between major nodes, such as New York to Washington. The private automobile with the internal combustion engine requires a distributed network of roads, with trunk lines for large volume interstate highways and local roads that service low volume traffic between any two given points. Transportation needs engines and fuels; control methods such as the whip, the steering wheel, and the brake; and navigation methods such as the compass, maps, and the modern GPS. The other major form of land transportation is a network of pipelines to carry water, oil, gas, and sewage.

The performance of transportation is measured by the speed, the weight carried, and the degree of security from natural storms and from human bandits and war. The record for a human runner in the Marathon is 20 kph over 44 km. In comparison, the record for a horse at the Kentucky Derby is 63 kph; and the top speed of a running cheetah is 120 kph. An automobile and a commuting train can easily do 100 kph for a very long time without becoming tired, but they are surpassed by the top speed of the magnetic levitation train, which travels at 600 kph without the use of wheels.

6.1.1 Wheel

It takes energy to move an object, which is the product of the force needed and the distance traveled. It takes power to delivery energy quickly, which is the product of the force needed and the velocity of travel. On the land, the force required to move an object sliding on the ground and creating friction, is equal to the weight of the object times a friction coefficient, which depends on the surface contact between the object and the ground. The friction factor is small when the surfaces are smooth, such as steel on oiled steel, and is large when the surfaces are rough, such as lumber on a gravel road. Friction is also harmful as it creates heat, and wears out the objects. An engineer strives for fast delivery of heavy objects by using the least power.

Before the invention of the wheel, sleds and rollers with small diameters were used to move heavy objects, such as for the stones for building the pyramids. The pottery wheel was an earlier invention for a different purpose, and might have been a source of inspiration for the transportation wheel. The transportation wheel came during the Bronze Age, and brought tremendous benefits, since a well-lubricated wheel can reduce the friction coefficient by a factor of 100 and also reduces wear and heat generation. The oldest wheels were simple wooden disks with holes for the axles, shown in many Sumerian pictures of clumsy chariots. A wheel has a sliding friction between the axle and the hub, but the axle and hub have smooth and hard surfaces that are kept clean and lubricated.

The wheel with spokes was invented later, which allowed for lighter and swifter vehicles. The Egyptian Pharaoh on a chariot was often depicted going to war, shooting arrows and trampling on the bodies of his enemies (see Fig. 6.1). This kind of light, two-wheeled cart with a single horse or a pair of horses is frequently seen on Greek vases and Chinese tomb paintings, and is suitable for speed and maneuver. Chariot races were described in the *Iliad* as part of the funeral games to commemorate the death of Patrocleus, and were also part of the racing events in the Roman Coliseum. But the heavy four-wheeled wagon is suitable for carrying heavy loads and people in everyday lives. Since wheels work best on smooth and hard roads, instead of paths with large rocks and quicksand, paved roads were built in

FIGURE 6.1 Pharaoh on chariot.

ancient Mesopotamia, Egypt, and Crete. The Romans built great road systems, suitable for conquests and for bringing resources.

The chariots and other wheeled cars were greatly improved with the inventions of the differential gear on the axle, and of ball bearings. When a vehicle with two wheels sharing a single solid axle enters into a turn, the two wheels must revolve at the same rate, but the outside wheel needs to travel a longer distance than the inside wheel, which creates sliding and friction. The differential gear allows the outside wheel to revolve faster than the inside wheel, and makes possible a smooth turn without sliding. Ball bearings were found in early Roman ships. The first patent for ball bearings was awarded in Paris in 1869. They change the sliding friction between two surfaces to the much smaller rolling friction of balls.

Railroads appeared in Greece as early as 600 BCE, in the form of 6 km of grooves in limestone to transport boats across the Corinth isthmus, using power from slaves. Wooden wagon ways were used in England by 1650 for transporting coal from mines to canal wharfs for boats. These roads were made of cast iron plates on top of wooden rails. The Surrey Iron Railway in London opened in 1803, and featured a horse-drawn public railway on cast iron, and later on wrought iron rails. When the steam engine arrived, the first steam trains were introduced in 1825 by George Stephenson. The connection of the North American continent from Atlantic to Pacific was celebrated in Promontory Point Utah in 1869, where the Union Pacific and Central Pacific railroads met and the Golden Spike was struck, to form the First Transcontinental Railroad.

The California Gold Rush began in 1848 when John Sutter discovered gold in a creek. The "Forty-Niners" were some 300,000 people who went from the East Coast of America to the West, perhaps half by land and half by sea. The land travelers covered 5000 km by an assortment of vehicles, including the Conestoga wagon that was 7 m long and could carry a load of 5 tons. It was drawn by six to eight horses or oxen, and could have a top speed of 25 km/day. So this journey from New York to San Francisco would take something like 200 days. The Forty-Niners might also have sailed south to the Cape Horn, and then north to San Francisco, covering 30,000 km in 5–8 months. A combination sea–land route was to sail to Panama, cross the isthmus on land, then take another ship to San Francisco. This slow travel was dramatically changed on June 4, 1876, when an express train called the Transcontinental Express left New York City and arrived in San Francisco in only 83 h and 39 min! The railroad opened the American West, and accelerated the population movement and economic growth. Steam locomotives require numerous workers to load coal and remove ashes, and to maintain the boilers and pistons; locomotives are sooty and particularly troublesome in crowded metropolitan cities. Diesel trains used diesel engines to generate electricity from kerosene, and replaced the steam trains after World War II. Electric trains are much cleaner, and were adopted in the cities as soon as they became available.

6.1.2 Automobile: Gasoline

Ever since the introduction of the modern mass-produced automobile, America has had a romance with the car that represents personal liberty, freedom, privacy, and the flexibility to go from any point to any other point. More than 90% of all households own at least one car, which ranks after the house as the second most expensive possession. For suburban and rural Americans, a romantic date requires a car to provide transportation as well as privacy.

The modern automobile made its debut as the "horseless carriage." A three-wheeled steam powered car was built in 1769 by Nicolas-Joseph Cugnot in France, which moved at walking speed. It was like a train with a combustion boiler, a separate power stroke cylinder with piston, and steam as the working fluid; the engine was bulky and dirty, and it took time to light the boiler, followed by a long wait for the steam to be ready. The British Red Flag Law of 1865 limited all mechanical road vehicles to a speed of 7 kph, and required a crew of two with a third man going ahead to give warning. The internal combustion engine was a great advance, as combustion and power stroke take place in the same cylinder, and the hot combustion gas is the working fluid. In 1807, the brothers Nicéphore and Claude Niépce in France created an internal combustion engine called the Pyréolophore, which was fueled by a mixture of coal dust and resin mixed with oil. A better engine was invented by Étienne Lenoir in 1860, which was much improved in 1876 by Nikolaus Otto in Germany. The Otto engine is a four-stroke engine with a cylinder containing a piston, an inlet valve, an outlet valve, and a spark plug (see Fig. 6.2). In the first stroke, the cylinder pulls down and draws in a mixture of gasoline and air into the cylinder; in the second stroke, the cylinder pushes up to compress the mixture; at the beginning of the third stroke, a spark creates an ignition to generate heat and pressure that pushes the cylinder down with great force; and the fourth stroke is exhaustion of the burnt mixture.

The first working automobile was built in 1885 by Karl Friedrich Benz. The car had a two-stroke engine, a small wheel in the front and two large wheels in the back, and it ran at a speed of 15 kph. The same year, Gottlieb Daimler began by working with Otto, and then started his own company to develop a much faster single-cylinder engine, and installed it on a two-wheeler motorcycle. Later Daimler

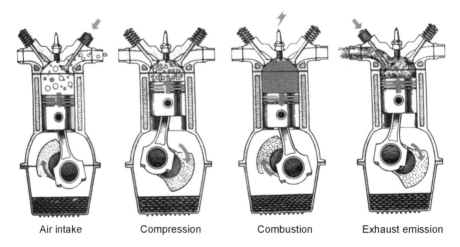

    Air intake   Compression   Combustion   Exhaust emission

FIGURE 6.2   Internal combustion engine.

installed his 0.8 kW engine on a stagecoach. The first Daimler automobile was built in 1889, and was too expensive for anyone but the wealthy aristocrats. When Henry Ford began the mass production of automobiles, the 1908 Model T was built for the mass market, and the price of a car fell from $850 to $260.

A fuel is required to propel the Otto engine, and there are a number of requirements, including the following

- High heat of combustion measured in calories/gram, or joules/gram.
- Being a liquid in operation and storage conditions, as solids are hard to move and gases take up too much volume.
- Volatile for easy start, but not too volatile to cause vapor lock.
- High octane number for engines with high compression and performance.
- Huge supply and low cost.

The heats of combustion for some commonly available fuels are methyl alcohol 5.3 kcal/g, ethanol 7.1, cotton seed oil 9.4, whale oil 9.5, and petroleum oil 10.3. Petroleum oil has the highest heat of combustion, as well as a large supply. There had been historic seepages of petroleum oil in the world, such as in the Middle East, and in Santa Barbara, California. The modern transportation system cannot depend on the random seepages from the ground for an ample and dependable supply. The supply of petroleum oil made a dramatic gain when the first oil well was drilled by Edwin Drake in 1859 at Oil City, Pennsylvania.

Petroleum is a mixture of many different types of molecules with different molecular weights, from the very light gases, such as methane, to the very heavy solids, such as asphalt. A modern refinery puts crude oil through a distillation unit, which separates the various fractions according to their volatility, or tendency to become vapor. The lightest fractions are gases with 1–4 carbon atoms per molecule, and they are methane, ethane, and the liquid petroleum gas (LPG) of propane and butane. Next comes gasoline with 5–11 carbons, and kerosene and jet fuel with 10–18 carbon atoms, and fuel oil with 16–24 carbons for home and business heating. The heaviest fractions are the residual oils and asphalts that are normally solid at room temperatures, which are used in ocean going ships and power plants, and for paving roads and roofs. The combustion of gasoline into water and carbon dioxide plus heat can be represented by the following formula:

$$C_8H_{18} + 12.5O_2 \rightarrow 9H_2O + 8CO_2 + \text{heat}$$

In 1860, kerosene for lamps was the most valuable product from petroleum, and gasoline was poured into creeks as useless by-products. Today, the most valuable petroleum products are the transportation fuels: gasoline for automobiles, and kerosene for diesel and jet airplanes. When the premium transportation fuels have been distilled from a barrel (42 gallons or 159 L) of crude oil, there is a great deal of light gases and heavy oils left over, which are of less commercial value. Many inventions were made to "crack" the heavier oils, and to "glue" the light gases, into medium sizes suitable for transportation. Thermal cracking was invented by William Burton in 1912, and catalytic cracking was invented by Eugene Houdry in 1927. Alkylation is the process to react small olefins with paraffins to make gasoline, and polymerization does the same by reacting two small olefins together.

The other important requirement for gasoline is the octane number, which measures the fuel's ability to resist knocking. This robs the engine of power during demanding situations such as climbing a hill under heavy load, and accelerating to enter a freeway. One solution to this problem was given in 1921 by Thomas Midgley, who introduced tetraethyl lead; and another solution was given in 1950 by Vladimir Haensel, who introduced platinum reforming, which transforms the low octane normal paraffins into the high octane branched paraffin and aromatics.

The transportation fuel of plants and animals can be compared to the transportation fuel for cars. Let us look at a table of combustion heats in kcal/g.

Substance	Heat (kcal/g)	Substance	Heat (kcal/g)
Hydrogen	29.2	Graphite	7.8
Methane	13.2	Ethyl alcohol	7.1
Petroleum	10.3	Bituminous coal	6.7
Whale oil	9.5	Methyl alcohol	5.3
Cotton seed oil	9.4	Oak wood	3.8
Cod liver oil	9.4	Glucose	3.7

Hydrogen has a much higher heat of combustion than carbon or graphite, and the better fuels have a high hydrogen/carbon ratio and little or no oxygen. A Canada goose weighs 9 kg, and eats grass rich in carbohydrates and protein. It converts carbohydrates and proteins (4.3 kcal/g) into fat (9.0 kcal/g). So the goose can pack 18,000 kcal either in 3.8 kg of carbohydrate or in 1.8 kg of fat. When the goose packs for travel, she converts starch into fat to save weight. The same consideration is involved when peanut and coconut plants pack their seeds for travel and dispersal, as they also turn carbohydrates into fat. When animal and plant fat are buried under the ground for millions of years, a chemical reaction took place to remove an acid group and petroleum is the result, which has an even higher heat of combustion of 10.3. A 20 gallon (76 L) tank of gasoline weighs 140 pounds (64 kg), and can be burned to create 700,000 kcal and drive 400 miles (670 km). This is an enormous amount of energy, especially when compared with how much human energy is required to push the same car for 1 km.

The thermal energy from the engine is used to move the car from standing still to a cruising speed of 60 mph (100 kph). To keep the car rolling at that speed, more energy is required to overcome the rolling friction of the wheel as well as the air drag. The air drag is not significant below 48 mph (80 kph), but becomes dominant at higher velocities. A blunt object creates more drag than a tear-drop shaped object with the rounded end in the front. The best streamlined design of an automobile requires many experiments in a wind tunnel. In most automobile designs, there can be a lift force that helps to support the weight of the car. In the opposite direction, the design of a racing car can have a negative lift that pushes the car down, which is important in preventing a rollover when going around a corner.

The Automobile Age transformed American society, and increased personal freedom and the development of the suburbs. Cities like Los Angeles and Houston grew up during the automobile age, and occupy far more land per inhabitant in low-rise houses in comparison with the older cities of New York and Boston. It is arguable that New York City is cleaner with a million cars and gaseous exhausts than it would be with a million horses and the associated solid and liquid wastes.

For an average family in the United States, the automobile is the second largest capital investment after the house. A driver's license is owned by two-thirds of all Americans, and an average car is driven 20,000 km each year, consuming 39 L/km. The current pattern for Americans going to work each day is 76% drive alone to work, 10% carpool, 5% take public transport, 4% work at home, 3% walk, and 2% use other means of transportation. In 2008, the 10 largest companies listed in the Fortune 500 included four oil and two automobile companies:

1. *ExxonMobil*
3. *Chevron*
4. *ConocoPhillips*
6. *General Motors*
7. *Ford Motors*
10. *Valero Energy*

We have also paid dearly for our love affair with the automobile. There are 6.8 million motor vehicle accidents per year, leading to 3.4 million injuries and 42,000 deaths. Motor vehicles are the fifth highest cause of death, behind only cardiovascular diseases, malignancies, pulmonary diseases, and diabetes. The automobile contributes significantly to the emission of carbon dioxide into the atmosphere and global warming. Air pollution in the Los Angeles basin is caused by the emission of unburned hydrocarbons and nitrogen oxides from automobiles, and leads to smog, especially during the summer. The Clean Air Act Extension of 1970 required the Environmental Protection Agency to develop and enforce regulations to protect the general public from exposures to airborne contaminants that are known to be hazardous to human health. It led to the development and use of the catalytic converter beginning in model year 1975, which is still legally required. The catalytic converter's active ingredients are the precious metals platinum and palladium, which must be imported from Russia or South Africa.

6.2 WATER TRANSPORTATION

Swimming was the earliest form of water transportation, which was supplemented with floating devices such as logs, bundles of papyrus reeds tied together in Egypt, and inflated animal skins in Mesopotamia. The floats were later joined together to increase stability and carrying capacity; hollow boats were the next development to keep passengers and cargo dry. Water travel on rivers and lakes is more hazardous than land travel, as winds and waves can cause greater damage to stability, and the vessels can capsize and sink. It is even more heroic to travel in the open sea or ocean beyond the sight of land, where there can be much greater waves and storms, and the danger of dashing against rocks or running aground. Ocean voyage beyond the sight of land requires navigation methods that primarily depend on the sun and the stars. The development of harbors and lighthouses was critical to the success of Alexandria as a great city in Egypt. The Polynesians managed to travel 20,000 km in their sailing canoes from Taiwan to Tahiti, Hawaii, and Easter Island over enormous, empty oceans. Their achievements required more courage and optimism and used less technology, in comparison with the later achievements of Portugal and Spain in the Age of Discovery.

There is a great deal of information on ancient ship building and sailing in Book 5 of *The Odyssey*, explaining how Odysseus built a ship to leave Calypso's island of Ogygia, where she had held him captive for 7 years. He was pining to return to his home in Ithaca, to his wife Penelope and his son Telemachos. Zeus sent Hermes to inform Calypso that it was time to let Odysseus go, and she consented to give him all that he needed. Let us look at the inventions and technologies at his disposal.

> She gave him a heavy bronze ax, with both blades well honed, with a fine olive haft lashed firm to its head. She gave him a polished smoothing adze as well, and led him to tall trees of alders, black poplars and firs that were seasoned and drying for years, ideal for easy floating. He fell twenty trees, trimmed them clean with his ax and split them deftly, trued them straight to the line. She then brought him drills, which he used to bore through all his planks and wedged them snugly, knocking them home together, locked with pegs and joints. He made a merchantman broad in the beam and bottom flat. He put up half-decks pinned to close-set ribs and a sweep of gunwales rounded off the sides. He fashioned the mast next and sank its yard in deep and added a steering-oar to hold her right on course, then he fenced her stem to stern with twigs and wicker, bulwark against the sea-surge, floored with heaps of brush. Then she gave him bolts of cloth to make the sail, and he rigged all fast on board, then eased her down with levers into the sunlit sea. He was done on the fourth day.
>
> On the fifth day she bathed and decked him out in fragrant clothes. Then she gave him two skins—one for dark wine and a larger one for water, and added a sack of rations, filled with her choicest meats to build his strength. The wind lifting his spirits high, Odysseus spread sail—gripping the tiller, seated astern—sleep never closing his eyes, forever scanning the stars, the Pleiades and the Plowman late to set and the Great Bear that mankind also calls the Wagon: she wheels on her axis always fixed, watching the Hunter, and she alone is denied a plunge in the Ocean's baths. Hers were the stars the lustrous goddess told him to keep hard to port as he cut across the sea. And seventeen days he sailed, making headway well.

It may be surprising that the nymph Calypso would have such a chest of tools and knowledge of navigation, and a warrior would have such detailed knowledge on ship building. Let us identify all the stars that were mentioned for navigation. Pleiades is a beautiful star cluster in Taurus the Bull, the Plowman is Bootes the Bear-Watcher with the bright star Arcturus, the Great Bear is the Big Dipper, and the Hunter is Orion. There was no mention of Polaris located at the true north, in the constellation of the Little Bear. In the latitude of the Mediterranean, the Bear is the only reliable group of stars that are always visible at any hour of the night or any season of the year. However, the Great Bear is not at the true north, as it lies between 50° and 61°N.

We can now use this information to speculate on where Ogygia was. If Odysseus kept the Bear to his left or port, he would be sailing roughly in an easterly direction for 17 days. If he had an average speed of 2 knots, he would have sailed 1500 km, so he began his trip from the islands of Majorca to Ithaca; a more optimistic speed of 4 knots would mean a starting point beyond the Pillars of Hercules or Gibraltar, into the Atlantic Ocean; a more pessimistic speed of 1 knot would mean he started from Sicily. On the 18th day, he was shipwrecked by Poseidon, and ended up on the island of Scheria where he met the princess Nausicaa. After banquets and storytelling, King Alcinous sent him home on a ship where the crew sat at oarlocks and rowed him home to nearby Ithaca.

Water transport is well suited for carrying large and heavy loads due to the buoyancy of water. It is a struggle for land vehicles to move a house from one location to another, but it is easier for water transportation as the floating power of water can support unlimited weight. Water transport is superior for carrying inexpensive and heavy commodities such as lumber, stone, grain, cement, and coal, as well as for removal of refuse. The largest oil tanker, Knock Nevis, has a length of 460 m, is taller than the Empire State Building, and can carry 550,000 tons of oil. Most of the modern metropolises are located near a seacoast with a port, such as London, New York, and Shanghai; or are located on a navigable river, lake, or canal, such as Paris and Chicago. Some ancient capitals such as Beijing, Mexico City, and Timbuktu were landlocked and depended mostly on land transport. The arrival of trains made possible the modern landlocked cities of Atlanta and Denver.

Canals connecting two bodies of water were another great advance in the ancient world. Previously isolated areas could be brought in touch with the world of knowledge and trade by digging navigable canals. The Grand Canals of China were more than 1200 km long and connected the grain producing areas south of the Yangtze River with the capitals and garrisons in Kaifeng and Beijing. The Erie Canal (1825) opened up a connection from the populated northeastern seacoast to the Great Lakes, and transformed the cities of Buffalo, Detroit, Chicago, and Minneapolis into cities of world commerce. The Suez Canal (1869) shortened the distance from Europe to the Far East from 20,000 to 12,000 km, and bypassed the Cape of Good Hope; the Panama Canals (1914) shortened the distance from New York to San Francisco from 26,000 to 10,000 km, and bypassed the Cape Horn.

6.2.1 Sailing

A canoe can be powered by pushing with bare hands, but a paddle is a great improvement; the oar requires oarlocks, and is a more advanced technology. Sailing is much less tiring than paddling, and is particularly suitable for heavy loads and long voyages. The most direct way to catch wind is with a square sail that hangs from a horizontal yard attached to the vertical mast, and is filled with wind from behind (see Fig. 6.3). A boat with a square sail can also sail up to a right angle to the wind, when a steering oar or rudder is used to keep the boat going straight. Egyptian ships were depicted in tomb paintings and scale models, designed for the calm Nile River with a

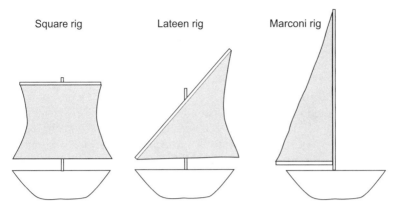

Square rig   Lateen rig   Marconi rig

FIGURE 6.3 Square and lateen sails.

single square sail and many oars, as well as two steering boards on both sides at the stern for changing direction. Greek ships had to brave the more turbulent Aegean Sea, and they began to have bottom keels, which made them much more stable and strong, with radiating ribs and planks to cover the sides. The Homeric poems distinguished between a fast ship-of-war, which was low, long and narrow, with a rowing crew of 50 and a ram in the front for attacking other ships; and the slow merchantman that was tall and broad with very few oars, and had much larger cargo spaces for shipping wine and olive oil. Several Viking ships have been discovered in the burial mounds of chieftains, such as the Gokstad Ship in the Oslo Museum. They were shallow boats that excelled in coastal waters, and had single square sails for running in front of the wind or rowing against unfavorable winds.

A sailing ship with a square sail cannot sail into the wind. When the wind blows from the north, the ship can "run" in front of this tail wind and travel south; when the ship desires to sail east, it can "reach" by rotating the yard and the ship can sail by 45°; but the ship cannot sail north into the head wind. When the wind is unfavorable or if the wind dies, a warship would drop the sail and row, but a merchantman could not afford many rowers and would be motionless or be "in irons." The triangular lateen sail was developed later for the Upper Nile, the Red Sea and the Indian Ocean. The slanted yard is attached to but not perpendicular to the mast, and the yard can be swiveled so that the sail can change from perpendicular to parallel to the ship length. It has the ability to sail within 45° of the wind, such as to the northeast or the northwest in a pattern called "close haul" or "beating." By executing a sequence of zigzag or alternating tacks of northeast and northwest, it can even sail directly into the wind! These "points of sail" are shown in Figure 6.4. Subsequent sailboats used both square and triangular sails, such as the Christopher Columbus flagship the Santa Maria, which had square sails on the foremast and mainmast, as well as a lateen sail on the third or mizzenmast, for varying weather conditions. Modern sailboats use the fore-and-aft or Marconi rigging where there is no yard; the top of the triangular mainsail is directly raised to the top of the mast, and the bottom is attached to a horizontal boom that can swivel from perpendicular

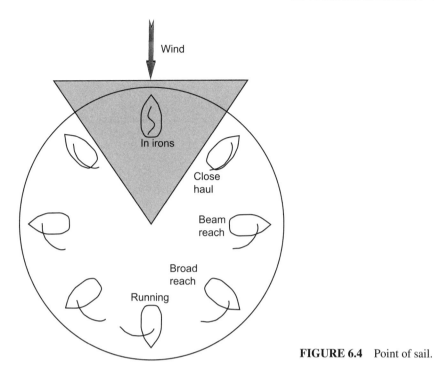

FIGURE 6.4   Point of sail.

to parallel with the ship length. The USA-17 boat, from the 2010 race of the America's Cup, can sail within 20° of the wind!

A taller sail can catch more wind and deliver more power. However, when a sail is taller than the length of the ship, there is a strong tendency to "heel," or to raise the windward side and lower the leeward side, so the ship may capsize; this heeling is more acute when sailing close to the wind, and a captain should move the crew to the windward side of the boat to maintain balance. One invention to counter this heeling is to place a deep and heavy keel at the bottom. This tipping can also be balanced by an invention from the Pacific Islands: a catamaran has an outrigger placed on one side of the sailing canoe, and a trimaran has two outriggers placed on both sides of the canoe.

The maximum sailing boat speed depends on the boat direction to the wind. When a sailboat is running in front of the wind, the maximum speed that it can execute must be slower than the wind, as the driving force is proportional to the speed difference. When the ship is heading toward 90° from the wind in a reach, and the sail is pivoted at around 45°, it is possible to run faster than the wind. The north wind creates a force on the pivoted sail, with a force pointing southeast to drive the boat forward, which generates two components: a force pointing east to drive the boat forward and a force pointing south that causes the boat to heel. In the 2010 America's Cup race, the boat USA-17 was able to reach a top speed of 22 knots (42 kph) when the wind speed was only 6–7 knots!

A ship plowing into the water creates bow waves, which often sets the limit on the top speed. A longer ship with a single hull can travel faster than a shorter ship. The feasible speed is proportional to the square root of the "Length of Water Level," which is the length of the wetted portion of the ship—usually shorter than the total boat length that includes projections above the water. The traveling ship creates a number of crests and troughs along the side of the ship. When the ship speed increases, the distances between the crests and troughs will increase, till there is only a single wave with the bow riding on the first crest and the stern on the second crest. This takes place according to the Froude analysis $V = 4.62\sqrt{\lambda}$, where V is the boat speed in kilometer/hour, and λ is the length of water level in meters. This works out to

Ship length (m)	Top speed (kph)	Top speed (knot)
3	8.0	4.2
10	14.6	7.7
30	25.3	13.3
100	46.2	24.3
300	80.0	42.1

When the top boat speed is matched with the boat length, both the bow and the stern of the boat are on top of crests. When the boat tries to travel faster, the bow rides on top of a crest but the stern drags below the next crest, so that the boat is climbing up which greatly increases the water resistance. The first voyage of Christopher Columbus from the Canary Islands to the Bahamas covered 6300 km over 35 days, for an average speed of 180 km/day or 7.5 kph. The Santa Maria was 21 m long and the Nina was 18 m long, so they could not possibly travel faster than 10–11 knots. The length of a wooden ship is limited by the availability of tall trees used to make its keel, and is seldom longer than 100 m or 24 knots. Steel ships can be much longer. The Titanic had a length of 270 m and a speed of 21 knots, which is much less than the top critical speed. It is not economical for a commercial ship to cruise close to the critical speed, which requires higher energy cost. But the aircraft carrier Enterprise has a length of 342 m and a speed of 33.6 knots.

For a ship that does not make bow waves, such as a submerged submarine, it is possible to travel faster than this Froude velocity. The nuclear submarine USS Triton in 1960 took the Magellan route to circumnavigate the globe while submerged for the entire cruise. The trip of 70,000 km was covered in 84 days and 19 h, for an average speed of 34.5 kph or 18 knots. Another method to exceed the Froude velocity is by a fast and light boat that skips over the water surface in a process called planing, such as the hovercrafts that ride on air jets and the multihull catamarans. Since the USA-17 is a trimaran, it can reach a speed of 22 knots (42 kph) even though the boat length is only 30 m; if it were a single hull boat, the boat speed would be limited to 13 knots. Donald Campbell achieved a speed of 460 kph on his Bluebird, which was a hydroplane with jet engines.

6.2.2 Navigation

When a traveler is in unfamiliar territory, such as a featureless desert or ocean, he or she needs to find a method for identifying the right direction to reach the destination. Celestial navigation depends on observing the positions of the sun and the stars. The sun is always in the south at noon. The sun rises in the east and sets in the west, but this is strictly accurate only during the equinoxes (March 21 and September 21). During the summer solstice (June 21), the sun actually rises in the northeast, and during winter solstice (December 21), it rises in the southeast. The further north the traveler is in latitude, the greater is the seasonal angular correction. It is simpler to rely on the Polaris star of the night sky, which is at the tip of the handle of the Little Dipper, and always at the north. Since Polaris is not visible in the southern hemisphere, the nearest substitute is the Southern Cross, which is inconveniently located at 60°S. However, should the traveler need directions at an inconvenient time of the day, in any season, or in cloudy weather, then these sky watch methods would fail.

In 585 BCE, Thales of Miletus discovered the property of lodestone of attracting iron and other lodestones, and named this phenomenon after Magnesia where the lodestones were found, but did not mention its application in navigation. The compass was first mentioned around 250 BCE in China, and is one of the greatest inventions to benefit travelers. According to legends, around 2600 BCE the Yellow Emperor had a battle in a thick fog, during which his troops were getting lost. He introduced the South Pointing Chariot, which always pointed to the south, and he won the battle. Historical records have been found from the sixth century, on the use of the South Pointing Chariot for travelers over boundless plains, with a figure on the roof that always pointed toward the south, no matter which way the chariot turned. This chariot made its way to Japan by the year 720.

The lodestone is a piece of iron oxide ore that has been magnetized, and has a tendency to rotate to the north–south direction and align with the earth's natural magnetic field. The rotation force is feeble, so the needle in the compass must be free to rotate without friction forces. In ancient China, the solution was to carve the lodestone into the shape of a spoon or ladle, and place it on a smooth polished surface to let it rotate freely (see Fig. 6.5). Then the handle of the spoon would point south, which gave it the name of zhi-nan-zhen or the south-pointed needle. The ladle shape was chosen to resemble the Bear or the Dipper. In the year 1086, Shen Gua described in his *Dream Pool Essays* that, "Alchemists rubbed the needle with a lodestone, making it capable of pointing to the south." He also pointed out that often the needle points slightly east of true south, which today is called magnetic declination. This correction was noticed by Christopher Columbus during his voyage. This correction is small near the equator, and can be enormous near the North or South Magnetic Poles. Shen Gua pointed out that this magnetic needle could be attached to a piece of wood, and left to float in a bowl of water; it could also be placed on a fingernail or on the rim of a bowl; but his favorite design involved suspending the needle from a thread in a place without wind.

The compass may have passed from China to the Middle East via the Silk Road, and then introduced to Europe. There is another theory that Europeans independently invented the compass, but its ability to point north–south was not discovered in Europe till Alexander Neckam in 1180, who wrote about magnetic directions that led to its use in navigation. The magnetic compass played a

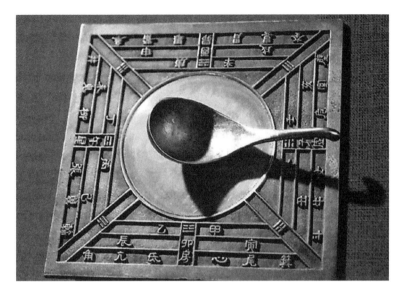

FIGURE 6.5   Chinese spoon compass.

crucial role in the voyages of Christopher Columbus, and in subsequent voyages of discovery.

In planning a trip on land, it is valuable to have a map of the terrain where you can see the starting and ending points, and the landmarks in between, such as mountains and rivers. The Lascaux Cave has the oldest known map of the sky that dates from 16,500 years ago, showing the three bright stars that light up the summer sky: Vega, Deneb, and Altair on the Milky Way. The earliest known terrestrial maps have been found in Babylon, which were followed by a sequence of increasingly more accurate descriptions of the world. In his book *Geographia*, Ptolemy of Alexandria included a map of the known world from the Atlantic Ocean to China, where each place was assigned a set of coordinates. Latitude is the north–south position on earth measured from the equator, which can be determined by measuring the elevation of the North Star Polaris, or the elevation of the sun at noon during the equinox. Ptolemy preferred to express latitude in terms of the maximum length of the summer day—which is 12 h at the equator and 24 h above the Arctic Circle. Longitude is the east–west position, and the current reference point is the Greenwich Observatory near London, which is set to zero. Ptolemy set the reference point for longitude 0 at the Canary Islands (modern longitude 16°W), in the Atlantic Ocean outside of Gibraltar.

A modern navigation chart gives not only accurate geographic information but also the prevailing direction and strength of wind and water currents. A captain with such maps and charts can reduce a good fraction of the time required to go from Seville in Spain to the Caribbean. The Gulf Stream that flows rapidly from Florida to Europe was noticed in 1513 by Ponce de Leon. In 1770, Benjamin Franklin had the Gulf Stream mapped accurately, and noted that its maximum speed was up to 9.3 kph. There is a prevailing tendency for the westerly winds above the latitude of 35°N to blow from the west, and the trade winds below 30°N to blow from the east. Thus it is advantageous to sail from Europe to America by the southern route with the trade winds, and to return by the northern route with the Gulf Stream. A ship should avoid the relatively windless 30°–35°N as the "horse latitude" where the wind often dies. This wind pattern is duplicated in the southern hemisphere, except for the westerly wind at 60°S that is called the Roaring Circumpolar Current as there is no land at that latitude. Captain Bligh tried to sail the HMS Bounty from east to west at the Cape Horn, and for a month he was unable to make headway against this current; finally he had to admit defeat and turned east to cross the Cape of Good Hope to reach Tahiti.

A traveler has several methods of navigation between two points, such as from Syracuse in Italy to Ithaca in Greece. Syracuse has latitude 37°30'N and longitude 15°30'E, Ithaca has latitude 38°21'N and longitude 20°43'E, and the distance between them is 521 km. The Pilotage method of navigation requires memorizing or having a detailed record of the coastlines and landmarks from Syracuse to Ithaca. The pilot should sail near the shore only in daylight, scanning the horizon and observing the landmarks to derive the present position. The Dead Reckoning method requires maps and instruments to measure the current position, and computations to project future positions based on current speed and direction. Ithaca is to the north of Syracuse by 51 min and to the east by 313 min, so the traveler should sail due east by a slight angle of 5.6°N. A traveler who plans to sail at a speed of 16.7 kph should arrive in 31.3 h. How can we constantly monitor the angle of sailing on the way? One method is to use a magnetic compass to check the direction, but we must remember that the magnetic compass points to the magnetic north instead of the true north, and that there is a correction factor involved that depends on local conditions. Our knowledge of the sailing speed and direction is only approximate, and may be affected by the winds and currents, so we have only an approximate idea of our path and arrival time.

We have previously discussed how celestial navigation depends on observations of the positions of the sun and a few selected navigational stars. It is relatively easy to find the latitude on a cloudless night. After sundown in the northern hemisphere when the weather is not too cloudy, the elevation angle of Polaris above the horizon is the latitude of the observer. Polaris is always overhead when you are in the northern hemisphere: if you live in New York, it is 40° high; when you are on the equator (such as at Singapore or the Galapagos Islands) it is barely visible on the horizon. The best instrument to measure the elevation of a star or the sun is the sextant, which compares the elevation of a star with the horizon. However, if you are in the southern hemisphere, Polaris is not visible and there are no bright stars near the South Pole, so you have to settle for the Southern Cross, which is really located at 60°S, thus off by 30°.

When you sail in daylight, you can find your elevation by the noon sun with a more complicated procedure. The sun runs around on a circuit in the sky called the ecliptic, which crosses the equator precisely twice a year at the spring and autumn equinox. If you live on the equator during the equinox, the sun will be directly overhead or 90° high zenith at noon. But if you live in New York, then the sun will be 50° above the horizon (or 40° from the zenith) at noon on these two days. What happens to the sun on other days? On the summer solstice the sun is 23.5° north of equator, and the New York sun is 73.5° high, which is higher and closer to the zenith; and on the winter solstice the New York sun is 26.5° high, which is lower and closer to the horizon. So using the sun for latitude is not as convenient, as you can do it only at noon and you have to know the season in order to make corrections.

It is a lot more difficult to find the longitude, and an additional piece of information is needed, such as the position of the moon or a star such as Spica. A more convenient method is based on the traveler carrying an accurate clock that is set to Greenwich Mean Time (GMT). If you sail west from Greenwich to New York, you will find that at noon in New York, your Greenwich clock will read 5 pm, or 5 h later. In general when the local clock is 1 h later than GMT, you are at 15°W; and when the local clock is 1 h earlier than GMT, you are at 15°E. Since New York clock is 5 h behind Greenwich time, it must be at longitude 75°W; Calcutta is 6 h ahead of Greenwich and is at longitude 90°E; Fiji is 12 h ahead of Greenwich and at longitude 120°E.

The GPS or global positioning system has made obsolete all other methods of determining positions on earth, and depends on a fleet of 24–32 synchronous satellites that orbit about 20,000 km above the earth's surface. From any point on earth, 8–10 satellites are within sight at all times. Radio signals come from each satellite carrying its position and the time of the signal, at the speed of light of 300,000 km/s, which take more than 65 ms to travel to earth. GPS has a very accurate atomic clock to measure this time delay, and uses it to calculate the distances to three satellites; since the positions of the three satellites at any given moment are known, this gives the GPS computer enough information to compute the longitude and latitude of your position.

According to the Archimedes Principle, the buoyancy force acting on a substance immersed in a fluid at rest is equal to the weight of the displaced fluid. Since the density of water is 1 g/mL, the buoyancy force on a ship is equal to the submerged volume V multiplied by the density of water. The ship has a "center of buoyancy" or CB, which is at the center of the volume of the hull below the water level, and a "center of gravity" or CG that depends on the weight distribution on the ship. The stability of the ship is improved by lowering the CG, such as by putting heavy ballasts at the bottom, so when the ship tilts it will have a natural tendency to return to the upright position. This is also the reason behind the advice that "you should not stand up in a canoe," which raises the CG and invites overturning and capsizing. The Swedish King Gustavus Adolphus wanted two rows of bronze cannons on the top deck of his warship Vasa to increase its firepower above all other enemy ships. The ship builder did not feel confident enough to warn the king about the threat to stability, so the CG was too high for stability. The tragic story was that Vasa tipped over on its maiden voyage at Stockholm Harbor in 1628, and stayed at the bottom till its recovery in 1959, and is now in a Stockholm museum.

A moving ship meets a number of resistances that require the expenditure of power to overcome. To drive a ship to a cruising speed requires the kinetic energy equal to $mV^2/2$, where m is the mass and V is the velocity. A moving ship encounters two main types of resistance: the friction drags and the surface wave drags.

When the velocity of an object submerged in water is low and the body is stream-lined, we have streamline flow: that is when streams of the flowing water part in front of the object and rejoin in the rear. When the velocity is high and the body is blunt, we have turbulent flow: when there is a separated and turbulent wake behind the object and much higher resistance. The design of a ship is helped by making scale models and putting them in towing tanks to measure their resistances. In more recent times, computational fluid mechanics took over much of the work, as it is much quicker and cheaper.

The conquest and colonization of Polynesia in the vast Pacific Ocean could not have taken place without the multihull sailing canoes of the Polynesians, as well as their methods of navigation out of sight of land. They may have originated from Southeast Asia, and reached many islands. They sailed from New Guinea to Fiji and Samoa in 1600–1200 BCE, over a distance of 4700 km in an empty ocean with very few islands. The precision of their navigation system must have been very great but is not well understood. They could not even rely on familiar stars, as Polaris is not visible south of the equator, and the Southern Cross is not visible North of the equator. The next leg of their epic journey came around 300 BCE to Cook Island, Tahiti and Marquesa; and they reached Hawaii in 500 and New Zealand in 1000. The distances among the Polynesian islands are contained in a triangle defined by New Zealand, Easter Island and Hawaii, and each side has a length of around 7500 km.

The great geographic discoveries of Christopher Columbus, Vasco da Gama, and Ferdinand Magellan depended on the availability of ocean going vessels and navigation methods out of sight of land. Such ships were not single acts of invention, but the result of a continuous series of improvements over many centuries. The China Clippers were built for speed to deliver tea from China to London. The Great Tea Race of 1866 took place between Fuzhou, China and London by nine ships, which took a route through the Sunda Strait and rounded the Cape of Good Hope, a distance of 25,000 km over 102 days, for an average speed of 245 km/day, or 10 kph. This speed is much below the top speed of 30 kph, as there were many days when the wind was not favorable. During the California Gold Rush of 1848–1850, a sailing boat for the "Forty-Niners" would take 5–8 months to sail from New York south to the Cape Horn, and then north again to San Francisco. Jules Verne was fascinated by this new world of technology and speed, and he wrote *Around the World in 80 Days* in 1873 where the trip was largely taken on steamships, except for three land segments that were taken on steamships. His steamer from Yokohama to San Francisco took 22 days, and New York to London took 9 days, while his train trip from San Francisco to New York took 7 days.

Water transport excels in the long distance transport of heavy and low cost commodities in bulk carriers. The dry carriers are used to ship coal, cement, iron ore, grain, and wood chips; the wet tankers are used to ship crude oil, liquefied petroleum gases, and chemicals. The container ships are designed to carry more valuable cargo such as furniture, toys, clothing, and computers; the roll-ons carry cars and buses; and the reefers are refrigerated to carry food items. The bulk of the globalization of the world is built on cheap water transport, so that the raw material of the less developed countries can be traded for the manufactured goods of the more developed countries, which leads to a better standard of living in both worlds. Singapore and Hong Kong have exports that are more than 200% of their GDP, and their continued prosperity depends heavily on the docks.

Many ships dispose of their wastes in the open ocean, which is still preferable to disposal in crowded harbors. Some of the ship cargo contains hazardous chemicals, and their leakage can cause a great deal of damage. The London Convention on the Prevention of Marine Pollution by Dumping of Wastes and Other Matters is an agreement signed by many nations. The United Nations Convention on the Law of the Sea was signed in 1982, and defined the rights and responsibilities of nations in their use of the world's oceans. In 1989, an Exxon oil tanker was delivering oil from Alaska to California when it hit a reef at Valdez in Alaska, and spilled an estimated 10.8 million gallons of crude oil, which was one of the largest ecological disasters in history. Thousands of animals died immediately, including seabirds, sea otters, harbor seals, bald eagles, and orcas. It also destroyed billions of salmon and herring eggs. Regulatory laws were strengthened in the aftermath, including the requirement to upgrade oil tankers to double-hull bottoms that are more likely to contain leakage after a collision.

6.3   AIR AND SPACE TRANSPORTATION

The freedom and speed of eagles have always inspired admiration and dreams of glory. The human conquest of air requires either a lighter-than-air apparatus, or a heavier-than-air apparatus with sufficient lift to counteract gravity. The Greeks had the legend of Daedalus escaping from the island of Crete with his son Icarus, by making wings glued to their backs with wax. But Icarus flew too close to the sun and the wax melted, so he plunged into the ocean. Chinese literature also carried many fantasies of flight and space travel: the poet Qu Yuan wrote *Li Sao* in 339 BCE, and described riding a blue dragon through the sky; the legend of Chang E described the lady who swallowed a pill of immortality and flew to the moon; and the Dunhuang Caves of 360 have numerous paintings of Fei Tian or immortals who fly through the air. Leonardo da Vinci made many sketches of flying machines, which he could never execute because human muscle power is not sufficient to lift a person.

The conquest with a lighter-than-air machine can be accomplished by a balloon lifted by hot air, or by hydrogen (discovered in 1766) or helium, all of which are lighter than air. Many people have discovered that a paper lantern lighted with a candle can lift up in the air, enhancing the night sky as a source of amusement at festivals. The brothers Joseph-Michel and Jacques-Etienne Montgolfier in France accomplished the feat of carrying humans in hot air balloons in a demonstration in Paris. They were inspired by embers lifting from a fire, while thinking about a new method to assault the English fortress of Gibraltar. They built a globe-shaped balloon of sackcloth held together with 1800 buttons, containing $790 \, \text{m}^3$ of air, and reinforced with a fish net of cords on the outside. In December 1782, when they lit the burner and pushed hot air into the balloon, it flew! For the first public demonstration in June 1783, their flight covered 2 km over 10 min, and achieved a height of 1800 m. By September 1783, they had constructed a balloon of taffeta coated with a varnish of alum for fireproofing, and attached a basket containing a sheep, a duck, and a rooster. It flew successfully at Versailles in front of King Louis XVI and Queen Marie Antoinette. In November 1783, the first free flight by humans was made on the outskirts of Paris, and the first passengers were Pilatre de Rozier, a physician and Francois Laurent, the Marquis d'Arlandes. Jacques Charles (1746–1823) was a French scientist and inventor, and achieved famed in establishing the Charles law, which states that under constant pressure, an ideal gas volume is proportional to its absolute temperature. In August 1783, he launched a hydrogen balloon based on the Cavendish method to generate hydrogen by adding sulfuric acid to iron or tin shavings.

The potential for the military use of flight was pursued from the beginning. During the Napoleonic Wars, the French army was unable to cross the English Channel, and a proposal was made to fly troops from Calais to Dover by an armada of balloons. Balloons go passively where the wind is blowing, but a lightweight and powerful engine was not available till much later. Count Ferdinand von Zeppelin began his research on balloons in 1890, after watching the use of balloons for signals in the American Civil War and the Franco-Prussian War. He built a series of rigid metal hydrogen balloons as long as 236 m, and the forward thrust was provided by Daimler internal combustion engines. With these balloons, regular transatlantic flights from Germany to North America were operated, with the Empire State Building as a dirigible terminal to dock. They were also used in World War I for patrols and bombing raids. These airships were large and light, and could be battered and damaged by strong winds. The flammability of hydrogen was the major detriment, and led to a number of spectacular disasters such as the Hindenburg disaster in 1937. Helium was discovered in 1895, but it was difficult to obtain in large enough quantities from mining operations, and it is heavier than hydrogen.

The legend of Daedalus has inspired a number of recent developments in human-powered aircraft (HPA), which is supposed to be powered entirely by human muscles, without the aid of jumping from cliffs or thermal updrafts. Such aircraft first became possible when very lightweight but strong material became available. In 1961, Derek Piggot achieved a flight of 650 m with an entirely human-powered take-off and landing. Paul MacCready built the Gossamer Condor in 1977, and bicyclist Bryan Allen flew it for 2172 m. Two years later, these two staged a flight from England to France over 35.8 km in 2 h and 49 min. The myth of Daedalus was finally realized in 1988, when the MIT Daedalus flew 123 km from Iraklion in Crete to the island of Santorini. Daedalus 88 was made of lightweight aluminum and Mylar film with a weight of 31 kg, and yet had a wingspan of 34 m and a top speed of 28 kph. It was entirely powered by the Olympic cyclist, Kanellos Kanellopoulos, who weighed 72 kg and could deliver 0.27 hp! The flight took a little under 4 h, ended in the water due to gusty winds, and Kanellopoulos swam 7 m to shore.

6.3.1 Airplane

Birds fly by flapping their wings downward, in order to push themselves up in the air. The avian anatomy includes a prominent chest to anchor powerful pectoral muscles to pull the wings downward. It would have been natural for the early inventors to design and build flapping wing devices. However, the kite with a fixed wing is the true ancestor of the modern airplane. George Cayley in England was a pioneering aeronautical engineer, who began experimenting in 1804 by flying fixed wing gliders from hilltops. He also made scientific measurements on the drag encountered by an object at different speeds and angles of attack. He found that stability and control were the most important factors for success, and deliberately place the center of gravity well below the wings. The German Otto Lilienthal in the 1890s made more than a thousand glides in his "hanging gliders," made experiments on birds, and gathered reliable aeronautical data. In 1896, he fell from a height of 17 m and broke his spine in a gliding accident, and was killed. The American, Samuel P. Langley, was the Secretary of the Smithsonian Institution in 1887. He began experimenting with rubber-band powered models and gliders. In 1896, he made an unpiloted model that flew more than 1500 m. He received grants from the War Department and the Smithsonian to develop a piloted airplane, and purchased an internal combustion engine of 50 hp, but he gave up the project after two crashes in 1903.

Human flight with heavier-than-air craft was finally realized by Wilbur and Orville Wright in 1903, who are shown in Figure 6.6. Wilbur (1867–1912) was the

FIGURE 6.6 Wright Brothers.

third child, and Orville (1871–1948) was the sixth child of Milton Wright, who was a bishop of the Church of the United Brethren in Christ, and of Susan Koerner, whose father built farm wagons and carriages. They moved in 1884 to Dayton, Ohio. When Wilbur was 11 and Orville was 7, they were given a small toy actuated by a rubber string, which would lift itself into the air. They built a number of copies and flew them successfully, and they struggled with scaling up to a larger version. Their father encouraged them to pursue such intellectual interests without any thought of profit. Wilbur had an accident in high school in 1886, when he was struck in the face by a hockey stick and lost some front teeth. He became depressed and detached, but later he blossomed and was comfortable conferring with lawyers and companies. Orville was mischievous and was even expelled from school for a while. He was more comfortable as a mechanic, loved to tinker and disliked management duties. The two brothers attended but never graduated from high school, and never married. In 1892, they opened the Wright Cycle Co. to sell, repair, and manufacture bicycles, and had a joint bank account. The bicycle is an inherently unstable machine, and the brothers developed a feel for stability and control.

In 1896, the brothers read in the daily papers about the glider experiments of Otto Lilienthal. They knew his data on the lift at various wind speeds, the demonstrations proving that curved surfaces are better than flat surfaces for the wings, and the critical importance of balancing the machine in air. The Wright brothers became avid readers of books on ornithology and aeronautics, and received reprints from the Smithsonian Institution on the works of Langley. They also read the wind tunnel data from George Cayley, which helped them to design the airplane.

In 1899, the Wright brothers began their experiments with airplanes. They were bachelors who lived in the same house, worked daily in the same shop, and could discuss problems all day together. They were convinced that wings and engines were sufficiently developed, and they concentrated on the unsolved problems of instability and the design for balance and control—which were responsible for the previous disasters and failures. The stability of an airplane involves three independent modes of rotation (see Fig. 6.7):

- *Pitch* when the airplane nose rises up in the air, or dives toward the ground.
- *Roll* when the airplane banks, so that the right wing goes up or down relative to the left wing, which is equivalent to heeling in sailing.
- *Yaw* when the airplane turns to face right or left.

They began their glider experiments in September 1900, by traveling to Kitty Hawk, North Carolina, which was listed by the Weather Bureau as the windiest place in the country. The Kill Devil Hill provides privacy from reporters, and has regular breezes and soft sandy landing surfaces. The Wright brothers designed their flyer as a biplane with two flat wings separated by struts, with a horizontal stabilizer in front. In 1903, they added power in the form of a specially built gasoline engine fabricated in their bicycle shop, with wooden propeller blades 2.4 m long made of laminations of glued spruce and driven by bicycle chains. Charles Taylor built the engine in 6 weeks, cast from aluminum, with 12 hp and weighing 77 kg. It had fuel

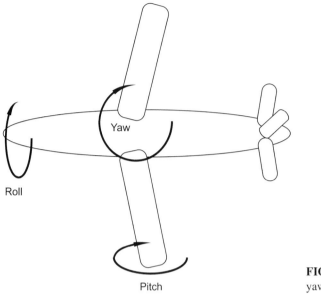

FIGURE 6.7 Roll, pitch, yaw.

injection with no carburetor, and gasoline was gravity fed into the crankcase through a rubber tube from the fuel tank, which was mounted on a wing strut.

The 1900 glider had a wingspan of 17′6″ (5.2 m), a length of 11′6″ (3.5 m), and weighed 24 kg. The 1902 glider became larger with a wingspan of 32′ (9.8 m), a length of 17′ (5.2 m), a weight of 112 pounds (50.1 kg), and was made of spruce wood. The engine and the pilot were placed behind the wings, and the pilot flew lying on his stomach on the lower wing, with his head toward the front. Their control for roll and yaw was by warping the rear of the wings to create lift on the two sides. In the modern design of yaw, this method of turning has been replaced by tail rudders that are similar to ship rudders, pitch is now controlled by the tail horizontal stabilizers, and roll by the ailerons or flaps on the rear of the wings.

The first manned flight of 1903 took place on the morning of December 17 at Kitty Hawk, flying into a head wind gusting to 45 kph. This flight was piloted by Orville who traveled for 37 m in 12 s. The next two flights covered 53 and 61 m, with an altitude of 3 m above the ground. After the fourth flight, a powerful gust of wind flipped it over several times. The airplane was severely damaged and never flew again. The brothers shipped it home, and in 1948, it was installed in the Smithsonian Institution in Washington DC. The historian, Fred C. Kelly, later asked Orville whether he foresaw the use of the airplane for war purposes. Orville replied, "At the time we first flew our power plane, we were not thinking of any practical uses at all. We just wanted to show that it was possible to fly." He also said, "I got more thrill out of flying before I had ever been in the air at all—while lying in bed thinking how exciting it would be to fly."

In 1906, they received US Patent 821,393 for a "Flying Machine," featuring the control mechanism by wing warping. In 1908, Wilbur did a public demonstration of flying at Le Mans, France, where he flew for 1 min and 45 s in a circle, and made banking turns and figure-of-eights. That same year, Orville did a demonstration for the US Army at Fort Myer, Virginia. His flight lasted 62 min, with Army Lieutenant Selfridge riding along as an official observer. However, at an altitude of 30 m, a propeller split and shattered, and the aircraft went out of control. Selfridge died that evening in an army hospital, while Orville was badly injured with a broken left leg and four broken ribs. In 1909, Orville and his sister Katharine joined Wilbur for a trip to France, and their flights were watched by the kings of England, Spain, and Italy.

The brothers formed the Wright Company in November 1909. Wilbur was president and Orville was vice president, with headquarters in New York City and a factory in Dayton. They sold a plane to the US Army for $30,000, which was a two-seater that could fly for 1 h at a speed of 67 kph. It was inevitable that a great success would attract imitators, and Glenn Curtiss was among the first to build airplanes but refuse to pay royalties. Wilbur took the lead in the patent struggle, but he died from typhoid fever in 1912 at age 45, and Orville reluctantly took over the company. A lawsuit was settled in the 1914 US Circuit Court in favor of the Wrights against the Curtiss Company. The legal struggles took so much time that the Wright airplanes did not develop rapidly and were surpassed by European makers. Orville was not happy to be a businessman, so he sold his company in 1915, stopped flying after 1918 and retired. His old injuries gave him much pain, and he was lonely and without family except for Katharine; but Katharine left him to marry in 1926 and then died from pneumonia in 1929. Subsequently Orville cooperated with Curtiss on airplane patent and priority claims, and helped to found the Curtiss–Wright company in 1929. He died of a heart attack in 1948.

An automobile travels on four wheels, and has one familiar rotational motion that is called yaw, that is controlled by the steering wheel, so that the automobile can turn left or right. In a ship or modern airplane, yaw is joined by the two additional and important rotational motions of roll and pitch. An airplane would be out of control without an effective "three-axis control." The roll on a ship is the familiar rocking motion when the left side goes down and the right side goes up, and then the motion is reversed. A multihull catamaran with two bottoms far apart has the advantage of a very much reduced roll. The roll or bank of an airplane is controlled by the two ailerons on the rear tips of the main wings, so if the left aileron is up and the right aileron is down, the plane will roll with left wing depressed and right wing raised. If the ship or plane is longer than it is wide, the pitch is normally a less severe rocking motion when the head goes up and the tail goes down, and then the motion is reversed. The pitch is controlled by the two elevators at the two tail fins, and they are designed to work together so that when they are up, the tail will be depressed and the plane will nose up.

SCIENCE MATTERS: DRAG AND LIFT

When a solid object is placed in a steady stream of air coming from the left, it will experience a drag force to the right, which is equal to the product of the wing surface area, the square of the velocity of travel, and a drag coefficient C_D. The drag force is a negative factor as it slows down the vehicle, and takes energy to overcome. When velocity is doubled, the drag force can increase not twice but four times. This solid object in air can also be designed to have a lift force that is vertical upward and can be used to lift the object against gravity; it can also be designed to push downward to stabilize a racing car rounding a corner. It depends on the same factors as the drag force, except the drag coefficient is replaced by a lift coefficient C_L. This lift force is a beneficial force, as it lifts the airplane into the air.

What is the origin of this lift force? When the wing or airfoil is designed properly, it has a blunt leading edge and a pointed trailing edge. It should be more flat at the bottom and more curved at the top, so the airflow along the upper wing surfaces will be faster than at the bottom. According to the Bernoulli equation, the sum of the pressure and half of the product of density with velocity squared should be a constant.

$$\rho V^2/2 + p = \text{constant}.$$

So the faster flow at the upper surface must be compensated with a lower pressure, and slower flow at the lower surface must be compensated with a higher pressure, which creates a net lift force. When you look at a parked airplane on the runway, you will notice that the wings are not horizontal but slightly tilted with the front higher than the back. This "angle of attack," α has an influence on the lift coefficient C_L. A good wing should have a high lift and a low drag coefficient, and the ratio C_L/C_D should be as high as possible. For modern wings, both coefficients increase with the angle, and the ratio reaches a maximum value at an attack angle of around $5°$. When the angle of attack is too high, the airplane will stall.

The speed of the first Wright brothers' flight in 1903 was 11.3 kph, achieved with a 12 hp gasoline engine. From then on, the engines became much more powerful and the speed rose. It reached 373 kph with Billy Mitchell in 1922, and 1117 kph with Chuck Yeager in 1947. The jet engine was invented by Frank Whittle in England and Hans von Ohain in Germany around 1937, and air speed greatly increased as a consequence. The speed of sound poses a significant obstacle in the race for higher speed, as air resistance increases dramatically around that speed; this speed is 1266 kph at 20°C and 1100 kph at the cruising altitude of commercial jets.

The Mach number is the ratio of speed divided by the speed of sound, so Mach 1 is at the speed of sound. In 1947, Chuck Yaeger was the first man to break the sound barrier at Mach 1. The supersonic turbojet Concorde first flew in 1969, and began regular transatlantic flights from London and Paris to New York and Washington for 27 years. It had a maximum speed of Mach 2.2, which is 2416 kph. Commercial jets may take 8 h to fly from New York to Paris, but the average supersonic flight time was under 3.5 h. However, it was not a commercial success and it was shut down after the recession of September 2001. The record air speed of Mach 10 was achieved in 2004 by an unmanned rocket plane.

The peaceful use of commercial aviation has greatly accelerated the globalization of the world. It is possible to reach very remote locations not connected by roads, and to land or airdrop goods or parachute passengers. An airplane can soar like the eagle, and survey larger territories that are not available otherwise. One of the famous discoveries made by airplanes was the Angel Falls in Venezuela, which is the tallest waterfall in the world; another famous discovery was finding ancient geoglyphs, such as the Nazca Lines in Peru. The role of airplanes in war was critical to their funding and rapid development. The bombing of cities brought war to civilians, no matter how far they were from the battlefronts. In the Battle of Midway in 1942, naval bombers dethroned the battleship, and replaced it with the aircraft carrier as the monarch of the ocean. The atomic bombs of Hiroshima and Nagasaki were carried and dropped by bombers.

6.3.2 Astronautics

The starry sky was thought to be the dwelling of immortals, and flying beyond the atmosphere of earth can be compared to achieving immortality. A fantasy to reach the moon has been found in many legends, which was thought to be a cold place with a perfect jade or marble palace. There is a Chinese legend about the King Yu who acquired the elixir of immortality, and his lady Chang E who stole and swallowed his elixir, then floated in the air till she reached the moon.

Most motions depend on pushing against a medium: the car pushes against the road, ships push against water, and airplanes against the air. But there is nothing to push against in the airless space between earth and moon. The physics of space travel depends on Newton's third law of motion, which decrees that for each action there is a reaction that is equal and in the opposite direction. If you sit in a wheelchair, and throw a heavy weight to the front, your chair will be propelled backward as a consequence. The principle of jet action is employed in the rocket and the jet airplane. The Chinese used rockets to carry flaming material in wars against the Mongols as early as 1212. Sultan Tippu in India used rockets in military attacks against the British. This so impressed William Congreve that he demonstrated a solid fuel rocket at the Royal Arsenal in 1805, which was subsequently used against the French in the War of 1812. This rocket attack is celebrated in the American national anthem when Francis Scott Key watched the 1814 British bombardment of Baltimore harbor, and wrote the words, "And the rocket's red glare, the bombs bursting in air, Gave proof through the night that our flag was still there."

SCIENCE MATTER: ESCAPE VELOCITY

The attractive force of an object with mass m_1 to earth with mass m_2, by gravity g at a distance of d, is given by

$$F = \frac{gm_1m_2}{d^2}$$

Here g is the gravity constant, $32.2 \, \text{ft/s}^2$ or $981 \, \text{cm/s}^2$. A mass of 1 kg on the earth surface, 6378 km from the core, experiences the force of 1 kg; but at the distance of 6,378 km above the earth, or $d = 12,756$ km, the same mass experiences the force of 1/4 kg. A ball tossed upward, or a bullet fired upward, would eventually slow down and fall back on earth. The escape velocity is the minimum speed required of an object to leave earth permanently, and is given by

$$V_{\text{es}} = \sqrt{\frac{2gm_2}{d}}$$

Notice that it does not depend on the mass of the object being thrown upward. The escape velocity on earth is 11.2 km/s or 7 miles/s. Since the earth rotates at a speed of 0.5 km/s at the equator, a launch to the east at Cape Canaveral near the equator requires the discounted speed of 10.8 km/s; but a launch from San Diego to the west would require the augmented speed of 11.6 km/s. The return trip from the moon is less demanding, since the moon is considerably lighter than earth; the return rocket from the surface of the moon requires only a speed of 2.4 km/s, or less than 5% of the rocket fuel.

Konstantin Tsiolkovsky in Russia published a paper in 1903 that showed how to achieve the required thrust to achieve escape velocity with liquid fuels, and suggested that it would be more efficient to use multistage rockets that could jettison the empty fuel tanks mid-way, to reduce weight and increase range. He also suggested guidance systems using a gyroscope, which is a rapidly spinning top that remains pointed in a fixed direction even when the vehicle is turning around. Robert Goddard produced two patents in 1914, one describing a multistage rocket, and the second describing a rocket fueled with gasoline and liquid nitrous oxide. He received support from the Smithsonian Institution, and published a book in 1919 on his theories of rocket flight. He experimented with liquid oxygen and gasoline engines, which were launched in 1926, and rose 13 m during a 2.5 s flight that ended in a cabbage field. His research attracted the interest of Charles Lindbergh and Daniel Guggenheim, the financier. Goddard went to New Mexico in 1934 and built a series of rockets that reached an altitude of 2.7 km. He held 214 patents for his inventions. In 1960, the US government agreed that the large military rocket engines for intercontinental ballistic missiles infringed on Goddard's patents, and paid his widow a million dollars in damages.

Hermann Oberth of Germany published a book in 1923 on rocketry into interplanetary space. One of his protégés was Werner von Braun, who began his own research on rockets in 1934. He used Goddard's plans, which had been published in various journals, to design his series of rockets at the village of Peenemunde on the Baltic Sea. The first V-2 rocket was launched toward London in September 1944. In May of 1945, von Braun surrendered to the Americans, and led a very successful research group in Huntsville, Alabama till 1970. He began to write about manned space stations in 1952.

In October 1957, the Russians launched Sputnik, which was 0.6 m in diameter and weighed 82 kg. It achieved an orbit around the earth with a period of 96 min, and was part of the effort to build intercontinental ballistic missiles. This set off the Space Race between the United States and the USSR, due to the potential military applications and to national pride. The "Missile Gap" became an important factor in presidential elections. America launched the first communications satellite in December 1958. The second Soviet launch carried a dog named Laika, but there were no plans to return it safely to earth.

Von Braun became the first director of the Marshall Space Flight Center in Huntsville at 1960, and he undertook the development of the giant Saturn rockets to carry heavy payloads into the earth orbit. In 1961, Yuri Gargarin became the first man in space, which was followed 23 days later by the American, Alan Shepherd. The American efforts accelerated past the Russian efforts after that.

On July 16, 1969, Neil Armstrong and Buzz Aldrin became the first humans to land on the surface of the moon. Their Apollo 11 was fired by a Saturn V rocket and had three modules for the journey: the service module containing life support systems, the landing module named Eagle that went to the moon and back, and the command module named Columbia that was the only one that returned to earth. On July 19, the three modules were parked on an orbit around the moon. The next day, Armstrong and Aldrin took the Eagle and landed on the moon, while Michael Collins stayed in the command module. Armstrong said, "Houston, Tranquility Base here. The *Eagle* has landed." They walked around, planted an American flag, collected moon rocks, and blasted off to return to the other two modules. They returned to earth on July 24, after a splashdown in the Pacific.

The rocket engine is the easiest engine to explain, as it is just like a firecracker. If you encase gunpowder in a cylinder where the wall has uniform properties, the explosion will push equally in all directions. But if the cylinder has a strong top and side, but a weak bottom, the explosion will escape to the bottom that pushes the rocket up—according to the third law of Newton that for each action, there is a reaction of equal strength and opposite direction. The rocket engine has a heat-resistant nozzle to restrict the flow, and channels the flow to be narrower and faster. We need a chemical reaction that produces a great deal of energy, but the fuel should not weigh too much. Since there is no air for the combustion, the rocket has to carry both the oxidizer and the fuel, just like a rocket filled with black powder. In modern solid fuel rocketry, which is loaded and ready to go, the oxidizer is usually ammonium perchlorate NH_4ClO_4, and the fuel is usually aluminum powder bound with a rubber cement. This combination has the advantage of being easy to store and handle, and easy to get ready at a moment's notice, which is great for submarines and remote locations. Unfortunately, it has a low thrust-to-weight ratio, so it is not good for long journeys; it is also costly as the storage cylinder is also the rocket engine, so the entire cylinder has to be built to withstand high pressure and temperature.

The liquid rocket, on the other hand, has separate storage tanks for the fuel and the oxidizer, which are pumped into the rocket engine, the only part of the rocket that sees high temperature and pressure. It has a much higher thrust-to-weight ratio. The oxidizer is usually liquid oxygen (boiling point of $-183°C$) or hydrogen peroxide (explosive compound), which are difficult to store and handle; and the fuel can be kerosene or liquid hydrogen (boiling point of $-273°C$) with added aluminum powder. When you see photos or movies of the space rockets, you usually see frost from the fuel tanks due to these low temperatures.

How do we describe and predict the orbits made by a spacecraft around the earth, as well as travel to the moon and other planets? The simplest case to explain is the motion of a single planet around the sun, which was done by Johannes Kepler (1571–1630). He was the assistant who analyzed the data of Tyco Brahe, the astronomer who made the most precise measurements. Kepler announced the three laws of planetary motion:

(1) The planet orbit around the sun is an ellipse, with the sun positioned on one of the two focuses (the earth orbit is only slightly eccentric, but enough to upset astronomers).

(2) The radius connecting the sun to the planet will sweep an equal area for an equal length of time (therefore the angular speed is faster when the planet is nearer the sun).

(3) The square of the period of rotation is proportional to the cube of the semi-major axis (so the inner planets from Mercury to Venus rotate much faster than the outer planets of Saturn and Neptune).

Later on, Isaac Newton (1643–1727) solved the Kepler equations of motion and found them to be consistent with his law of universal gravity. Since we are talking about universal laws of gravity, this set of Kepler planetary motion applies to the sun and any individual planet—as long as there are no other heavy objects nearby. But to describe the motion of the moon around the earth, these equations would have to be modified as there are now three bodies involved—the sun, the earth, and the moon. This is the notorious "three-body problem" that has no exact solution in differential equations, and one has to do numerical approximations that are very tedious and slow. When fast computers became available, these motions of celestial mechanics could be tackled readily. Now we can calculate even fancy many-body problems such as the "gravity-assisted" Planetary Grand Tour by Voyager 2. It was launched on August 20, 1977 and visited Jupiter, Saturn, Uranus, and Neptune in succession over a period of 12 years, and then went into interstellar space. The remarkable thing to remember is that once Voyager 2 was launched, the entire journey was programmed for the next 30 years, as there are neither rockets nor fuel to do corrections.

The conquest of space began with national prestige and military applications in mind, as the same technology could be used to deliver atomic bombs to enemy cities and arsenals. There were also some very practical applications. The earth satellites are responsible for constant monitoring of the earth, resulting in accurate weather predictions, mapping the earth, monitoring its various resources, relaying electronic signals, and GPS, which is responsible for guidance of motorists and hikers. Space conquest is also responsible for the space station observatories, including the Hubble Space Telescope that was launched in 1990, that are probing the universe with unparalleled clarity and depth. It also gives mankind the option of escaping earth for other habitats, if for some reason the earth became hostile to life.

REFERENCES

Batchelor, G. K. "An Introduction to Fluid Dynamics", Cambridge University Press, Cambridge, 1970.

Berger, M. L. "The Automobile in America: History and Culture", Greenwood Press, Westport, Connecticut, 2001.

Bond, B. "The Handbook of Sailing", Alfred A. Knopf, New York, 1980.

Boyd, J. E. "The Science and Spectacle of the First Balloon Flights, 1783", Chemical Heritage, pp. 32–37, September 2009.

Chatfield, C. H. "The Airplane and Its Engine", McGraw-Hill, New York, 1940.

Constable, G. and B. Somerville. "A Century of Innovations: Twenty Engineering Achievements that Transformed Our Lives", Joseph Henry Press, Washington DC, 2003.

Damon, T. D. "Introduction to Space", Krieger Publishing Company, Malabar, Florida, 1995.

Feodosiev V. I. and Siniarev G. B. "Introduction to Rocket Technology", Academic Press, New York, 1959.

Gillespie, C. C. "The Montgolfier Brothers and the Invention of Aviation", Princeton University Press, Princeton, NJ, 1983.

Guthrie, V. B. editor, "Petroleum Products Handbook", McGraw-Hill, New York, 1960.

Hamlin, C. "Preliminary Design of Boats and Ships", Cornell Maritime Press, Centreville, Maryland, 1989.

Howard, F. "Wilbur and Orville", Alfred A. Knopf, New York, 1987.

Kayton, M. editor, "Navigation: Land, Sea, Air and Space", IEEE Press, New York, 1990.

Kelly, F. C. "The Wright Brothers", Harcourt Brace, New York, 1943.

Kelly, F. C. editor, "Miracle at Kitty Hawk: Letters of Wilbur and Orville Wright", Da Capo Press, New York, 1996.

Landels, J. G. "Engineering in the Ancient World", University of California Press, Berkeley & Los Angeles, 1978.

Lattimore, O. "The Desert Road to Turkestan", Methuen, London, 1928.

Marchaj, C. A. "Sailing Theory and Practice", Dodd Mead, New York, 1964.

New York Times, "Agile and Fast, US Boat Wins First Race Easily", D6, February 13, 2010. "BMW Oracle Wins the America's Cup", D10, February 15, 2010.

Paine, L.P. "Ships of Discovery and Exploration", Houghton Mifflin, Boston, 2000.

Prandtl, L. "The Essentials of Fluid Dynamics", Blackie & Son, London, 1952.

Riper, V. and A. Bowdoin. "Rockets and Missiles", Greenwood Press, Westport, Connecticut, 2004.

Smits, A. J. "A Physical Introduction to Fluid Mechanics", Wiley, New York, 2000.

Taggart, R. editor, "Ship Design and Construction", Society of Naval Architects and Marine Engineers, New York, 1980.

Tupper, E. C. "Introduction to Naval Architecture", Elsevier, Amsterdam, 2004.

Wilcove, D. S. "No Way Home: The Decline of the Worlds Great Animal Migrations", Island Press, Washington DC, 2008.

Wilford, J. N. "On Crete New Evidence of Very Ancient Mariners". *New York Times* D1, February 16, 2010.

Wright, O. "How We Invented the Airplane", David McKay, New York, 1953.

CHAPTER 7

INFORMATION

We observe our environment and take appropriate actions for our benefit and security. Our eyes and ears are the most important organs to receive signals, which are transmitted to the brain for evaluations and actions, such as advancing toward food and fleeting from a predator. Selected items are stored in the memory for future use, and we can communicate to others about the direction of available food and approaching enemies. Our senses cannot detect some important signals and our brains are not perfect in making records and in recalling. We have needs to communicate to others who may be far away, or will come along later in time. We have invented many special tools for information reception, recording, and communications.

7.1 OBSERVATION

Animals have five senses to receive signals from the environment. The eyes are sensitive to light, and the ears are sensitive to sound; the skin is sensitive to texture, heat and pain; the tongue and the nose analyze tastes and odors. There are many lights and sounds that are too faint, and others are out of the range of our eyes and ears. We also need sensors that can monitor in dangerous and inaccessible environments, and continue to function when we are asleep. In a more primitive time, it was enough to know that a grown man would be taller than a child, and that a larger field of wheat would need more water; but in more complex societies, we need quantitative information for a tailor to measure for a suit, and for a builder to build irrigation systems.

7.1.1 Enhanced Sensing

There are many signals from our surroundings that we cannot see nor hear, which carry very important messages concerning our welfare, and we need sensitive instruments to detect them. There are many reasons why we cannot observe these light or sound waves:

- *Weak Signal:* A distant ship is too far to see, and requires a powerful telescope to make the image brighter and larger; or a sound is too faint to hear, and requires an amplifier to make the sound clearer and louder.

- *Extreme Size:* Bacteria are too small to see without a microscope to make the images larger; or a forest is so large that one has to get the perspective from a distance such as from an airplane or a satellite.

Great Inventions that Changed the World, First Edition. James Wei.
© 2012 John Wiley & Sons, Inc. Published 2012 by John Wiley & Sons, Inc.

- *Beyond Range for Sense Organ:* The light has a frequency range that is infrared or ultraviolet, or the sound is infrasonic or ultrasonic, so that they are not detected by the eyes or the ears.
- *Bad Location:* The object is inside a womb, inside a nuclear reactor, or inside a terrorist hideout.

To see objects that are too far away, we use telescopes to make the objects brighter and larger, particularly important in navigation, battlefield observations, and astronomy. To see objects that are too small, we use microscopes to make the image size much larger than the objects; it is particularly important to see microbes that can injure our health. The beginning of these technologies came from the refractive properties of light in a transparent material such as water or glass, and from the reflective properties of light from a mirror.

Light travels in vacuum at the speed of 300 million meters per second (Mmps), and slightly slower in air. But light slows down to a speed of 245 Mmps in water and to 190 Mmps in ordinary glass. A beam of light that travels perpendicular to the surface of water or glass would continue in the same direction but with a slower speed. However, a beam of light that travels at an angle to the surface would be bent to a smaller angle, in a process called refraction. The refractive index of a material is the ratio of light speed in a vacuum divided by light speed in the material, which is 1.33 for water and 1.52 for ordinary glass. The Greeks and Romans knew that a glass globe filled with water could be used to read letters that were small and dim. Alhazen (965–1040) was a mathematician in Iraq and Egypt, and he wrote a famous *Book of Optics*, which described the magnifying power of a lens. Roger Bacon described the properties of magnifying glasses, which was clearly influenced by Alhazen. The craft of making eyeglasses began in Venice and Florence in the thirteenth century, and expanded to the Netherlands and Germany.

A single convex lens has a focal length such as a meter; this is the distance where the sunrays can be focused to a single point through a magnifying glass, suitable for starting a fire. The highest magnification is obtained by putting the lens very close to the eye, and varying the distance to the object for the best focus. Bacteria are from 0.5 to 5.0 μm (1 m equals one million microns) in diameter, and can be seen with a light microscope. In 1675, Antoni van Leeuwenhoek pioneered the use of microscopes to see microorganisms; he used tiny fused glass beads, and obtained the highest magnification of 275 times by using beads with the smallest radius and focal length. Later developments called compound microscopes use multiple lenses: the objective lens to collect light from the object, and the eyepiece to focus into the eye or camera; they can reach a 1500× magnification with a theoretical limit resolution of 0.2 μm.

Visible light belongs to the family of electromagnetic radiations covering a range in wavelengths from 10^{-14} to 10^8 m, from the short-wave cosmic rays to the long-wave radio waves. The human eye can see only visible light that forms a tiny fraction of all kinds of light, falling between the wavelengths of 0.3–0.7 μm. Viruses are even smaller, on the order of 0.10–0.30 μm, thus they cannot be seen with a light microscope and need an electron microscope that has much more energetic beams and shorter wavelengths. A single molecule such as hydrogen has the size of about 0.0002 μm, and requires the services of a scanning tunneling microscope (STM). We have found many uses for shorter wavelength radiations; ultraviolet, X-ray, and cosmic ray have great penetrating power, and can be used to probe objects deeply buried, such as fractured bones in the body, and flaws in concrete and steel.

The telescope that brings distant objects closer requires two lenses: the objective lens, which is closer to the object being observed, and the eyepiece. There is a legend that in 1608, the Dutch lens maker, Hans Lippershey, observed two children playing with lenses in his shop, and commented that they could make a distant weathervane seem closer when looking at it through two lenses. His original telescopes had either two convex (bulging) lenses with an inverted image, or a convex objective lens with a concave (hollow) eyepiece to produce an upright image. Lippershey applied for a patent for his instrument "for seeing things far away as if they were nearby," but he was not successful. In Italy, Galileo heard about this invention, and improved upon the design for his revolutionary work in astronomy and the rise of modern science. He experimented with many versions, and was able to improve his magnification from 3× to 30×. Galileo was able to persuade the Venetian Doge and Senate for support, since his telescope could detect the approach of enemy warships hours before they were visible to the naked eye. He used his telescope to discover the satellites of Jupiter, the ring of Saturn, and the mountains and craters on the moon.

When you gather light with an objective lens with a diameter of 50 mm, and focus it on an image with a diameter of 10 mm, the brightness can be increased by a factor of 25. A defect of the refracting telescope is the "chromatic aberration," which means that the lens bends blue lights more than red lights, thus separating the colors in an image. This defect was later corrected by using lenses of different refractive power.

Another type of telescope is based on the reflective properties of light. A flat mirror reflects an image that has the same size as the object. A concave mirror gathers light toward an approximate focal point, and gives an image that looks bigger by zooming in; a convex mirror diverts light and makes an image look smaller, which is good for zooming out or panoramic viewing. However, a spherical mirror can focus rays only onto a small area, but a parabolic mirror can do a precise job of gathering rays onto a single focal point. Isaac Newton invented the reflector telescope in 1668, which is not subject to the chromatic aberrations of the refracting telescope that bends blue lights more than red lights. The largest conventional telescope was the 200 in. (5.1 m) Hale telescope on Palomar Mountain, which uses a single piece of borosilicate glass as the mirror. It is very difficult to make even bigger mirrors, and the larger optical mirrors are made of a number of small mirrors arranged on a parabolic frame made of steel.

The radio telescope was invented by Karl Jansky of Bell Telephone in 1931, which detects radio waves from space instead of detecting visible light. For astronomers observing light from distant objects, the atmosphere of the earth absorbs most of electromagnetic waves except for the "optical window" of visible light from 0.3 to 0.7 nm, some infrared light up to 10 μm, and the "radio window" of radio waves from 1 cm to 11 m. The atmosphere also has dust and turbulence, which make the images blurry. Thus, the only way to observe all the waves from the universe with clarity is to send a telescope into space. The Hubble Space Telescope, which was sent into a low orbit of 559 km in 1990, can collect not only the visible lights but also the ultraviolet and near-infrared. The details in some of the photos taken by this telescope, such as the Horseshoe Nebula, are spectacular. They are providing us with information about the origin of the universe, and the mysteries of dark matter and dark energy.

The longer infrared waves are heat rays that you can feel at a distance from a hot stove. They have been used in a night vision device (NVD) to identify warm objects such as a deer or a person in a cold and dark environment. This device has often been used by military, law enforcement agencies, and nature observers since 1939. To obtain stronger signals, some NVD send an infrared light to illuminate targets, and detect the returning infrared beams by electronic methods, which are invisible to the people and animals under surveillance.

Microwaves and radio waves are useful in telecommunications, and can see through clouds and mist. Radar was developed in 1941 using electromagnetic waves to identify the range, altitude, direction and speed of aircraft, ships, and weather formations. It is often cited as one of the decisive factors for the Allied victory in World War II. In 1887, Heinrich Hertz experimented with radio waves, and found that they could be transmitted through some types of material and reflected by others. Guglielmo Marconi pointed out in 1899 the possibility of using such a system for lighthouses and lightships, to enable vessels in foggy weather to locate dangerous points around the coasts. Many people in Europe and America began to explore this potential.

Robert Watson-Watt (1892–1973) was a Scottish engineer descended from James Watt, and considered by many to be the inventor of radar. He was initially interested in using radio waves for the detection of lightning that creates radio waves, to warn pilots of approaching thunderstorms. He used a directional antenna that could be manually turned to point directly toward a storm, in order to maximize the signal strength. He used a cathode-ray oscilloscope with a long-lasting phosphor, so that fleeting lightning could be recorded. At the start of World War II, Watson-Watt worked with the Air Ministry to advance air defense against German bombers from airfields that were only minutes away across the English Channel, creating the need to launch intercepting fighters quickly. In 1935, he made a demonstration with two receiving antennas that were 10 km apart, and he used the difference in arrival times to determine the direction of the approaching bomber. By the end of the year, the range was up to 100 km, and the Air Ministry set up five stations covering the approaches to London.

The Battle of Britain began in the summer of 1940, when Germany launched aerial bombings to destroy Britain's air force and military installations, and to force Britain to negotiate an armistice or surrender. Watson-Watt turned from the technical side of radar to building up a layered organization of people to run the network of radar machines, which tracked large numbers of aircraft and directed their defenses. The first indications of incoming air raids in Britain were received by the radar system located at Chain Home, which could detect formations of German aircraft when they began to organize themselves over their own airfields in

France or Belgium, sometimes only 20 min away from London. Thus, the Royal Air Force was constantly informed when the attacks were forming, and knew about their strength and direction even in cloudy and rainy weather. The information from visual observer corps were added to the radar results, and plotted by the Women's Auxiliary Air Force (WAAF) that received information via a telephone system. The British squadrons could take off quickly from the ground without the need to stay in the air; they were dispatched only when warned to intercept raids, and their interception rates were as high as 80%. The British commanders could see the entire battlefront and were able to direct fighters to where they were needed the most, but the Germans did not have an equivalent system. As a consequence, the German losses in planes and experienced pilots were too high, and the raids could not be sustained. By December 1940, the Battle for Britain over the skies was over, which convinced the world that Britain would survive the battle with Germany.

Watson-Watt traveled to the United States in 1941 to give advice on how to set up radar, especially in view of the inadequate US air defense against Japan during the Pearl Harbor attack. He was knighted in 1942, and was rewarded 50,000 pounds by the British government. Radar has become widely adopted in many other applications, particularly for civilian air traffic control when the weather is cloudy and the airfields may become invisible. Another important use is in road traffic, to measure the speed of cars in violation of speed limits. Watson-Watt later moved to Canada, and was pulled over for speeding by a radar policeman. His remark was, *Had I known what you were going to do with it, I would never have invented it!* At the age of 72, he married 67-year-old Dame Katherine Trefusis-Forbes, who was the founding Air Commander of the Women's Auxiliary Air Force, which had supplied the radar operators. They settled in Scotland for the rest of their lives.

Sound is mechanical vibration of the air, and the human ear can hear frequencies from only 20 to 20,000 Hz, or vibrations per second. We cannot hear sound that is too soft, but we can compensate with microphones, amplifiers, and loudspeakers. The loudness of sound is measured from 0 dB or decibel that is at the barely audible "threshold of hearing," to the unbearable 140 dB of the "threshold of pain." Touch is the main sense involved in grooming and embrace, which are the primary methods of bonding and mating between loved ones. Pain is a signal of distress that needs appropriate action, such as retreat or repair. We can use our hand to feel whether a child is feverish, but a thermometer gives a more precise quantitative reading. We can smell spice and perfume, and taste saltiness and sweetness. We have many inventions for the chemical analysis of vapor and liquid, such as spectroscopy and chromatography; they greatly extended our ability to discover and quantify the chemicals in our surroundings as well as inside our bodies, for safety and health. We can measure the concentration of CFC in the atmosphere to parts per billion, which may be damaging to the ozone hole.

One of the earliest sound sensing instruments in history was the first seismograph of Zhang Heng (78–139), a court official in the Eastern Han Dynasty who excelled in mathematics, inventions, poetry, and statesmanship. The ratio of the circumference of a circle to its radius, or π, was initially assigned the value of 3, but he managed to improve that to 3.162. Zhang invented a seismograph to measure the tremors of the earth that were faint and distant. It was a bronze vessel mounted with eight dragons pointed at the cardinal directions, with bronze balls in their mouths. The frame was attached firmly to the ground; but a pendulum at the center was free to swing, and was connected by arms to the dragon heads. When the earth moved in an earthquake, the frame moved with it, but the loose balls with their inertia moved much less; the difference in motion would cause the ball closest to the direction of the quake to come loose and fall into the mouth of a bronze frog, so the astronomer could detect an earthquake and its direction as far as 500 km away. We have much more accurate seismographic instruments now, to measure direction and distance. The famous Richter Scale measures the strength of the shaking amplitude of the waves; it works on a logarithmic scale so that a scale 7 quake is 10 times greater in amplitude than a scale 6 quake, but the energy released is 31.6 times greater.

Sounds beyond the human range of hearing can be used for detection and diagnosis: very low frequencies are called infrasonic, and extremely high frequencies are called ultrasonic. Dolphins and bats can emit ultrasonic sounds, and use the return waves for ecolocation to navigate and to catch prey. The use of sonic waves to detect submarines, or sonar, was developed by Paul Langevin in 1917. He made an apparatus that sends a beam of sound into the water, which bounces off an object and returns to the sender to be analyzed. The time it takes for the beam of sound to travel can be used to compute the distance between the object and the sender. When there are two or more senders/receivers in action, the distance of the object to different senders would be sufficient to compute both the distance and the direction of the object. This is particularly important in submarine warfare. A similar principle is used in ultrasound analysis of the human body; it is particularly important for viewing a fetus in the womb, to evaluate the health and gender of the baby, and to detect twins. Sonic waves are also used in industry to measure the thickness of objects, and to find flaws in concrete, cement, and metals.

7.1.2 Quantitative Measurements

Quantitative information is needed for planning and taking appropriate actions. Methods and standards are controlled by the National Institute of Standards and Technology (NIST) in the United States, and by the Bureau International des Poids et Mesures for many international communities.

The measurement of *length* was probably the first to be developed into a quantitative form. The most natural units would be a person's own limbs, such as the distance from the elbow to the tip of the middle finger, usually ranging from 16 to 21 in. (0.41–0.83 m), which was named "cubit" in Egyptian hieroglyphic signs. For shorter distances, the width of a "palm" without the thumb is about 3 in. (76 mm), which can be further divided into 4 "fingers" of 3/4 in. (19 mm) width. Each person has limbs of a different size than those of other persons, but the variability can be diminished by a standard such as the royal cubit based on the limb of the king. However, a young king would have a shorter limb than the fully mature king. The Romans introduced the "foot," which is 16 digits or 12 in. (0.305 m). For longer distances, the Greeks used the "stade," which is about 600 ft (182 m). For even longer distances, the more convenient concepts of "days of marching" for roads, and "days of sailing" were used at the start. The Roman "mile" is 1000 paces or 2000 steps, which is 1479 m and slightly shorter than an English mile. For the important Roman Appian Way, the earliest known milestones were erected to measure the distance traveled. In 1889, an international method for a standard meter was adopted by keeping a platinum–iridium bar at the International Bureau of Weights and Measures, to be measured at the melting point of ice. In 1960 there was an improved standard meter, based on the wavelength of the orange-red emission line of krypton-86 in vacuum, which was then multiplied by 1,650,763.73.

We have invented many powerful instruments to quantify measurements, even the diameter of the observable universe, which is at least 93 billion light years, or 8.80×10^{26} m. We can also measure down to the radius of an atom that can be as small as 30 pm or 10^{-12} m; the radius of the atomic nucleus is 10,000 times smaller than the atom, as small as 1.5×10^{-15} m. This range of lengths from universe to atomic nucleus encompasses a range of more than 40 orders of magnitude! The associated concept of area is measured in square meters, and the concept of volume is measured in cubic meters.

The measurement of *weight* is important for goldsmiths and grocers, and the balance with two equal arms was a very early invention. The balance with two pans of equal size, and a set of standard weights, was shown prominently in Egyptian tomb figures, which also became a symbol of impartial justice. In the Egyptian afterlife, the balance was used to weigh the heart of a dead person against a feather, to determine whether the person deserved eternal bliss in the afterlife. The Egyptian "shekel" is around 8 g; and their larger unit is the "mina," which is close to a pound (0.45 kg), and of greater interest in the market of meat and grains; the even larger unit is the "talent," which is more than 130 pounds (59 kg), comparable to the weight of a person. Spring scales do not use standard weights, but use a spring with known spring constants according to Hooke's law, where the displacement of the spring from the rest position is proportional to the weight. The electronic versions of balance depend on the change of electrical resistances from a change of length, and can be much more accurate.

How would you weigh an elephant, with a weight that is up to 5000 kg? That was the challenge to the child Cao Chong, the eldest son of the ruler in northern China in 220. His ingenious solution was to load the elephant on a boat in a river, and mark the water level on the boat. Then the elephant was replaced with a number of rocks till the boat sank to the same level. It is then easy to sum the weights of all the rocks. Unfortunately, the boy died at age 13 and did not become a wise ruler or eminent scientist. How do we weigh the mass of the earth, which is around 6×10^{24} kg? The moon rotates around the earth in a stable orbit, so the gravitational pull must be exactly balanced by the centrifugal force created by the velocity of the moon; thus the mass of the earth must equal the square of the moon's velocity, multiplied by the radius of the orbit and divided by the gravity constant, $m = v^2 r/g$. (We measure the gravity constant g by dropping rocks from towers, and compare heights with times of travel.)

The passage of *time* in a day controls human activities on when to wake up and when to eat. The Sumerians around 2000 BCE introduced the sexagesimal system based on the number 60, and used it to count slaves, fish, grain, and time.

This system survives today only in our measurement of time, as well as longitude and latitude of the earth and sky. Thus, we have 60 s to a minute, 60 min to an hour, and 24 h/day of 86,400 s. Over the centuries, there have been many instruments for measuring time. The sundial uses a gnomon or a thin rod to cast a shadow on a set of markings to mark the hours, but does not function on cloudy days or at night. The earliest water clock was found in Egyptian tombs from around 1500 BCE, and could measure time even at night when the sun was not shining. The hourglass uses the flow of sand instead of water to measure the flow of time. The Chinese invented the first mechanical clocks that were driven by an escapement mechanism. The most accurate devices now are the cesium atomic clocks; there is one in the National Institute of Science and Technology that can track time as accurately as a second over a billion years. The accuracy of clocks is also vital to determine the longitude of an object, which is essential in GPS, used for military, commercial, and personal devices such as automobile navigation and outdoor trekking.

The seasons of the year are important for the growing of plants and the mating and migration of animals; thus they are important for the hunter-gatherers to gather plants and to find animals, as well as for the farmers to plant crops. The positions of the sun and stars have always been important to determine the seasons, and the *calendar* was an invention to keep track of the change of seasons. The annual rise of the Nile coincided with the first reappearance at dawn in the east of the brightest fixed star Sirius, which was "the bringer of the Nile" and the start of inundation by floods, suitable for planting crops.

The equinox happens twice a year, in spring and in autumn, when the day is just as long as the night. The "tropical year" measures the time from one spring equinox to the next spring equinox. Such a calendar can be used to predict the seasons. In northern land near the Arctic Circle, the length of the day in the winter is much shorter than the length of the night, which brings the end of growth and plenty. Stonehenge was carefully aligned toward the position of the sunset on the winter solstice, which marks the day when the sun begins to return to the north, and an important time for festivals and celebrating rebirth.

The Sumerian year has 360 days, and the Egyptian calendar has 365 days. But the actual length of the tropical year is 365.242199 days, so that the spring equinox is later than the predictions on the Egyptian year by 0.242 days/year; this difference is insignificant for 1 year, but after a thousand years the error is 242 days, which makes the Egyptian calendar useless for keeping track of the seasons.

The Julian calendar was adopted in the year 46 BCE, which introduced the notion of a leap year of 366 days where every year divisible by 4 would be a leap year. Since the length of the average Julian year was $365^1/_4$ days, the error declined to only 7.8 days per thousand years. After 1600 years, this error became significant again, and the current Gregorian calendar was proclaimed in 1582. This time, the years divisible by 4 (such as 1896) are still leap years, with the exception of the years divisible by 100 but not by 400 (such as 1900) that remain as ordinary years with 365 days. So the average Gregorian year is now 365 97/400 days long, and the error has declined further to a more acceptable 0.3 days per thousand years. What should we do after a few thousand years?

Instead of predicting the annual seasons, lunar calendars were designed to predict the phases of the moon, which is important in religious affairs and in the heights of tides. In Mesopotamia, the beginning of a new month was determined by the actual observation of the waxing crescent moon in the western sky at twilight, which was also used to mark the beginning of Ramadan in the Islamic calendar. The Islamic calendar is based on cycles of the moon, and a year always has 12 lunar months. An average "synodic lunar month" from new moon to new moon has a length of 29.531 days, so one compromise solution would be to have the month alternate between 29 and 30 days. But an average year has 12.369 full moons, therefore the Islamic year is accurate on lunar phases, and does not predict the seasons. That is why the sacred month of Ramadan can occur in the summer or in the winter. The Chinese calendar is based on the moon phase as well as on the seasons, and the new moon is always on the first day of the month. The compromise solution in China is to throw in a leap month when needed to adjust for the seasons, so that each year may have 12 or 13 months. In our Gregorian calendar, some months have two full moons at a frequency of about once every 3 years—and the extra one is called the Blue Moon.

The weather and the health of individuals are measured by temperatures and pressures. The thermometer that gives quantitative measurements of *temperature* is not a single invention, but gradually progressed over many centuries. Hero of Alexandria knew of the principle that substances such as air expand and contract with heat, which could be measured with a closed tube that is partially filled with air and dipped in a container of water. When the temperature rises, the air in the tube expands and pushes the water level down. Avicenna of Persia built such a device around 1000. Galileo built a sealed glass cylinder containing clear water and a number of glass bulbs containing different proportions of air with colored liquids and metal weights, thus containing different densities that would not depend on temperature. When the temperature rises, the clear water expands and has less density, and the set of glass bulbs with different fixed densities are designed to sink in sequence as the liquid is warmed. The Galilean thermometer measures temperature qualitatively by the number of sunken bulbs without a continuous scale, and responds to temperature changes slowly, as it takes time to warm up a column of water from the surrounding air. It is also difficult to compare the measurements taken by two different Galilean thermometers.

In 1654, Ferdinando II de' Medici, Grand Duke of Tuscany, had a sealed tube with a bulb and a stem, and partly filled with alcohol, which was the first modern-style thermometer with a scale. A universal standard was needed to compare the results from different thermometers. In 1665, Christiaan Huygens suggested using the melting and boiling points of water as standards. In 1724, Daniel Fahrenheit produced a temperature scale that bears his name, with the peculiar values of 32 and 212 as the melting and boiling points of water, and using mercury as the fluid because it has a high coefficient of expansion. In 1742, Anders Celsius proposed a more convenient scale with 0 and 100 as the melting and boiling points of water. The normal temperature of a healthy person is 37°C, which can change due to menstrual cycles or fever.

The earliest pressure measurements were also by liquid columns. Evangelista Torricelli is often credited with inventing the barometer in 1643. A tube sealed at one end is completely filled with a liquid, immersed in a pool of liquid and inverted so that the sealed end is up. The liquid level will fall and create a vacuum, and the height of the liquid column is a measure of the atmospheric pressure. Another version involves a liquid filled U-tube with one end open to a reference pressure, and the other end open to the object being measured. The difference in the heights of these two sides, divided by the gravity constant and the density of the liquid, measures the difference in their pressures. When the reference side is a vacuum and the other side is atmosphere, the difference in liquid height would be 34 ft (16 m) for water or 30 in. (760 mm) for mercury. As one climbs a mountain, the atmospheric pressure will drop, so that at the peak of Mount Everest at 13,700 m, the air pressure is only 1/3 of sea level pressure. The normal blood pressure of a healthy person is systolic 90–120 mmHg, and diastolic 60–79 mmHg.

7.2 RECORDS

The brain selects pieces of information to store as memories for future use, such as where food is plentiful, which animals are dangerous, and what are the best ways to catch a fish. The brain as a recording device has many defects: it has a fragmentary recall of details, it lacks objectivity, and the memory records fade with time and perish upon death. The oldest human records may be the cave paintings of animals and hunting that have been found in many parts of the world; the oldest ones were found in the Chauvet Cave in France and date from 32,000 years ago. Excavations in caves and burial grounds have yielded many portable arts—carved horns and bones representing animals and people, such as the Venus of Willendorf. Writing came much later, around 3500 BCE, which marks the beginning of history. There are no recordings of ancient sounds and motion, which requires technologies that did not exist until the 1800s.

7.2.1 Text, Writing, and Printing

Excavations from as early as 8000 years ago found protowriting items such as knots and tokens used for recording inventories and accounting for debts owed and paid. The first problem to solve in the development of writing is the creation of a method of representation or encoding, an accepted convention of which symbol represents what object or idea—that is, a language. The use of picture symbols to represent objects is called the *logogram*, and the use of sound symbols to represent speech is called the *phonogram*. History began with writing, which makes permanent records of human achievements and experiences in addition to mundane accounting. The oldest writings, from the Mesopotamian city of Uruk, were cut into wet clay with a stylus in a form called the cuneiform, and were symbols representing sounds. These earliest clay tablets of Sumerian cities date from 3500 BCE, and are records of temple income kept by priests, with pictures of cow heads, ears of corn, or fish, together with numerals (see Fig. 7.1). Egyptian hieroglyphics date from 3000 BCE, which were written on papyrus made from the reed, and were a mixture of pictorial representations and symbols of sounds (see Fig. 7.2). Chinese

FIGURE 7.1 Sumerian cuneiform.

writing dates back to 1300 BCE, carved at first onto turtle shells and oxen bones to record royal divination results; it was based mainly on picture symbols enriched with sound symbols. A shell from the Shang dynasty (1600 BCE) is among the oldest, shown in Figure 7.3; the inscription shows a question posed by royalty on the success of the next harvest, and the name of the diviner is Wei—possibly an

FIGURE 7.2 Egyptian hieroglyphics.

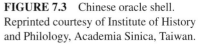

FIGURE 7.3 Chinese oracle shell. Reprinted courtesy of Institute of History and Philology, Academia Sinica, Taiwan.

ancestor of this author. The writing symbols or characters were greatly modified over the centuries so that one can compare them to the modern characters, and can barely discover the shapes for the sun and the horse. Most modern principal writing systems were based on sound symbols representing speech, including the Greek and Latin alphabets and the Japanese kana.

Before the invention of alphabets, it took thousands of symbols to record the many objects and sounds needed for everyday lives, and the art of writing and reading was limited largely to the priests and professional scribes who could invest the time to learn and memorize them. The earliest alphabet came during the reign of King Ahiram of the Phoenicians, around 1600 BCE. It had 22 letters, all consonants, and was written from the right to the left. To read this text aloud, the reader had to provide the appropriate vowels, which required more memorized knowledge and created ambiguities for later generations. Nevertheless, the Phoenician alphabet spread rapidly to many neighbors, and was used to write the Aramaic, Hebrew, and Arabic letters in the East, and the Greek, Etruscan, and Roman letters in the West. The Greek language added vowels to the consonants, and was written from the left to the right. The alphabet made possible the use of a much smaller set of symbols that are easier to memorize, and makes writing and reading much easier. A modern reader of Chinese needs to memorize about three thousand characters in order to read newspapers and simple texts. A modern reader of Japanese needs to master an alphabet of 51 letters, in addition to the Chinese characters.

The Epic of Gilgamesh dating from 2150 BCE is generally regarded as the oldest literature of men. The Code of Hammurabi was written in 2000 BCE, and was the first recorded body of law by which a government operated. The spread of Buddhism was greatly helped by the writing of scriptures in 500 BCE. Writing led to a profound change in society, as contracts could be written down and enforced, without relying on memories that were unreliable. Religious and civic laws could be written down and enforced uniformly over different places and time, and could be more objective. The history of a people can be written down, about their journeys and wars, their rulers and invaders, their joys and sorrows, and about practical ideas on how to farm and herd animals. However, it was difficult and expensive for an ordinary person to find the time and material to learn reading and writing, which were reserved for the scribes and priests. Even the elite children of upper class families may have preferred the active life of hunting and warfare to the difficult and scholarly tasks of reading and writing.

There were demands for numerous copies of important texts, such as religious and legal documents for different provinces of an empire, which created the need for large scale copying. Each text had to be copied by hand, which was slow, expensive, and difficult, especially when copying a book with thousands of words without making some mistakes. When a copy was taken to another place and read at a subsequent time, the reader far from the source could not be sure how accurate the copy was. Other concerns were the high costs of buying the medium for writing, hiring the script to copy, and storing the scrolls in libraries. There were very few books available, and they were kept in restricted libraries that were accessible only to the elite. Even a learned lay person who could read might not have had access to many books, and could not know the contents in the sacred books of religion and law of the land. In classrooms, the teachers read to the students from the few books or notes available, and the students did not have access to books. Even today, the associate professors at English universities are called Readers, in honor of the fact that they had access to books.

A handwritten parchment book was an expensive investment. In 1424, the Cambridge University library owned only 122 books, each of which had a value equal to a farm or vineyard. Society was divided into a few ruling literate people who were the priests and the scribes, and the common illiterate people who were farmers and trade people. The economy rapidly expanded during the Renaissance, which created increased demand for books, and Europe was ready for a revolution in book reproduction. Two critical inventions were needed: an inexpensive durable writing medium, and rapid and accurate reproduction.

Many types of materials have been used for writing throughout history. The clay tablets of Mesopotamia were inexpensive, but they were heavy to move and difficult to store. The Egyptians soaked the pith of papyrus plants, cut them into strips, laid them across each other at right angles, pressed and dried them, and gummed them together to form a scroll. The only source for papyrus in the ancient world was the reeds from Egyptian marshes. In Greece and Rome, parchment was made from untanned leather, especially from calves, goats, and sheep. It was washed and soaked in lime, dried on a stretch frame, shaved with a knife and smoothed to produce writing surfaces. For lengthy documents, rectangular pages were sewn together in long strips and then rolled up. In the second century at Pergamum, the book was invented to replace the scrolls that took shelf space; the sheets were bound together with a cover to form a codex or a book. A book of 200 quarto pages (about 250 mm tall) of parchment would require the skins of no less than 12 sheep, so the material cost was a significant part of books and libraries.

The oldest forms of writing in China, from 1350 BCE, were on the plastrons of turtle shells and shoulder bones of oxen, which were used by kings for divination about rituals, war, hunting, crops, and births. At the same time, bronze vessels were used for rituals and ceremonies, and were cast with words engraved on the sides. Stone tablets called stele were also carved for memorials. For long narratives and texts of history, as well as military messages from the king to the general, vertical strips of bamboo and wood were strung together to make scrolls. This may have led to the Chinese custom of writing from top to bottom.

Paper was the great invention that swept all previous media off the main stream. Cai Lun, the eunuch and court official of the Han emperor, was recognized as the inventor of paper in the year 105, although he might have taken the credit from the ingenuity of his staff. His official biography said that, *In ancient times writings and inscriptions were generally made on tablets of bamboo which is heavy, or on pieces of silk which is costly. Cai Lun initiated the idea of making paper from the bark of trees, remnants of hemp, rags of cloth, and fishing nets. He submitted the process to the emperor in the first year of Yuan-Hsing and received praise for his ability.* The process involved placing shredded fibers from various sources into water to soak and boil for many days to remove nonfibrous material such as lignin, leaving behind only fibers to make a suspension with added binders. Then a sieve mould was immersed into the suspension, and lifted to remove a layer of the matted fiber on the mould. The fibrous mat was removed and compressed to remove as much water as possible, and then left to dry to form the paper. Paper could be made in large quantities from cheap material, which made writing and copying a great deal more affordable.

The secret of papermaking was passed to the Arabs in 751 during the battle of Talas River near Samarkand, when several Chinese artisans were captured. The secret of papermaking moved to Europe after that. Wood pulp is the most important source of fibers in modern paper, which consists of about 60% cellulose bonded in a matrix of lignin, which should be removed to prevent yellowing and brittleness with time. Special paper can be made adding opaque fillers to make it possible to write on both sides, or by coating with polymers to make photographic papers. Besides writing, there are many other uses of paper, including bags and wrapping, playing cards, religious charms, and money. It is fortunate that at the time of the invention of printing, cheap paper was available. Otherwise, it is hard to imagine the number of sheep that would have been needed to sustain the rising demand for books in

Renaissance Europe. The invention of paper must be counted as being of equal importance to the invention of printing.

There were many technologies for reproducing text before printing was invented. Sumerian seals featured carved images of gods and demons, and were pressed onto wet clay to make impressions. Seal rings with the owner's names were also used to impress on wax to authenticate a document. An early form of printing took place in China around 700, which involved carving official decrees on a stone tablet, followed by rubbing the surface with ink and pressing a piece of paper onto the inked stone tablet to transfer the words. An important ingredient of printing in China was the ink, which was made from boiled linseed oil and pine resin smoke. The oldest printed text that we have was made in the year 868 in China, called the Diamond Sutra. It was found in a cave in Dunhuang together with many other Buddhist writings, which were executed from carved wood blocks. Since each page of a book had to be carved onto a separate piece of wood, the reproduction of a book of many pages was difficult and slow, so it was reserved only for very important material with a wide readership—such as religious pamphlets and banners. The carved woodblock of each page could be saved for possible future uses.

The movable-type printing is based on carving a single letter or character on a small block, and assembling numerous small blocks for a page. The earliest mention of movable-type printing was by Shen Gua (also known as Shen Kuo) in China, from his *Dream Pool Essays* written in 1088. He gave credit for the invention to Bi Sheng (990–1051). He wrote

> During the reign of Chingli, [1041–1048] Bi Sheng, a man with no official position, made movable type. His method was as follows: he took sticky clay and cut in it characters as thin as the edge of a coin. Each character formed a single type. He baked them in the fire to make them hard. He had previously prepared an iron plate and covered his plate with a mixture of pine resin, wax, and paper ashes. When he wished to print, he took an iron frame and set it on the iron plate. In this he placed the types, set close together. When the frame was full, the whole made one solid block of type. He then placed it near the fire to warm it. When the paste [at the back] was slightly melted, he took a smooth board and pressed it over the surface, so that the block of type became as even as a whetstone.

A page of the movable type is composed by assembling the blocks of individual characters or letters into a frame, which is used for printing. After the page is printed a sufficient number of times, it can be taken apart into individual letters or characters, and reassembled to form another page. An early movable type appears to have been made in Korea around 1100. For a language based on an alphabet with less than 50 letters, this is a very advantageous method.

Johannes Gutenberg (1398–1468) was a goldsmith in Mainz, Germany, and the first European to use movable-type printing in 1439, as well as inventing the mechanical printing press. It was a revolutionary improvement upon handwritten manuscripts and woodblock printing, and transformed European bookmaking. Gutenberg's invention may have been independent of the earlier inventions from the Far East. He is often regarded as one of the most influential inventors of the millennium in Western history.

Gutenberg was born in Mainz around 1398, and his father was a goldsmith who taught him much about working with metals and coins. The family was poor and was forced to move from Mainz; by 1434, Gutenberg was working in Strasbourg as a goldsmith. It was in 1440 in Strasbourg that Gutenberg perfected and unveiled the secret of printing based on his research, and in 1448 he was back in Mainz where he took out a loan from his brother-in-law to build a printing press. By 1450, the press was in operation and he borrowed 800 guilders from a moneylender named Johann Fust. In 1455, Gutenberg published his elegant 42-line Bible, and produced about 180 copies, mostly on paper and some on vellum. The Gutenberg Bible is one of the most prized books in the world, and there are about 48 complete copies today. Gutenberg sold his Bible for 30 florins each, which was roughly three years' wage for an average clerk, but it was much cheaper and more accurate than a handwritten Bible that could take a scribe over a year to make, with numerous errors in copying.

Later, Fust sued Gutenberg about the loan, which had risen to 2000 guilders, and the court favored Fust by giving him control over the Bible printing workshop and half of all printed Bibles. In 1465, Gutenberg's contributions were recognized, and he was given the title of "Hoffmann" by Archbishop von Nassau of Mainz, which included a stipend, an annual court outfit, 2180 L of grain and 2000 L of wine. He died in 1468 in Mainz, and the location of his grave has been lost.

There were no written descriptions of the Gutenberg invention, and his exact methods are under dispute among experts. The movable-type printing press begins with a type for each letter, which Gutenberg made from an alloy of lead, tin and antimony. The current method is to use a hard metal punch with the letter carved on it, which is hammered into a soft copper sheet to create a mould or matrix, to create hundreds of identical letters. This is then placed into a holder, and cast by filling with a hot-type metal, a softer alloy of lead, tin, and antimony, which is cooled down to create types. Subsequently the letters are arranged according to a text on a rack and inked, then pressed against a piece of paper to make the print. Some experts believe that instead of the copper sheet, Gutenberg used sand that made a less precise type.

The phenomenal success of this invention also depended on the availability of three other affordable inventions: paper, oil-based ink, and printing press. We have already discussed the low cost of paper compared with the parchment involving many sheep. The ink used by European scribes for writing was an aqueous solution of gum with lampblack as the pigment, which worked well on semiporous paper surfaces. But a water-based paint does not wet the hard surfaces of metal types, but pools in globules and does not spread evenly. A new ink was developed by Gutenberg, which was based on linseed oil with the grounding of lampblack or charcoal, and forms thin films on type surfaces. He also needed a device that would push the inked typefaces firmly against the paper, to achieve clear transfer. The solution was the printing press with a screw to provide pressure that may have been inspired by the wine or olive presses that squeeze grapes for wine and olives for oil.

When books became cheap and easily available, there was a prodigious increase in the number of printed books. It has been claimed that eight million books were printed between 1450 and 1500, which was more than all the scribes of Europe had produced in over 1000 years, since the year 330. The entire classical canon became reprinted and widely circulated in Europe. More people had access to accurate knowledge, so more people became learned and could discuss these works. Students could have their own books, and could learn by solitary reading in addition to attending lectures. The teachers did not have to waste time on reading the texts aloud to the students, so more time could be spent in discussing the meaning and interpretation of the texts. It became possible to cross reference two or more texts to compare consistency and to detect the corruption of scribes. The laymen could read the Bible without interpretations from priests, and could read the law decrees without relying on lawyers. This broke the monopoly of priests, lawyers, and professors, who were previously the only ones with access to the few available hand-copied books. Printing played a factor in the Protestant Reformation, as the 95 theses of Martin Luther posted on the door of his church were printed and circulated widely. Luther also issued broadsheets outlining his anti-indulgences position, which evolved into newspapers and mass media. Literacy soared in all advanced nations, so that it was soon expected that all adults should be literate.

Before Gutenberg's invention, the medieval churches had used stained glass windows and statues to illustrate the stories from the Bible, as the congregation was illiterate and could only understand pictures. Printing led to a movement away from relying on spoken words and images for information, toward a culture of relying on printed words for information. The printed word also helped to unify and standardize the spelling and syntax of national languages, to decrease their regional variability within one nation. Daniel Defoe wrote that, "Preaching of sermons is speaking to a few of mankind, printing books is talking to the whole world." The other early items that were printed included edicts, money, playing cards, and religious charms. Newspapers were made possible by printing, and books no longer had to be about religion or laws, since people could afford to read cookbooks, detective stories, comic books, gossip magazines, and even pornography.

Electronic books, or eBooks, have become immensely popular since the introduction of the Kindle ereader by the giant bookseller Amazon in 2007. Although the device is smaller than a book, it can hold several thousand books without illustrations. The cost of reproduction can be exceedingly small, as there is no paper, ink, or printing press involved. In 2011, Amazon announced that it had crossed the threshold for selling more eBooks than printed books.

7.2.2 Picture and Photography

Ancient drawings and paintings of animals and humans are much older and more vivid than the ancient writing of texts. People from another culture cannot decipher Egyptian hieroglyphics or a foreign alphabet, but they can all appreciate a horse painting from the Chauvet Grotto in France. To capture the likeness of a horse, the Chauvet artist needed great skill and experience, to represent in a few strokes all the details that he considered important, while leaving out numerous other details. The camera obscura is a method to capture an image in great detail. It was described by Mo Ti in 420 BCE, and was known to Aristotle. Al-Hazen (965–1039) proposed its use for observing solar eclipses. In his *Dream Pool Essays*, Shen Gua (1031–1095) mentioned the reverse images formed by a small hole. He had used a dark box equipped with a pinhole, held at a variable distance from the receiving surface. Focusing was accomplished by varying the distance between the pinhole and the receiving surface, till the focus was sharp. A pinhole is easy to make, but it admits very little light so the resulting picture is quite dim. A later improvement came to camera obscura by replacing the pinhole with a single convex glass lens, which admitted more light but had the unfortunate property of chromatic aberration described earlier. Johannes Vermeer might have used the camera obscura to project a scene onto a piece of paper, and then sketch the outlines with charcoal to produce a highly accurate perspective.

The steps involved in making a photograph may be described as: (a) gather and focus an image on a surface for inspection, (b) record and amplify the image and make it permanent, (c) display the image, and (d) make numerous copies of the image. The fleeting camera obscura image on the surface is not permanent, and has to be fixed or recorded before we have a permanent image. Many chemical substances are light sensitive, and change their chemical composition when exposed to light. The ability of the sun to darken silver salts was known as early as 1727.

In 1816, Joseph Nicéphore Niépce in France experimented with a number of light-sensitive substances. He dissolved bitumen of Judea in lavender oil and used the solution to coat a pewter plate, which hardened under light. After 8 h of exposure, he developed the picture by washing with a mixture of oil of lavender and white petroleum, which dissolved the unexposed bitumen that had not been hardened by light. After rinsing and drying, the picture became permanent, as the already hardened bitumen was no longer sensitive to more light. The exposure time was very long and the images were not sharp. In 1829, Niépce took in a partner, L.J.M. Daguerre, who was a scene designer and painter for the Paris Opera. In 1837, Daguerre used silver halide as the light-sensitive material, and mercury vapor as the developer, and was able to make clear and permanent photographs with a reasonably short exposure of 20 min. Two years later, he developed a silvered copper plate, and exposed the plate to iodine vapor to form a layer of silver iodide that was also sensitive to light. The unchanged silver iodide was washed away with sodium thiosulfate, leaving the polished silver to form the shades. This product was called the daguerreotype.

An ideal photographic recording should have a portable surface covered with a fine image material that is sensitive to light, so even a small amount of light would produce a microscopic latent image. The image material should have a dynamic range so that more light would produce a stronger latent image, and should also have a developer that amplifies the invisible microscopic latent images by a million times into visible images. Finally, there should be a fixer to wash away all the remaining unexposed image material so that the film is no longer sensitive to light for preservation. Earlier photographs used silver halide on a glass plate. Celluloid was invented in 1873 by John Wesley Hyatt, and was used for photographic film by George Eastman and the Kodak Company.

The celluloid film with silver halide has been supplanted by modern electronics that are much more sensitive to light, and do not involve a "wet" processing. The "photomultiplier" is a photoelectric device that converts light into electrical signals. It was originally a vacuum tube, and can detect light and multiply the current by as much as 100 million times; the vacuum tubes have been replaced by semiconductor devices. Willard S. Boyle and George E. Smith of the Bell Laboratories invented the charge coupled device (CCD) around 1970, which is the most sensitive technology today for light and digital imaging, and is used in all the digital cameras. They used a capacitor to convert light into electric charges in proportion to light intensity, and then arranged many capacitors into a linear array. This array is then transferred to an amplifier and processed for recording. A two-dimensional image requires many such linear arrays of capacitors. Boyle and Smith shared a Nobel Prize in Physics in 2009, which was also won by Charles Kao for the development of fiber optics.

Our photographs are among our most cherished possessions, for memorable people and events. We preserve and display our pictures of weddings, holidays, children growing up, playing sports and music, sights and tourism. We remember the images of historic events such as the atomic bomb at Nagasaki and marines raising the flag on Iwo Jima. We also use photographs to show the faces of historical persons, to teach the science of nature, and to explain the principles of technologies.

7.2.3 Sound and Phonograph

We do not have the sounds of our ancestors talking or singing, but some fragments of their music have survived to this day, due to the invention of music notation and symbols, which can be found in cuneiforms from as early as 2000 BCE. They are the equivalent to written texts, and they require special knowledge to decipher the pitch and duration of each note, and whether the music included the harmony of many notes together. The most familiar musical scales repeat the notes at each octave, and the most important scales are the pentatonic (5 notes), the diatonic (7 notes), and the chromatic (12 notes).

Sound involves pressure waves in the air, and the human ear can hear frequencies between 20 and 40,000 Hz, or vibrations per second. An instrument such as the flute has a relatively pure sound, where one note consists of mainly one frequency and the loudness of the sound produced depends on the amplitude of the vibrations. An instrument such as the oboe produces a complex sound that is the mixture of many frequencies. A sound recording device must faithfully capture and record all these instantaneous vibrations by frequency and amplitude as they change with time, and be able to play them back with amplification.

The key inventions in the phonograph involve the ability to (a) gather and magnify sound waves, (b) record the vibrations on a durable medium, (c) play back the sound, and (d) reproduce many copies. Thomas Edison invented and developed the carbon microphone, which consisted of two metal plates with a pack of carbon particles between them. When the sound wave hit the front metal plate that acted as a vibrating diaphragm, the carbon particles became alternately compressed and loosened, so their electrical conductivity alternately became higher and lower. He also developed the first successful phonograph in 1877, by gathering the sound from a large area with a large trumpet, and focusing them onto a small stylus. His recording device was a hand-cranked cylinder wrapped with a tinfoil sheet, and he used the stylus to cut the vibrations onto grooves on the rotating tinfoil. The frequency and depth of the cutting was controlled by the sound vibrations. The playback was on the same machine, with a needle in the grooves to feel the varying frequency and depths, and to translate them back into sound to be amplified by the trumpet. It was not a commercial success because the grooves wear out after only a few plays, the tinfoil cylinder was difficult to store, and there were no good methods for copying and mass-producing cylinders for the listening public. Many other materials were used for the cylinder later, including a wax-coated cylinder and hard plastic cylinders.

Flat disk recordings became the winning technology, despite the disadvantage that a disk had to run at a constant number of revolutions per minute, so the outside grooves moved faster than the inside grooves. Later in 1877, Emile Berliner came out with his gramophone that had a recording surface made of a flat zinc disk coated with beeswax containing spiral grooves, into which the sound signals were cut at varying widths while the depth remained constant. Berliner's disk could be copied and mass produced by stamping, and was easier to store than the cylinder. This method led to a separation of the professional recording instruments from the consumer listening devices.

An entirely different method of recording was created by Valdemar Poulsen of Denmark, who made a magnetic wire recording in 1898. Each small domain on the wire could be magnetized to point north or south, and this alternation could be made to code the sound. This led to the later development of recording on the more economical magnetic film and tape. Memory devices were also made with electronic tubes and transistors, which have so many advantages that digital recording has replaced analog recording. Today, one of the most efficient methods of recording is the optical Compact Disk, where the sound signals are cut into a plastic and aluminum disk as digital dots and dashes.

Playback of sound recordings was originally done by a large trumpet, which only focused the sound in a specific direction without making it louder, which would be needed for a large audience in a concert hall or public gathering place. Amplifiers were developed to magnify the strength of electrical signals, so that even faint electrical signals could be converted into signals that are 100 million times stronger. Loudspeakers came in to convert electrical signals back into sound waves, often by driving a diaphragm connected to a horn. Reproducing sound faithfully over a large range of frequencies can be more an art than a science, and it is usually necessary to build large horns for the low frequencies, called "woofers," and small horns for the high frequencies, called the "tweeters."

In 1901, Eldridge Johnson founded the Victor Talking Machine Company and began making many famous recordings including the Red Seal records of tenor Enrico Caruso, violinist Jascha Heifetz, and pianist Sergei Rachmaninoff. The world was ready for the synchronized recording of cinema with sound, which became the "talkies" that swept away the silent movies.

7.2.4 Motion Pictures

Attempts to depict motion might be as old as Paleolithic cave paintings, where animals in motion were sometimes shown with extra legs superimposed. The 1912, Marcel Duchamp painting of *Nude Descending a Staircase No. 2* was a continuation of this tradition of superposition to give a sense of motion. Most children have seen the common "flip book," featuring a series of pictures that vary gradually from one page to the next, so when the pages are turned rapidly, the pictures appear to be animated. Its success depends on the human physiology of "persistence of vision," so that a rapid succession of pictures, preferably more than 20 per second, fuses into an appearance of continuous motion. Zoetrope is a wheel-of-life or magic lantern device that consists of a cylinder with many slits cut vertically in the sides, and gives glimpses of a sequence of drawings painted on the inside. When the cylinder rotates, such as by a stream of hot air from a candle underneath, an illusion of motion is produced. When a sequence of photographs taken in rapid succession is available, it can be displayed by the linear flip book or the cylindrical zoetrope as a one-second motion picture. The critical invention needed for longer motion recording was a mechanism to record the sequence of a greater number of images, and another mechanism to project longer sequences.

In 1872, Leland Stanford asked Eadweard Muybridge, a noted photographer, to settle a question about a galloping horse—whether all four legs leave the ground at the same time, and whether at that moment the four legs are splayed far apart or bunched together under the belly. Muybridge developed a scheme for instantaneous motion picture with up to 30 photographic cameras placed 0.3 m apart, all facing a white wall. Between the wall and cameras was a series of strings, each of which triggered an electromagnetic shutter of the camera. With this elaborate setup, he recorded a horse galloping. After reviewing all these photos, singly and with a machine, he showed that, indeed, all four hooves were off the ground and bunched together under the belly. He subsequently produced 11 volumes of publications containing 20,000 photographs of animals and humans in motion.

A satisfactory motion picture needs to record action at a minimum of 30 pictures/s, so that a movie that is only 1 min long would need 1800 pictures. If the individual pictures are only 25 mm long, the total length of these pictures glued together would be 45 m. In 1891, Thomas Edison patented the "Kinetoscope camera," or a peep-hole viewer, which was installed in penny arcades for people to watch simple short films. The first motion picture was called Cinematographe, produced in 1892 by the brothers August and Louis Lumiere, and was shown in Paris in 1895. They introduced the iconic film perforations on the side of a film strip as a means of advancing the film through the camera and projector. The first movie film was 17 m long, which was hand cranked through a projector and ran for 50 s.

Many companies were formed to make motion pictures of a longer length, which would remain silent for the next 30 years. Director D.W. Griffith was the first to make a motion picture in Hollywood in 1910. Sometimes live musicians and sound effects were employed to liven up the silent motion pictures. It was difficult to synchronize the picture being projected on the screen with the sound on disks from the disk machine. In 1919, Lee De Forest was given a patent for the sound-on-film technique, which photographically recorded the soundtrack onto the side of the film strip so they were totally synchronized. The first successful "talkie" was The Jazz Singer starring Al Jolson in 1927.

Combining several art forms can create extra pleasure. The Tang poet Wang Wei was skilled in both poetry and painting, and critics said that, "He has painting in his poem, and he has poetry in his painting." Richard Wagner was a pioneer in synthesizing poetry, music, visual art, and drama into the single art form of opera as musical drama. Alexander Scriabin wrote a symphonic poem, *Prometheus: Poem of Fire* in 1915, which featured a "keyboard of lights" that was built like a piano but had flashing lights as well as music when played. In 1932, Aldous Huxley published his novel *Brave New World*, in which he introduced a satiric form of entertainment inspired by the "talkies," and called the "Feelies." This fanciful future entertainment included the feel of a bear rug on the skin, in addition to sight and sound, but Huxley did not mention how to record the feelings on the skin, or how to play this back for the audience. In 2009, the New York Times reported a performance of "Green Aria: A Scent Opera" at the Guggenheim Museum, which had music coordinated with 30 named fragrances including subtle, pungent, intoxicating, and stinking! For instance, smoke was issued when the story theme was fire. The producers realized that if they were to broadcast scents from the stage using fans, it would take about 50 min to clear the auditorium of the first scent to become ready for the next scent. Thus, they provided each seat with its own adjustable tube that sent out small amounts of scent to each nose.

7.3   COMMUNICATION

The flowering plants have bright colors and scents to advertise to insects or birds about the availability of honey and pollen. Animals send messages by gestures, sounds, touch, and scent to members of the family or to a herd, as well as to potential enemies. Male peacocks show their big ornamented tails, and the male bowerbirds build colorful nests to attract potential mates. The chirps of crickets and cicadas, the songs of birds and of the humpback whales, and the roar of lions are all means of communication.

People also communicate by gestures and sounds, which can be private and intended for only one person, or can be an oration or a broadcast intended for many people. Personal messages without amplification can be delivered to only those in the immediate neighborhood, as the voice can reach only a limited distance. We have invented simple solutions for distant communication, such as flashing mirrors and smoke signals at daytime, and flares at night to warn about the approach of enemies.

7.3.1 Text Messages

Before the invention of writing, prearranged signals such as knots or markings on a piece of wood could be delivered by a courier to convey a message. Since the invention of writing, a letter can convey much more precise and complex information. To keep such information confidential, sealed envelopes were developed; and to authenticate the sender, signatures and engraved seals have been used.

The first great invention in rapid text messages over long distances was the telegraph, invented by Samuel Morse (1791–1872), who was a student at Yale and became a very successful painter of prominent people. In 1825 when he was away from home to paint a portrait of Lafayette in Washington, his father sent a messenger on horseback with the sad news that Samuel's wife had died in New Haven. As a consequence of this delay in news from home, he changed his career and became interested in rapid long distance communications. He met Charles Jackson in 1832, who was schooled in electromagnetism, and he conceived of the idea of a single-wire telegraph. The sender is connected to the receiver by a loop of electric wire powered by a battery; the sender has an electric key and the receiver has a buzzer. When the sender pushes down the electric key, the loop is connected and the battery causes electricity to flow, which causes the buzzer to sound for the receiver. A single battery has the power to carry a signal over only a few hundred yards of copper wire.

Morse managed to obtain financial support to build more elaborate equipment, and increased the distance by relay stations with batteries, so he could send a message through 10 miles of wire. He then lobbied the US Congress for more support to build a long distance telegraph, and in 1843, obtained a sum of $30,000 to construct an experimental 63 km telegraph line between Washington DC and Baltimore, along the right-of-way of a railroad. This involved new methods to make copper electric wires of great length, to insulate the wires from leakage to the surroundings, and to support the wires on telegraph poles at regular intervals; it also required many relay stations with batteries to boost the signal strength. An impressive demonstration occurred on May 1, 1844, when news of the nomination of Henry Clay for US President was telegraphed from the convention in Baltimore to the Capitol Building in Washington. On May 24, 1844, Morse's company and its telegraph lines were officially opened for business, and he sent the famous words "What hath God wrought!"

He was also an inventor of the Morse code with Alfred Vail, based on earlier codes. It used a sequence of dots and dashes to encode each letter of the alphabet: a dash is required to be three times as long as the dot, and the pause between letters is also required to be three times as long as a dot. A single dot represents the letter E, and a single dash represents the letter T. It takes two symbols to represent the letters A, I, N, and M; it takes three symbols to represent 8 more letters, and four symbols to represent the 12 remaining letters. He designed it so that the more frequently used letters used the shorter signals. The numerals from 0 to 9 are represented by five symbols.

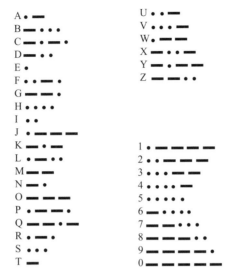

The first trans-Atlantic cable was laid in 1866, which was a tremendous financial investment and feat of engineering, as the wires had to be insulated and had to function under heavy pressure, and because the relay stations could not be built in the middle of the ocean. In 1872, Australia was connected to England. The most famous message with the Morse code is the distress signal "SOS" that is transmitted as " . . . --- . . . ". Telegraphy has had great military and diplomatic significance, as it can transmit information at the speed of light of 300,000 km/s, so that a telegraph can reach anywhere on earth in less than 1 s. The Battle of New Orleans was fought on January 15, 1815 and was won by General Andrew Jackson over the British, resulting in 2000 casualties and the death of General Edward Pakenham. The Peace Treaty of Ghent had already been signed on December 24, 1814, but the news of this peace treaty did not reach New Orleans till February. This costly battle would have been avoided if the telegraph had been available at that time.

The telegraph sends a message through a copper wire by the flow of electrons. A radio message is delivered without any wires, and it goes through air or even a vacuum as a beam of light. In fact, it involves a form of electromagnetic waves with wavelengths that cover an enormous range, from 100×10^9 m for extremely low-frequency radio to 10^{-12} m gamma rays; the most important subset is the visible light of 300 to 700×10^{-9} m. The product of wavelength and frequency equals the velocity of light.

Name	Frequency (Hz)	Wavelength (m)
Gamma ray	300×10^{18}	10^{-12}
X-ray	300×10^{15}	10^{-9}
Ultraviolet	30×10^{15}	10^{-8}
Visible	3×10^{15}	10^{-7}
Infrared	300×10^{12}	10^{-6}
Microwave	300×10^9	10^{-3}
Very high frequency	300×10^6	1
Medium frequency	300×10^3	10^3
Super low frequency	300	10^6

The invention of the radio involved many critical stages and outstanding inventors, and the principal inventors were Joseph Henry, James Maxwell, Heinrich Hertz, and Guglielmo Marconi.

Joseph Henry (1797–1878) came from a humble family in Albany, New York. At the age of 13, he apprenticed as a watchmaker, and did not attend any college. His life took a turn when he was 16 and read a book on scientific topics, which so inspired him to study nature and engineering that he enrolled in an academy. He was interested in magnetism and did many experiments, including winding insulated wires tightly around an iron core to make a more powerful electromagnet. He demonstrated that he could cause an electromagnetic effect over a distance of 200 ft without a connecting wire, and speculated on the existence of electromagnetic waves. He also did research in aeronautics and in acoustics. At 1832, this genius without a college degree became a professor at Princeton University, and in 1846, he was appointed the first secretary of the Smithsonian Institution.

James Clerk Maxwell (1831–1879) of Scotland developed his theory of electromagnetism in 1864, which ranks with Newton's laws of motion and Einstein's relativity as one of the foundations of physics. Maxwell explained all the electric and magnetic phenomena known in his day by four differential equations, and predicted the existence of waves of oscillating electric and magnetic fields that travel through space at the constant speed of light. At that time, all the known waves traveled in a material medium such as air, water, or a solid; he postulated a medium called aether that permeates all space including even a vacuum. But he did not work on the practical application of using these waves to carry messages at the speed of light to distant points. He also worked in many other fields, including the optics of color vision, and suggested that three black-and-white photos could be taken through red, green, and violet filters; if transparent prints of these images were projected onto a screen using three projectors with similar filters, the result would be seen by the human eye as a picture with complete reproduction of all the colors.

Heinrich Rudolf Hertz (1857–1894) of Germany was able to demonstrate the Maxwell electromagnetic theory of light in 1886, by building many apparatus to produce and receive high-frequency radio waves. His radio wave transmitter consisted of a high-voltage induction coil, a condenser, and a spark gap to create an oscillating spark. The frequency of the oscillation was determined by the configurations of the induction coil and the capacitor. Then he made a receiver by bending a copper wire into a circular shape with a brass ball at one end and a sharp point at the other; the distance between them could be adjusted with a screw so that it had a natural oscillating frequency to match that of the transmitter. A spark from the transmitter could travel a long distance without a wire, reaching the receiver and causing it to spark. Hertz also studied the velocity of electromagnetic radiation and found it to be the same as the velocity of light as predicted by Maxwell. However, the strength of the signal decreases as the reciprocal of the square of the distance traveled—in other words, if the signal strength is 1 at 1 mile from the source, then it drops to 0.25 at 2 miles, and to 0.01 at 10 miles. Long distance communication requires powerful methods of magnification. Hertz was skeptical about the practical impact of his work, and his promising career ended early as he died at the age of 36. The unit of hertz or Hz is named after him, which measures frequency in cycles per second. By the time of his death, we had the theory and the experimental proof for wireless telegraph and radio, and an entrepreneurial and resourceful inventor was needed to pull all this together.

Guglielmo Marconi (1874–1937) was born to a wealthy Italian landowner father and an Irish mother. He did not discover new theories or invent powerful components, but he was a very dedicated researcher who assembled components and improved many systems; he was also a very good entrepreneur and had the social connections to raise the financial support needed to build a large and practical system, and had great vision on the most important applications of telecommunications. He was inspired by the work of Hertz and Nikola Tesla, and he set out to develop practical wireless telegraphy.

Marconi based his invention on the Hertz oscillator and receiver, and the Morse telegraph key and code. By 1895, he was able to send and receive a radio message over 2.5 km. A great deal of capital was required to go from ideas to a practical radiotelegraph, but Marconi did not find enough interest and support in Italy. He traveled to England and benefited by his fluency in speaking English, so he was able to attract investors and funds, including the British Post Office and Navy. The great challenges were to increase the distance of the signal, and to make the transmission stable in all weather conditions. He did many experiments to determine that the two significant factors were the power of the signal and the height of the transmitting towers. The signal could also be distorted or interrupted by electrical storms and by solar flares. If radio waves acted strictly like beams of light, it would be impossible to reach great distances due to the curvature of the earth. Nevertheless, Marconi's experiments showed that much greater distances could be reached, which was later explained as ground hugging transport and refraction of the waves from the atmosphere. He adopted the carbon granular rectifier for reception, which was a great deal more sensitive than sparks. He took out a patent in 1896, formed the Marconi Company in 1897 after sending a signal 57 km over land from Salisbury to Bath. He recognized the importance of ship-to-shore and ship-to-ship communications, and sailed to the United States to cover the reporting of the America's Cup in international yacht races. By 1899, Marconi had sent a signal 15 km over water across the Bristol Channel to Wales, and later 52 km across the English Channel to France.

Now Marconi's big challenge was to send a signal across the Atlantic Ocean. In 1901, he built a sending station at Poldhu in Cornwall, England, and sent a message to Signal Hill in Saint Johns, Canada. The tower was 61 m high, and the distance traveled was 3400 km, a feat that stunned many people. However, this system was not stable and could be affected by many adverse atmospheric conditions. He found that the transmission of messages was generally better at night than at daytime, and he worked to make the transmission more stable and dependable. In 1903, he was finally able to send a message from America to Europe, from a station in South Wellfleet, Massachusetts. This first trans-Atlantic signal was also a great publicity success, as it was a message of greetings from President Theodore Roosevelt to King Edward VII of United Kingdom. It opened the door to a rapidly developing wireless industry.

Marconi also benefited from the publicity over the sinking and rescue of the Titanic, which struck an iceberg near midnight and went down in the Atlantic

Ocean on April 15, 1912. It was carrying 2223 passengers and crew on board, with 20 lifeboats that could hold only 1178 persons. Two wireless operators were working for the Marconi Company on the Titanic, and they immediately sent out distress signals that reached four nearby ships in the ocean. The ship Carpathian was 93 km away, and changed course to arrive in 4 h to save 706 survivors in the freezing water. The ship Californian was only 32 km away, but the radio was turned off as the operator had gone to sleep, so the Californian did not participate in the rescue. The practical utility of wireless telegraph for daily lives became well recognized. From then on, the radio became mandatory in ship operations, and was required to be kept on and attended at all times.

Marconi continued to refine and expand upon his inventions over the next few years, and turned increasingly toward the business aspect of his work. In 1909, he won the Nobel Prize in Physics, which he shared with Karl Ferdinand Braun, whose modifications to Marconi's transmitters significantly increased their range and usefulness. Honors continued to roll in, and he was made a Senator in the Italian Senate, and King Victor Emmanuel III made him a noble Marquess. He also joined the Fascist party in 1923, and was made a member of the Fascist Party Grand Council. At his second marriage in 1927, his best man was Benito Mussolini. In 1931, Marconi arranged the first radio broadcast of Pope Pius XI, and he said

> With the help of God, who places so many mysterious forces of nature at man's disposal, I have been able to prepare this instrument which will give to the faithful of the entire world the joy of listening to the voice of the Holy Father.

He died in Rome in 1937 at the age of 63. A state funeral was given for him in Italy, and all the radio stations throughout the world observed 2 min of silence.

The passage of electromagnetic waves through the atmosphere is subject to absorption by gases and dust, and subject to manmade or natural electromagnetic disturbances. Fiber optics are currently the best method of transmission over longer distances, which can handle a higher data rate (bandwidth), and are immune to electromagnetic interferences. A pure glass fiber can carry light over a long distance with little loss, even when it is coiled. The light can be kept inside the fiber core in a phenomenon called "total internal reflection" when a cladding material is used on the outside with a lower refractive index than the core. This concept was promoted by Charles Kuen Kao of British Standard Telephones and Cables, and he set the goal of loss to less than 20 dB/km. This goal was reached by Corning Glass Works in 1970, by doping silica glass with titanium. This loss was later lowered to 4 dB/km by using germanium dioxide as the dope. A typical optical fiber has a diameter of 125 μm, and a core of 100 μm. Kao was diagnosed with Alzheimer's disease in 2004; when he won the Noble Prize in Physics in 2009, his acceptance speech was delivered by his wife Gwen.

7.3.2 Picture

A picture can be sent over a wire, after being processed by a line scan from top to bottom, to produce a number of lines of varying black-and-white or gray scales. A low-resolution picture may be made with 10 lines/in. (0.39 line/mm), and a much finer picture is made with 100 lines/in. This is the principle behind the facsimile machine, commonly known as a fax machine. An early form was the Pantelegram of Giovanni Caselli in the 1860s. An insulating ink was used to write a message on a metal plate, and a pendulum with a stylus was used to scan the surface, which would then produce an electrical current when the pendulum was over the surface where there was no ink. This electric current could be transmitted to a receiver, which had a similar stylus over a paper soaked with potassium ferricyanide, which would darken when an electric current passed through. The first Pantelegram was sent from Lyons to Paris in 1862. It was reported that a sheet of paper 111 mm × 27 mm with 25 handwritten words took 108 s to transmit. The resolution of the picture depends on the spacing of the scan lines, and a finer resolution can be made at the cost of a slower speed of transmission. The most important use of this machine was for signature verification in banking. A wireless method of transmission was invented by Richard Ranger of the Radio Corporation of America in 1923, which was the forerunner of today's fax machines.

7.3.3 Sound

A string or wire connected to the bottoms of two tin cans or paper cups can be used to carry speech between two people. The vibrations created by talking into the cup cause vibrations in the bottom of the cup, which is transferred to the wire and travels to the bottom of the receiving cup acting as a diaphragm. Without magnification, this signal fades over distance and cannot be used for long distance communications.

A spark carries only the simplest on–off information. The human voice and music consist of vibrations over a range of frequencies and volume, and carry much more complex information. Around 1850, many people began to experiment with enhancement methods to extend the range and fidelity of sound transmission. Elisha Gray of Chicago devised a tone telegraph, with several vibrating steel reeds tuned to different frequencies, which modified the flow of an electric current. Several sets of vibrations at different frequencies could be sent on the same wire at the same time without interference, and Gray received a patent in 1875 for using the electric telegraph to transmit musical tones.

The first practical telephone was the invention of Alexander Graham Bell (1847–1922) in Boston. He was born in Edinburgh, Scotland, and made his first invention when he was 12, which was a machine to de-husk wheat for a flour mill. His father and grandfather were teachers of elocution, who wrote books explaining how to instruct deaf people to articulate words and to read other people's lip movements. He was also deeply affected by his mother's gradual deafness, and learned to use a manual finger language to tap out conversations. In 1870, the family moved to Canada and lived in Ontario where Bell built a workshop. Later he became a professor of vocal physiology and elocution at Boston University. His research on hearing and speech led him to experiment with hearing devices, and was given the first US patent for the telephone in 1876. He received financial support for his research, and hired an assistant, Thomas A. Watson, who was an experienced electrical designer and mechanic. Bell used a microphone with a vibrating reed or diaphragm to gather the voice, and the compression waves in a magnet were changed into fluctuating electric currents. When this fluctuating electric current was transmitted over a wire to the receiver, the process was reversed to create mechanical vibrations to reproduce the sound. He said, "Mr. Watson, come here, I want to see you," and Watson responded. Bell made many public demonstrations of his invention, including one at the Centennial Exhibition in Philadelphia. The Bell Telephone Company was created in 1877, and by 1886, around 150,000 people in the US owned telephones. In 1879, the Bell Company acquired Edison's patents for the carbon microphone, and made the telephone practical for long distances. In 1915, Bell made the first transcontinental telephone call from New York City to Thomas Watson in San Francisco over a 5500 km wire.

When radio waves were created by sparks, an uncontrolled range of frequencies were involved and they could carry no more information than dots and dashes for Morse code. When radio waves were created as a continuous stream of controlled frequencies and amplitudes, they could be used to carry much more information, such as voice and music. The method of amplitude modulation (AM) was introduced by Reginald Fessendon in 1906 to use radio to transmit speech and sound. Remember that human perception of sound is between 20 and 40 kHz. Fessendon started with a continuous radio carrier wave with a fixed frequency higher than audible sound, and set the amplitude to be variable and to conform to the sound signal to produce the AM wave, which is broadcast. This sound signal is shown in Figure 7.4, and the AM wave has a fixed but much higher frequency, and has higher amplitude when the sound signal is stronger. In the receiver, the fixed frequency is removed and the wave amplitude is recovered and magnified to be heard. In the United States, the frequencies assigned to AM broadcasting are in the range between 520 and 1610 kHz. The receiver needs an antenna to intercept as much of the radio signal as possible, followed by a tuner to filter out all other frequencies and to retain only the frequency of interest; this is followed by a detector to recover the signal, an amplifier to increase the signal

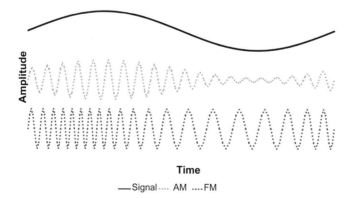

Time

—— Signal ···· AM ····FM

FIGURE 7.4 Sound by amplitude modulation and by frequency modulation.

strength, and a speaker to produce the sound. Amplitude modulation has low fidelity and is subjected to interference from random atmospheric and electrical background noises, and is now mainly used for speech.

The method of frequency modulation or FM was introduced by Edwin Armstrong in 1933. He used a signal system of fixed amplitude, but the frequency was made variable according to the sound wave signal. As a result, a broadcast station has to be assigned a frequency range, instead of a single frequency. FM is much less affected than AM by random noises in the environment, and gives much better music fidelity. In the United States, the frequencies assigned to FM are much higher and take up more frequencies, ranging from 87.7 to 108.1 MHz. Speech and sound carry much more information and intimacy than the dot-dash of the radiotelegraph, and multimedia communications have become revolutionary in the way we communicate. Franklin D. Roosevelt was president during the Great Depression, and inaugurated a series of 30 evening radio speeches between 1933 and 1944, known as his Fireside Chats. These Chats served to inform and assure the public directly of Roosevelt's views on the falling economy and the coming war, increased his bond and direct communication with the public, and were effective in pressuring conservative senators and congressmen to pass his New Deal measures.

7.3.4   Motion

After the simultaneous transmission of pictures and sound had been achieved, the challenge of the transmission of motion required a solution to the problem of synchronization, so that when a person opened the mouth, speech would come out at the same time. This is similar to the challenge of the "talkies" in motion pictures.

The basis of recording and transmitting a picture is scanning one line after another from the top to the bottom of a page, comparable to a farm tractor plowing a field. The scanning converts the picture into many horizontal lines of variable shades, which are then transmitted consecutively. The facsimile machine of the late 1880s was the precursor to television, which sent only a single picture. Paul Niepkow of Germany invented a mechanical television, with a mechanical system of scanning by perforating a spiral of holes in a spinning disk, then the light that goes through the holes hits a selenium sensor that is sensitive to light. The mechanical televisions never became popular, and were totally replaced by the electronic television, which became practical with the invention of the vacuum tubes.

Philo Farnsworth (1906–1971) made the first working television, which he publicly demonstrated to news media on September 1, 1928. He was born to a Mormon family in Utah, and later moved to Idaho. Farnsworth developed an early interest in electronics, and excelled in chemistry and physics at high school. He discovered that he could use a cathode ray tube (CRT) to generate an electrical television signal without a spinning mechanical device. The CRT is a vacuum tube where a beam of electrons from a hot cathode (negatively charged) is accelerated by voltage from an anode (positively charged) toward a screen equipped with a fluorescent coating that emits light when struck by an electron. The CRT also has a set of electric fields to deflect the electron beam from left to right so as to scan a single horizontal line, and then to move down a small distance to scan the second line, and so forth. He needed a system to convert an image into a beam of electrons, so he used an "image dissector," which focuses a photographic image onto a layer of cesium oxide that emits electrons proportional to the intensity of the light. The image dissector is not very efficient, because the bulk of the electrons produced are discarded, so a very bright light is needed.

In 1930 Vladimir Zworykin was recruited by David Sarnoff of RCA, and visited Farnsworth's laboratory when he copied the dissector. David Sarnoff tried to recruit Farnsworth as well, who declined the offer and joined Philco in Philadelphia instead. Vladimir Zworykin invented the "iconoscope," which is superior to the "image dissector" as it has photosensitive material to capture all the electrons. The commercial broadcast of electronic television began in 1936 in Germany, the United Kingdom, and the United States. A lawsuit over infringement between Zworykin and Farnsworth was resolved in 1939, when RCA agreed to pay Farnsworth 1 million dollars over 10 years, in addition to license payments to use Farnsworth's patents. Farnsworth left Philco in 1934, and died in Utah in 1971. The commercial broadcast of color television began in the United States in 1950, and required three guns for the colors red, green, and blue.

The dominance of the CRT gave way to advances in the LCD, or liquid crystal display, which can be much thinner. In this system, millions of LCD elements are arranged in a grid, and each act as a shutter that can be opened or closed to control light coming through, plus a filter to produce red, green, and blue lights. The shutters are made of liquid crystals to create polarized filters, and the alignment of two polarized filters determine whether light comes through or not. The advantage of the LCD is in a thinner and lighter display; however it is not very efficient, as most of the light generated in the screen is blocked by the polarized light filters. A more advanced method is the LED, or light emitting diodes. They are semiconductor diode devices that produce colored light directly without filters, and have the advantage of very low energy consumption and long lifetime.

The closed-circuit television is a point-to-point transmission of sight and sound for only the intended audiences and not for broadcast. It is particularly important for work in the military, crime control, banks, casinos and airports for surveillance, and to monitor factory assembly lines for quality control.

7.4 INFORMATION TOOLS

The great advances in information acquisition, storage and communication have depended on many tools that were developed: some of them are "hardware" or devices, and others are "software" or methods.

7.4.1 Digital Storage and Information Theory

The earlier methods of information storage have been described as analog, or nondigital. Texts were stored as inked lines on paper, pictures were stored in granular salt crystals in emulsions, and sound as grooves in tin foil. Digital methods of storage have captured a major share of the market, due to its versatility and ease in manipulation. It has turned out that a string of 0s and 1s is the most convenient way to store information, in a number system based on 2, or binary arithmetic.

First we convert all the decimal numbers into binary numbers. We represent the first few numbers as

Decimal	0	1	2	3	4	5	6	7
Binary	0	1	10	11	100	101	110	111
Decimal	8	9	10	11	12	13	14	15
Binary	1000	1001	1010	1011	1100	1101	1110	1111

Thus, the first digit on the right stands for 1, the second digit from the right stands for 2, the third digit stands for 4, and so forth. A string of two-digit can code up to 4 numbers, from 0 to 3; a string of three-digit can code up to 8 numbers, and a string of four-digit can code up to 16 numbers. In general, an n-digit string can code up to 2^n numbers, therefore a number as high as 2^n is regarded as n bits of information.

Bits	1	2	3	4	5	6	7	8
Numbers	2	4	8	16	32	64	128	256
Bits	9	10	11	12	13	14	15	16
Numbers	512	1024	2048	4096	8192	16384	32768	65536

For the human operators, it is often more convenient and compact to write in the hexadecimal system where each number from 0 to 15 is represented as a single symbol; it is the convention to use $A = 10$, $B = 11$, $C = 12$, $D = 13$, $E = 14$ and $F = 15$. For instance the number 15 in the decimal system is 111 in the binary system and is F in the hexadecimal system.

How does a computer store text information? The simplest method today for English is the ASCII method (American Standard Code for Information Interchange), which uses 7-bits to store the 128 most commonly used "glyph" such as

Number	0	1	2	3	4	5	6	7	8	9
ASCII	48	49	50	51	52	53	54	55	56	57
Letter	A	B	C	D	E	F	G	H	I	J
ASCII	65	66	67	68	69	70	71	72	73	74
Letter	a	b	c	d	e	f	g	h	i	j
ASCII	97	98	99	100	101	102	103	104	105	106

So a string of 8 or an 8-bit (also called a byte) can be used to code $2^8 = 256$ characters: more than enough for the 10 Arabic numerals, plus the 26 lower case letters and the 26 upper case letters, plus a number of symbols such as ?, @, #, and so on. To encode all the characters of the world's living languages would require a much larger code, such as the Unicode Transformation Format, called UTF-16, which is the standard used for Windows. This 16-bit word will code $2^{16} = 65,536$ different characters. It has room for all the Modern European, Greek, Cyrillic,

Arabic, Hebrew, Cuneiform from Mesopotamia, and Linear B from Crete. For instance ✳ is 00E8, • is 00E9, ● is 00EA, and ○ is 00EB. The largest fraction of available character space is taken by the CJK, or Chinese–Japanese–Korean characters. In this UTF system, each character requires 16 bits, or two bytes. A text with a thousand words, each with an average of five characters, would require 80,000 bits or 10 KB of memory for coding.

It is said that a picture is worth a thousand words. What is the information content and the storage capacity needed for a picture, and how do they compare to a thousand words? That certainly depends on the size of the picture, the resolution or detail, the number of shades from white through gray to black, and whether it is in black-and-white or full-color. We are in a digital age where all the information of text, sound, pictures, and movies are stored in binary digits of 0 and 1, and we can readily compare the requirements of storing different types of information.

How do we digitize an 8 in. × 10 in. (203 mm × 254 mm) photo for storage? We divide it into a rectangular grid of dots or pixels, where the resolution is determined by the number of dots per inch (dpi). A coarse-grained picture may have 72 dpi (2.8 dots/mm), and a fine-grained picture may have 300 dpi (11.8 dots/mm). If we have an 8 in. × 10 in. photo at a medium resolution of 150 dpi, then the total number of pixels would be calculated as $1200 \times 1500 = 1.8$ million pixels. Next we consider the depth, or the dynamic range from pure black through various shades of gray, to pure white. In 8-bit coding, black is coded as 0, white is coded as 255, and a medium gray is 127. If this same image is a color photo, each pixel has to be coded for the three colors of red–green–blue, so there are 5.4 million bits of information. Therefore, the memory required for this photo is now 43.2 million bits, or 5.4 MB. In comparison with the 1B needed for each character in ASCII and 2B for UTF, a thousand words of five characters each would indeed be equivalent to a picture. Fortunately for managing these millions of bit, we have a technology of data compression, which gives us the flexibility to give up some detail to reduce the memory required, by as much as a factor of 10.

A picture for digital storage and analysis is divided into numerous space intervals, but the digital storage of sound involves dividing a continuous sound wave into a number of discrete time intervals, and measuring the strength of the signal only at the time intervals (see Fig. 7.5). The quality of the digitization depends on the length of time intervals between samplings (sampling rate) as well as the volume or grades of intensity (depth). Since the audible frequency of the human ear is about 20–40,000 Hz, the lower notes do not need a fast sampling rate, but the higher notes need fast sampling rates. The rule is that the sampling rate should be at least twice the frequency that you want to hear clearly. A rate of 8 kbit/s is considered good enough for telephone conversation, and 96 is adequate for FM radio quality, but 224 is for CD or the highest quality of recording. If the sampling rate is 96,000 per second and the depth is 16 bits, then 1 min of recording would require 100 million bits, or 12 MB.

Naturally, there has been demand to store more information in a given storage space, which also speeds up the transmission rate over the Internet. It turns out that most information coding methods are not very efficient, so there is a significant amount of redundancy or fat that can be squeezed out without significant loss of

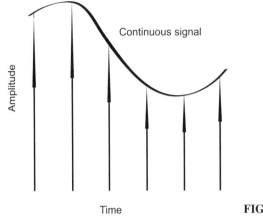

Continuous signal

Amplitude

Time

FIGURE 7.5 Digital signal.

fidelity, which is called lossless data compression. This is the method favored by medical applications and scientific research where fine distinctions can determine the outcome of important questions. The problem of the information content of a message, and thus how much compression you can make without loss, was solved by Claude Shannon in 1948 with the establishment of the subject of Information Theory. Shannon established a limit, and the best that a computer programmer can do is to come close to that limit. It is also possible to compress the data further and lose more fidelity, while still leaving enough detail for enjoyment; this is called lossey data compression, which can achieve a compression of greater than 10 to 1. This is favored for entertainment and amateur photography, where fine distinctions are of less importance. Some examples of the current methods are

Compression	Image	Audio	Video
Lossless	GIF, TIFF	WMA-lossless	CorePNG
Lossey	JPG	MP3	MPEG

The storage of digital information can take place on a number of media, such as writing 0 and 1 on a piece of paper, which is not easy for retrieval by a computer sensor. There have been many developments to increase reading speed and storage density, including punched cards, magnetic wires, magnetic tapes, electronic tubes, transistors, and optical disks. The older 3.5 in. magnetic floppy disks had a storage capacity of 1.4 MB. A standard optical CD has the storage capacity of around 700 MB, so you can store 432 small photos or 27 large photos. The optical disk is a thin polycarbonate disk with a layer of aluminum on top. The data is imprinted on the aluminum layer by a laser beam to create a number of "pits" of dots and dashes. In the early versions, each pit was 100 nm deep by 500 nm wide, with a length from 850 to 3500 nm. The signals are read by a 780 nm wavelength semiconductor laser through the bottom of the layer. The newer Blu-ray Disk can hold 25 gigabytes, and uses a blue laser beam with a shorter wavelength, and consequently a higher resolution. If you want to send a photo to a friend over a slow modem with a speed of

14.4 kbps or 14.4 thousand bits per second, the smaller photo will take almost 2 min, but the larger photo will take 30 min. Fortunately, there are faster connections, such as an ADSL line that has a message capacity of greater than 1 Mbps, or one million bits per second, so even the large photo would take only 26 s.

It is possible to take a message and determine its information content quantitatively in terms of a unit called *entropy*, which can be considered as the minimum number of yes–no questions that need answers for the complete information. Shannon's modern mathematical information theory had a profound influence on messaging, communications, encoding, cryptography, and many other topics. Information theory quantifies the basic information needed for the efficient allocation of storage capacity and communication channel capacity, as well as for the efficiency of packaging for compression.

When you toss a fair coin, the outcome is either heads or tails with equal probability, and the information content is an entropy of $H = \log(2) = 1$. Information theory uses only the logarithm based on 2, with the following equivalences: $\log_2(x) = 3.3222^*\log_{10}(x)$. So the outcome of the coin toss is settled after one question, "Is it heads?" If the coin is biased, so that the probability of getting heads is p, and tail is $(1 - p)$, then the entropy is $H = -p^*\log(p) - (1 - p)^*\log(1 - p)$, which is less than a fair coin of 1, and more than a biased coin with two heads where you already know the answer before any toss.

If you pull out a letter randomly from a jar containing the four letters A, B, C, and D, we can determine the outcome by asking only two questions. The first question should be, "Is it A or B?" If the answer to the first question is yes, the second follow-up question would be, "Is it A?" If the answer to the first question is no, the second question should be, "Is it C?" Therefore, the entropy is 2 for 4 equally probable outcomes. In general if we have n equally probable outcomes, we have the entropy given by $H = \log_2(n)$.

n	1	2	4	8	16	32	64	128	256
H	0	1	2	3	4	5	6	7	8

For a fair dice, where there are six equally likely outcomes from 1 to 6, the entropy is between 2 and 3, and the formula gives $H = 2.585$. This reflects the need to ask between two and three yes–no questions, depending on your luck. You may ask first, "Is it 1 or 2?" and if the answer is yes, you need to ask only one more question; but if the answer is no, you have four outcomes left and you have to ask two more questions. In a game of "Twenty Questions," you can ask no more than 20 yes–no questions while trying to arrive at the identity of an object that can be animal, vegetable, or mineral. You can, in principle, settle a question that has an entropy of 20, so the possible number of outcomes is $n = 2^{20} = 1,048,576$.

What is the information content of the Declaration of Independence?

> *When in the Course of human events, it becomes necessary for one people to dissolve the political bands which have connected them with another, and to assume among the powers of the earth, the separate and equal station to which the Laws of Nature and of Nature's God entitle them, a decent respect to the opinions of mankind requires that they should declare the causes which impel them to the separation.*

The first paragraph has 71 words, 336 characters, and 70 spaces between words. An English word is made from a 26-letter alphabet, so each character can carry $H(26) = 4.701$ bits of information, thus the entire paragraph can have an entropy of $4.701 \times 336 = 1579$ bits of information. However, the letter E is the most frequently used letter in the English language, and the letter Z is the least used. The frequency of letter usage in everyday English has been measured many times, and they are summarized below:

E	T	A	O	I	N	S	H	R	D	L	C	U
0.127	0.091	0.082	0.075	0.070	0.068	0.063	0.061	0.060	0.043	0.040	0.028	0.028
M	W	F	G	Y	P	B	V	K	J	X	Q	Z
0.024	0.024	0.022	0.020	0.019	0.015	0.010	0.010	0.007	0.002	0.002	0.001	0.001

For each 1000 letters used in the common English language, E is used 127 times and Z is used less than once. There is a better than 51% probability that a random letter grabbed from all 26 would be among the top six of {E T A O I N}. Using the customary frequencies of the letters, and the Shannon formula of

$$H(X) = - \sum_{i=1}^{n} p_i \log_2(p_i)$$

we arrive at the result that a single English letter has entropy of 4.129, which is 88% of the maximum entropy of 4.701 for uniform frequency. However, two consecutive letters are not necessarily independent of each other; for instance, the letter Q is always followed by the letter U in English, unless we encounter a foreign name. If you are shown a sequence of letters taken from an English text, such as {*When in the Cours**}, then a dictionary will show that the next letter can only be chosen from the set {e, i}. So the entropy of this character is much reduced. The five vowels, AEIOU, have a collective probability of 38.1% in the English language. But out of 44 presidents of the United States, only 5 of them have last names that start with a vowel, or an 11.4% probability: John Adams, John Quincy Adams, Chester Arthur, Dwight D. Eisenhower, and Barack H. Obama.

When you have a message to decipher, you need to consider the context of the message before you use the ETAOIN frequencies. The information content of a letter in a sentence depends on the all the letters that come before, but it is often on the order of 1.0 to 1.5 bits per letter. This is to say that the English language is more than 73% redundant, and that it is possible to write it much more compactly. However, some redundancy is often a good thing, as it can become the key to recovering a partly damaged message.

7.4.2   Secret Coding: Cryptography

One if by land, and two if by sea, wrote Henry Wadsworth Longfellow about the midnight ride of Paul Revere and the signal Revere had arranged with his friends to hang lanterns in the Old North Church if the British began an invasion in 1775.

How do we send a coded message to an intended recipient, which would be cryptic to an interceptor? There has to be an agreement ahead of time on the code, on which signal means what message. When Theseus sailed to Crete to meet with King Minos and the Minotaur, he promised his father that if he was successful, he would return to Athens with a white sail; but if he failed, his shipmates would raise a black sail. He was successful but forgot to replace his black sail with a white sail, which caused the despair and death of his father who watched from the shore. The smoke signal was on the Great Wall by the Chinese for 2000 years to warn the capital that enemies were coming, and the smoke from one beacon to another beacon was faster than runners or horsemen. The Native Americans often sent signals by shiny objects and by smoke. Naval ships signal each other with flags, and the International Code of Signals assigns a specific flag to each letter.

Sending secret messages can have many problems. Consider the following: A sender transmits an encoded message for one or more intended receivers. During transmission, the message becomes corrupted by noise, or may be garbled, or the message gets partly erased so that only a fragment remains. To ensure a successful message, although the sender would like to send the message in the shortest form, it should have some redundancy so a slightly corrupted message can still be read. Another solution would be for the sender to conceal the message, or encrypt the message with a key, which only the intended receiver has. This key is necessary to decipher the message, and a hostile spy without the key would not be able to do so.

There is a very long history of sending and intercepting encrypted messages. The simplest method is the Shift Cipher, where each letter in the text is substituted with another letter by a shift in the order in the alphabet. The Caesar Cipher was used by Julius Caesar during a siege in the Gallic War, where each letter in the message is shifted by three places in alphabetical order, such as

Order	1	2	3	4	5	6	7	8	9	10	11	12	13
TEXT	A	B	C	D	E	F	G	H	I	J	K	L	M
Cipher	d	e	f	g	h	i	j	k	l	m	n	o	p
Order	14	15	16	17	18	19	20	21	22	23	24	25	26
TEXT	N	O	P	Q	R	S	T	U	V	W	X	Y	Z
Cipher	q	r	s	t	u	v	w	x	y	z	a	b	c

We can see that A = 1 is shifted to d = 1, and B = 2 is shifted to e = 2, and W = 23 is shifted to z = 26. When X = 24 is shifted to 27 beyond the range of 26, we wrap it back to a = 24. This is call modulo mathematics based on 26, where 27 = 1, 28 = 2, and 29 = 3. A message such as "ATTACKATDAWN" is encrypted as "dwwdfndwgdzq." The intended recipient knows the key is 3, and can decode and reverse the encrypted message into the original message. It is more secure to delete the spaces between words, as they call attention to short words such as "I" and "at" that are easier to identify. The shift cipher is a very simple encryption, so it would not be difficult to break the code, especially if the spy knows that this is a shift cipher written in English, but does not know the key $n = 3$. One can break the code by the exhaustive attack, first trying $n = 1$, then $n = 2$, to exhaust all 26 different shifts and see which of the results make sense. But the message may be coded in another language such as Spanish or a truly obscure language like Navajo; then this cipher would require the much harder double-exhaustive attack on both key value and language type. The Navajo Code Talkers were used by the US Marines to send secret messages in the Navajo language over radio and telephone in World War II. All native speakers of Navajo are from the United States and there is no written Navajo literature, so the Japanese navy could not break the code, as they had no knowledge of the Navajo language.

A more sophisticated method is the Substitution Cipher, where each letter is replaced by an arbitrary letter or symbol. The key now consists of 26 pieces of information on each letter, instead of a single number. Mary Queen of Scots was imprisoned by Queen Elizabeth, and she was involved in the "Babington Plot" to assassinate Elizabeth. Messages between Mary and Babington were encoded by substitution letters and smuggled in beer barrel stoppers. The messages were intercepted and decoded based on frequency analysis. Six conspirators were hanged, drawn, and quartered. Mary was beheaded in 1587.

Edgar Allen Poe wrote a short story, *The Gold-Bug*, about the pirate treasure of Captain Kidd in 1843, featuring a substitution cipher that was solved by letter frequency. In the *Return of Sherlock Holmes*, written in 1905 by Arthur Conan Doyle, there is a story about the Adventure of the Dancing Men, which is an encryption of childish looking dancing figures, where each letter is replaced by one of these dancing men. The first task to break this code is to compile a list of the number of different dancing figures, and see if it is only a few dozen, which would means that it is an alphabetical language. If the list is hundreds or thousands, we may have an intractable nonalphabetical language such as Chinese. In this Sherlock Holmes story, it turns out that the language is English with 26 letters. The first symbol in this message can be a letter from A to Z, so if we assume that it is A, the second symbol can be a letter from B to Z. Since there are $52! = 8 \times 10^{67}$ possible permutations of 26 letters, an all out exhaustion attack would take too long to accomplish. Sherlock Holmes was able to crack the code by using the letter frequency of the normal English language. He waited till he had collected enough messages of this form sufficient to do some statistics, and decided that the figure that appeared most frequently had to be the letter E, and the next most frequent letters must be from the set of {T A O I N}. After some trial and error, he cracked the code and caught the bad guy.

The German Enigma machine used in World War II employed a wheel with the 26 letters that could be set at arbitrary positions to do substitution cipher, and the settings could be changed every day. The Germans never sent long messages to prevent deciphering through the letter frequency method. Eventually, their more fiendish invention was to use three wheels in sequence, and ultimately eight wheels. The cracking of this complex method was done through the work of Alan Turing and involved the invention of modern computers, which will be described later.

An even more sophisticated encryption method is the Vigenere Cipher, which involves a keyword such as "GAUL." As $G = 7$, $A = 1$, $U = 21$, and $L = 12$ in the order of the alphabet, we take a plain message and shift the first letter by 7, the second letter by 1, the third letter by 21, and the fourth letter by 12; then we repeat by shifting the fifth letter by 7, and so on.

Text	I	L	O	V	E	Y	O	U
Order	9	12	15	22	5	25	15	21
Add	7	1	21	12	7	1	21	12
Shift	16	13	36	34	12	26	36	33
Cipher	p	m	j	h	l	z	j	g

When we shift O at 15 by 21 and get 36, it is equivalent to $36 - 26 = 10$ that is the letter j. If you know the key is the four-letter word "GAUL," you can easily decode and reverse the procedure and obtain the original text. But if you know only that $n = 4$, but not the keyword, an exhaustive attack would mean not simply 26 guesses, but the binomial number $(26, 4) = 14,950$ trials. Even letter frequencies would not be that helpful, as the most frequent letter E would be mapped in turns to L, F, Z, or Q.

We mention here another form of encryption called the Permutation Cipher, where the letters are not changed but their positions in a sentence is changed. For instance if we use the key of routing:

Plain	1	2	3	4
Cipher	2	4	1	3

then a message can be permuted according to the key such as

Plain	I	A	M	G	O	I	N	G	F	O	R	A	W	A	L	K
Key	2	4	1	3	2	4	1	3	2	4	1	3	2	4	1	3
Cipher	m	i	g	a	n	o	g	i	r	f	a	o	l	w	k	a

To break this code, you need to discover both the length of the key and its permutation.

Currently, the most difficult code to break is the RSA code that stands for the inventors Rivest, Samir, and Adleman. It was invented in 1978 based on the multiplication of very large prime numbers. Most numbers have divisors such as $12 = 2 \times 2 \times 3$. A prime number does not have any divisor, such as 3, 7, and 13. There are 5 primes smaller than 10 (1, 2, 3, 5, 7), which is 50% of numbers smaller than 10; there are 26 primes smaller than 100, which is 26%; there are 168 primes smaller than 1000, which is 16.8%; and there are 78,498 primes smaller than a million, which is 7.8%. A google is 10^{100}, and 0.43% of the numbers less than a google are primes. In the RSA system, one chooses two prime numbers p and q, preferably more than 100 digits each, so their product $n = p \times q$ would have two hundred digits! After some computation, these two numbers give rise to two numbers: the "private key" d, known only to the intended recipient, and the "public key" e, which is published. A message is encoded with the public e key and sent, and the intended receiver will use the private key d to decipher and recover the message. A spy does not know the private key, which can be obtained if he/she can "factor" n into the product of multiplying p with q, which is a notoriously difficult computational problem. Many of the commercial and government documents now use the RSA coding method, and numerous hackers have sought methods to crack this challenging code, which is helped by ever faster computers. This is an ongoing race between the coders and the hackers, and the coders can increase the difficulty of cracking the code by increasing the prime numbers p and q to 200 digits each, or more!

The breaking of a code can changed the course of history. During World War II, the German Enigma machine was broken by the British with the help of code breakers at Bletchley Park with fast computers. After the code was broken, the British knew the deployment of German submarine packs sent to intercept British shipments of food and weapons from across the Atlantic, and were able to organize large convoys protected by warships to avoid crippling losses. This led to the defeat of the U-boats in the Battle of the Atlantic.

In 1942, the Japanese navy had a string of victories in the Pacific, and had planned to lure the remaining US aircraft carriers into a trap, and to occupy the Midway Atoll. The Japanese Naval Purple Code was broken by the Americans who then knew about the invasion, when they were coming, and that the numerically superior Japanese navy had been divided into four task forces so the four carriers had only limited escort. The United States set up an ambush of its own, the US Navy stopped the invasion and sank four Japanese aircraft carriers and a heavy cruiser, which decisively weakened the Japanese Navy. The Battle of Midway was called "the turning point" of the War in the Pacific. In 1943, when the Japanese commander of the Imperial Navy, Admiral Yamamoto, visited the Solomon Islands, the detailed arrival time and location, as well as the size of the squadron (two bombers and six fighters) had already been deciphered. President Roosevelt ordered the navy to "Get Yamamoto." Eighteen fighters were assigned for this mission to kill Yamamoto, and knew precisely when and where to find him. Both bombers were shot down by US fighters, and he was killed.

7.4.3 Electronics

Information needs to be received, processed to derive useful information, stored and retrieved, and communicated to others. With digital information, electronic signals may need magnification when they are received from a great distance, and when they should be presented by public address to a large audience. When there is a large volume of signals and symbols to handle, it becomes necessary to have information processing machines to handle them. Modern information equipment, computers, and networks are built on electronic components and devices.

The beginning of electronics was based on the observation that a hot metal filament has the tendency to eject electrons, especially toward a positively charged anode, in a phenomenon that is called "thermionic emission of electrons." When the filament is placed in air, it has a tendency to burn out after a while, and the ejected electrons meet resistance from air molecules. When the hot filament is encased in a vacuum tube, it is called the cathode tube and can have a longer life, and the electrons flow as a straight line to the anode, which is a cool metallic plate. John Ambrose Fleming invented the first vacuum tube in 1904, which is now called the diode. Electrons can pass from the heated filament cathode to the cool plate anode, but the flow of electrons is not reversed when the voltage difference is reversed, as the plate is not heated and cannot emit electrons. Thus, the diode can be used for a number of functions, such as "rectification" to convert alternating currents AC into direct currents DC. It can also be used as a logic gate to do logical and arithmetic operations.

The more elaborate triode was invented by Lee de Forest in 1907, who added a wire grid between the cathode and the anode. The flow of electrons from the negatively charged cathode to the positively charged anode can be controlled by adjusting the voltage on the grid. When the grid is negatively charged, some of the electron flow from the filament would be repelled and diminished; a positive charge on the grid would lead to a higher flow. So if an alternating signal is placed on the grid, the flow of electrons from the cathode to the anode would be an amplified alternating signal. The triode found usage in switching a signal on or off, as well as in amplifying signals in music for microphones and loudspeakers, and in the long distance transmission of signals. The initial applications were driven by military and government needs, followed by consumer needs when simpler devices that cost much less became available. In 1944, the ENIAC computer operated with 17,000 vacuum tubes and weighed 27,000 kg; the electric power requirements and periodic replacement of burnt tubes made it unsuitable for many practical applications. A great incentive for miniaturization was created.

The semiconductor arrived in 1947, and was based on entirely different principles than hot wires emitting electrons in vacuum tubes. The electrical properties of most materials can be classified as either conductors like copper, or as insulators like rubber. There is a class of elements with conductivities that depend on impurities and external conditions. The middle of the periodic table has these elements:

Group	III	IV	V
First row	Boron	Carbon	Nitrogen
Second row	Aluminum	Silicon	Phosphorous
Third row	Gallium	Germanium	Arsenic
Fourth row	Indium	Tin	Antimony

The group IV elements carbon, silicon, and germanium are insulators when they are very pure, but they become conductive when they contain a small amount of impurities. Germanium with contaminants from group V are donors of electrons, which become mobile and carry electricity; but contaminants from group III are donors of holes or deficiencies of electrons, which can also be mobile and carry electricity.

The invention of the transistor was made at Bell Laboratories by John Bardeen, William Shockley, and Walter Brattain in 1947, which was based on the element germanium. Bardeen and Shockley added small amounts of antimony to germanium at the level of 10 ppm, and discovered that the conductivity of germanium increased 200 times. This is called "doping," and the addition of electron-rich group V elements give the N-type semiconductor, and the addition of electron-poor group III elements give the P-type semiconductor. The conductivity is also very dependent on the surrounding electrical field, which gives rise to the method called "field effect" to control conductivity. The team also discovered the point-contact method, by using "cat's whisker," which is a wire contact touching the germanium surface that injects electrons or holes, and thus controls electron flow. It is now possible to do most of the work formerly done by the vacuum tube with the transistor, which is much smaller in size and weight, uses lower voltage and power, can be kept at room temperature, has a longer life, and can be mass produced in a factory.

The transistor is often mentioned as the greatest invention of the twentieth century, for which Bardeen, Shockley, and Brattain won the 1956 Nobel Prize in Physics. In 1972, Bardeen became the only person to win two Nobel Prizes in Physics when he shared the prize with Cooper and Schrieffer for a theory on superconductivity. The first silicon transistor was the next important invention, made by Gordon Teal in 1954, based on the inexhaustible and inexpensive element of silicon. This was the beginning of the phenomenon called "Silicon Valley," of start-up companies creating irresistible products and young pioneers becoming billionaires.

The task of amplifying sound needs an electronic circuit, which consists of vacuum tubes or transistors connected to the input and the output signals, as well as many components including resistors, inductors, and capacitors. Resistors are passive elements that diminish electric flow, and are often made of carbon powder mixed with ceramic powder and resin; capacitors are also passive elements that store electricity, and are made of metal plates sandwiched with a dielectric insulator such as paper and glass; and the connecting wires are made of copper. The manufacturing of an electronic circuit requires the assembly of hundreds of thousands of tubes or transistors as well as passive elements, which requires a great deal of hand labor and time, and is subject to errors in an assembly line.

The integrated circuit, or IC, was invented independently in 1958 by Jack Kilby of Texas Instruments based on germanium, and by Robert Noyce of Fairchild Semiconductor based on silicon. Instead of using copper as conductors and ceramics as insulators, the integrated circuit is made entirely of a single block or chip of silicon that has different regions doped by different N- or P-type material. The electronic circuit is designed on a piece of paper, and then photolithographed or printed onto the silicon surface like a photograph. The idea was a compromise: silicon is not as good a conductor as copper, and not as bad an insulator as ceramic, but the benefits of integration were so great that it overcame these objections. Speed increased and power consumption dropped since the elements are very small and are located very near each other. A chip can be a few square millimeters in area, with up to a million transistors per square millimeter. Robert Noyce went on to found the Intel Corporation together with Gordon Moore, but Noyce died in 1990. Jack Kilby received the 2000 Nobel Prize in Physics, and he noted the crucial role of Robert Noyce in his acceptance speech.

VLSI is very large-scale integration, which combines thousands of transistors into a single chip. The microprocessor incorporates most or all of the functions of a central processing unit (CPU) in a computer, placed on a single integrated circuit. They were first made by Noyce and Moore in 1971 for electronic calculators. Forty years later, the company Nvidia has a computer that uses 1.4 billion transistors for logic. Progress was so rapid that it inspired Moore's Law: that the long-term trend of progress in electronics is to double the capabilities of electronic devices every 18–24 months, including processing speed, memory capacity, and cost per unit of memory. There are many forecasts that there is a limit to Moore's Law, and that sooner or later this torrid pace will have to slow down.

7.4.4   Computers

The word computer was originally used to describe a simple device used for doing the arithmetic of addition and multiplication. The word has evolved to mean a

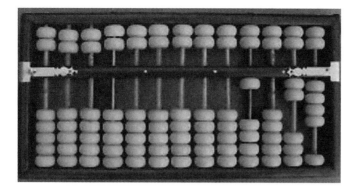

FIGURE 7.6   Abacus.

complex machinery that fetches input information, processes the information according to instructions or a program, stores important information, delivers output information, and takes appropriate actions. In many ways, the modern computer can be compared to the central nervous system of the human spinal cord, which receives information from the skin and muscles and sends out simple reflex motor commands; and the brain for the more complex functions of analysis and actions. Computers are now involved in searching for information from around the globe, in analyzing and reorganizing information, in storing and communicating information, and in games and entertainment.

The abacus was used in Sumer as early as 2700 BCE for addition and subtraction, and was also used in Egypt, Iran, and Greece (see Fig. 7.6). The number shown in this figure is 1024, represented as four beads moved up to the horizontal divider on the first groove from the right, two beads on the second groove, and one bead on the fourth groove. Each bead above the horizontal divider counts as five. The second groove to the left counts as 10, and the third groove counts as 100, and so forth. The abacus is particularly good for addition and subtraction. When it appeared in China around 1400, the counters were pierced wooden beads that slid along wood dowels or metallic wires. John Napier was responsible for both the calculating rods known as "Napier bones" and for the slide rule, which is based on his observation of the logarithm of numbers in 1614. The formula $\log(a * b) = \log(a) + \log(b)$ converts the more difficult operations of multiplication and division into the easier operations of addition and subtraction. In the generation of scientists and engineers that worked up to the 1970s, the slide rule was the most important workhorse in computation, as well as the symbol of the profession.

The modern electronic digital computer is the result of an immense series of discoveries, developments, and detours by a creative sequence of scientists and engineers, with perhaps many more exciting decades of improvements yet to come. There are many inventors, and it is possible only to point out some of the most outstanding pioneers and their contributions to this mighty forest. The modern desktop or laptop computers can be used for addition and multiplication, but they have become *universal information machines*, where the information can be numbers, digitized words, pictures, sounds, and motion.

The machine control of mechanical operations had a strong start in 1801 when Joseph-Marie Jacquard invented a loom for weaving carpets. It was controlled by punched cards, which marked the beginning of programming or a set of instructions on what operations a machine should do. In 1880, Herman Hollerith used punched cards to process the large quantity of information generated by the US population census, and he was reportedly inspired by seeing a railroad train operator punching tickets to record the passenger characteristics. The Swiss music box has a rotating cylinder with programmed pins, which pluck a row of metal strips to play the tune. The player piano is an automatic piano playing machine controlled by a roll of paper with perforations.

There were many computer developments in the twentieth century. The Atanasoff-Berry computer was among the first electronic digital binary computing devices, built in 1937 by professors from Iowa State College. Konrad Zuse invented a program-controlled computer in 1941, and employed magnetic storage in 1955. The British built the Colossus computer in 1943 to break the German wartime codes.

John von Neumann (1903–1957) was said to be the most brilliant person in his generation. His picture is shown in Figure 7.7. It is said that, "He was indeed from Mars, but he had made a thorough detailed study of human beings, and could

FIGURE 7.7 John von Neumann.

imitate them perfectly." He was a member of the generation of brilliant Hungarian scientists including Theodore von Karman, Leo Szilard, Eugene Wigner, Edward Teller, and Michael Polanyi. When Wigner was asked why there were so many Hungarian geniuses, he said that he did not understand the question, as there was only one genius—John von Neumann. He was relaxed and treated his creativity with humor; he was also friendly and not condescending, had a sense of humor and liked parties. Laura Fermi said, "Dr. von Neumann is one of the very few men about whom I have not heard a single critical remark. It is astonishing that so much equanimity and so much intelligence could be concentrated in a man of not extraordinary appearance." He never insulted people who were less brilliant, and never engaged in political debates. He had a perfect memory for mathematics, but he could not remember faces. A graduate student explained, "When Dr. von Neumann came out of a room, he would be besieged by groups of people who were stuck by some calculations. He would walk down the corridor with these people around him. By the time he walked into a door for the next meeting, he would likely have suggested either the answer or the best shortcut to get them."

He was born in 1903 in Budapest, Hungary, to Max and Margaret Neumann, who were nonpracticing Jews. He was named Neumann Jancsi, and he was the oldest of three brothers. His father was a very prosperous banker and influential in politics; the family had servants, nursemaids, governesses, private tutors for five foreign languages, as well as fencing masters and piano teachers. His father also became a prominent economic advisor to the government, and was elevated to hereditary nobility in 1913, so he called himself von Neumann in German.

In 1914, he entered the Lutheran Gymnasium high school, a year behind Eugene Wigner who would later receive the Nobel Prize in Physics. The mathematics teacher thought that high school would be a waste of time for the young pupil, so he was turned over to tutors from the University. He started publishing research papers in mathematics, but continued to attend classes with all the other students. His grades were excellent in everything except handwriting, physical exercise, and music. A career in science for the insecure Jewish minority was desirable as it could be practiced anywhere in Europe; it was more secure than banking, which depended on the local politics that could turn against Jewish minorities at any moment. In the postwar turmoil, the family bank suffered during the Communist government of Bela Kun. A year earlier, Eugene Wigner had been sent to the University of Berlin to study chemical engineering because it was considered to be a more useful and secure profession. Then, von Neumann decided to carry on his undergraduate and graduate education simultaneously in two distinct disciplines and in three cities several hundred miles apart. He completed his doctorate in mathematics at Budapest, and his chemical engineering degree at Berlin and later at the ETH at Zurich.

In 1930, he married Mariette Kovesi who was from a rich family that had converted to Catholicism. He published a landmark book, *Mathematical Foundations of Quantum Mechanics*, in 1932 in German. He came to America to join the mathematics department at Princeton, and then joined the newly founded Institute for Advanced Study in 1933, with a salary of $10,000 per year, which was very high at that time. He divorced in 1937, and the next year married Klari Dan, who was a childhood friend and recently divorced from her second husband.

He turned his attention to artillery ballistics, of predicting the landing of a cannon ball after being launched from a gun. This is a very complex problem as air friction plays a major role in slowing down the ball by creating shock waves and turbulence. The air becomes thinner at higher elevations, so that solutions to the relevant nonlinear differential equations were needed to produce practical and accurate firing tables. It is even more difficult to hit a moving object that might be turning and accelerating as well. Von Neumann worked with the Aberdeen Proving Ground in Maryland, and in 1939 even sat for the examination to become a lieutenant of the US Army, but was turned down for an officer's commission because he was over 35. Nevertheless, in 1940 he became a member of the Scientific Advisory Board, serving the Army Ordinance at Aberdeen Proving Ground and the Naval Ordnance at Silver Spring. In September 1943, he arrived at Los Alamos for atomic bomb research, to work on the engineering of the implosion plutonium bomb. He would be involved with the atomic bomb for the rest of his life. Using a punched card sorter from IBM for computation, he designed the implosion lens, and determined the optimal height from which to drop the bomb for maximum impact.

In a totally different area, he published his economic masterpiece, *Theory of Games and Economic Behavior* with Oskar Morgenstern in 1944. This provided an enduring method to explain the behavior of people and corporations in economics. It led to a new academic discipline and a new method to understand social behavior. The Nobel Prize in Economics for John Nash in 1994, and for Thomas Schelling in 2005, were both based on the game theory of von Neumann.

He returned to computers in 1944, when he met Herman Goldstine on the railroad platform at Aberdeen while waiting for the train to Philadelphia. Goldstine took him to see the computer development in Philadelphia, which included the ENIAC project of the Electrical Engineering Department at the Moore School. Von Neumann was interested in the architecture of the computer, and in 1945 wrote *First Draft of a Report on the EDVAC*. He established the principle of a stored program in the computer, which is similar to the way that data is stored. He described a system of computer architecture with

   a. an arithmetical and logical operations unit,

   b. a central control unit to provide sequences of operations,

   c. memory for data as well as the program of instructions, and

   d. input and output devices.

This von Neumann architecture is shown in Figure 7.8. He decided to build his own computer at the Institute for Advanced Studies, which would be named the MANIAC. Weather forecasting became the project goal, so he recruited meteorologist Jules Charney who came in 1948. This move later led to the Princeton Geophysical Fluid Dynamics Laboratory, which has made important contributions to the science behind global warming.

Von Neumann returned to working on the bomb in 1946 for the Bikini nuclear test, as he believed that deterrence was needed in world politics. He returned to Los Alamos to make the bomb lighter and smaller, so that it could be easily carried. In January 1950, Truman announced that the Atomic Energy

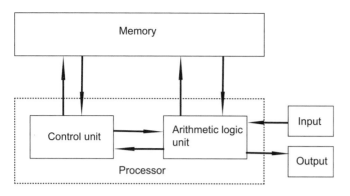

FIGURE 7.8 Von Neumann machine. Reprinted courtesy of Institute for Advanced Study, Princeton, NJ, USA . Alan Richards photographer. From The Shelby White and Leon Levy Archives Center.

Commission would work on the hydrogen bomb. John von Neumann made calculations on the ENIAC and the IAS computers, which showed that Stanislaw Ulam was right that the Teller design would not work. Subsequently, Ulam had a breakthrough in 1951, ensuring that the newly designed hydrogen bomb would work. The first hydrogen bomb explosion took place in November 1952, on the Pacific island of Eniwetok. Exposure to radioactivity at that test may have led to von Neumann's death from bone or pancreatic cancer in 1957, at the Walter Reed Hospital in Washington DC.

John von Neumann led a life of wealth, genius, and fame. In contrast, the life of Alan Turing (1912–1954) was much more modest and middle class. His picture is shown in Figure 7.9. He was born in London. His father, Julius Turing, studied at Oxford and entered the Indian Civil Service in 1896; he specialized in Indian law and the Tamil language, and was posted in Madras. Julius worked for 10 years in India, and wrote reports on agriculture, sanitation, and irrigation. He returned to England in 1907, met and married Ethel Stoney, who was the daughter of an engineer in the Indian railroad. They had a son, John, in 1908 and a second son, Alan, in 1912. Julius and Ethel left for India in 1913, and their two boys were left in the care of a retired army couple, Colonel and Mrs. Ward, who lived by the sea with four daughters. During the World War, Ethel returned to England. In the Turings' social middle-class circles, they were not allowed to play with working-class children, and the appropriate professions for them were usually army, navy, church, medicine, or law.

In 1931, Alan Turing failed to get a scholarship to Trinity College, and entered his second choice of King's College at Cambridge. He loved to run as exercise, which he did without much speed or grace, but with much staying power. Turing was interested in pure mathematics, but also in applied mathematics for industry, economics, or useful arts. In 1935, he became a Fellow of King's College, one of 46 selected, which came with a stipend of £300 per annum for 3 years, with no explicit duties. He published a paper about an improvement to a paper by von Neumann.

FIGURE 7.9 Alan Turing. Reprinted with permission © National Portrait Gallery, London.

In 1936, Turing published a famous paper, *On Computable Numbers*. He proposed the Universal Turing Machine, which worked by a set of instructions or algorithms, and proved that any automatically programmed computer was equivalent to a Turing machine. The Turing machine has the following parts:

(1) A tape divided into cells that is infinitely long to the left and the right.

(2) A head that is centered on one cell at a time, and can read and write symbols on it.

(3) A table of instructions for each cell, such as to erase, to write a symbol, and to move left or right by one cell.

(4) A state register to remember the steps taken.

All automatically programmed computers today can be considered special forms of the Turing machine.

That same year, Turing went to Princeton University to study with Alonzo Church, and met John von Neumann at the Institute for Advanced Study. He received his doctorate in 1938 and returned to Cambridge. In 1939 he moved to Bletchley Park, which was the cryptography unit of the British government. The Germans were using a machine called Enigma to code their messages, and the

British were working to break the code. It had a number of wheels with gears, with the 26 letters of the alphabet engraved on it. Since each wheel could be set in 26 possible positions, three wheels could be set in a combination of $26^3 = 17,576$ possible configurations. The naval version used eventually 8 rotors, which increases the complexity to $26^8 = 209$ billion combinations. Bletchley Park had a lucky break when the intelligence people in Poland turned over secret documents intercepted from the Germans by a lucky accident. Turing helped to design a machine called the bombe that was instrumental in 1940 to solve the naval Enigma system.

In 1941, he was engaged to Joan Clarke who was a fellow worker at Bletchley Park, but later it was broken by mutual consent. In 1942 he suggested the design of Colossus, which was to be the world's first programmable digital electronic computer. He went to the National Physical Laboratory in 1945, and worked on the design of Automatic Computing Engine. The following year, he designed a stored-program computer, and in 1948, moved to the University of Manchester.

Turing's life ended tragically in 1952, after a homosexual affair with 19-year-old Arnold Murray, which was considered a criminal offence at that time. He confessed to the affair, and was convicted in the same manner as Oscar Wilde had been 50 years earlier. He was given a choice of imprisonment or the punishment of hormone therapy to reduce libido. He accepted the latter, which amounted to chemical castration for a year. His security clearance was removed, and he was barred from cryptographic work. His death in 1954 was ruled a suicide, possibly from an apple laced with cyanide.

Since 1966, the Turing Award has been the highest recognition for contributions to computer science, sponsored by the Association of Computing Machinery. The play, *Breaking the Code*, about Turing's life and work, played in London and New York, and was also made into a BBC television production where Derek Jacobi played the part of Turing. In September 2009, the British Prime Minister Gordon Brown formally apologized to Turing, "On behalf of the British government, and all those who live freely thanks to Alan's work, I am very proud to say: We're sorry. You deserved so much better."

7.4.5 Network

There are many ways that a message can be sent and received for the distance of a few feet or all the way across the globe. Copper wires and optical fibers can carry messages that are critical to national security, as well as for everyday affairs. Wireless methods of sending use electromagnetic waves that can reach places without cables, even distant space satellites. However, wireless transmissions are vulnerable to enemy attacks by severing main trunk lines, and by natural disturbances such as solar flares. A network of communications linkages would be a lot less vulnerable, and could act as a web or net that could be cut in a few places, while still allowing the message to be routed around and reach their destinations. Many governments and private organizations have their own private networks to serve private purposes, and these organizations would gain very much in stability and convenience if their private networks were integrated. The problem for integration is the great variety of computers and softwares used by different organizations, with different methods to encode and decode messages, as well as different protocol or rules on how to send and route millions of transmissions. For example, should all the users agree to use only English and code them in ASCII, and give their addresses in the form of "username@organization.com.uk", to be managed by a central organization to avoid duplication and confusion? A great deal of leadership is required for an organization to propose such a system of communication protocols, as well as a cooperative spirit from many organizations to give up their own systems and adopt a universal system for the common good.

The greatest inventions in communications since the telegraph are the electronic mail or email (connecting two parties) and the Internet (connecting many parties); these inventions have completely revolutionized the way that we communicate and have become connected to the entire world. The Internet is a standardized global system of interconnected computer networks that use the standardized Internet Protocol Suite, called TCP/IP that stands for Transmission Control Protocol/Internet Protocol. It connects millions of academic and business entities, individuals, and government agencies by copper wires, fiber optics, wireless connections, and satellite communications. The Internet supports electronic mail, in addition to video, online shopping, gaming, and social networking. It began around 1960 as an effort of the Department of Defense Advanced Research Project Agency or ARPA. Its objective was to improve communications and make it less vulnerable to disruptions; it employed a "packet switching" method that bundles information and data into "packets," and sends each packet to one of the nodes of the network. Then the packets are routed to other nodes according to the protocols TCP/IP, written in 1972 by Robert Kahn and Vinton Cerf. They are used in the World Wide Web, which was proposed by Tim Berners-Lee in 1989. The underlying concept of the www is the hypertext, which makes it possible to encode and send tables, images, sounds, and motions.

The Internet has totally changed the way that we send and receive messages to and from others, search for information from various organizations and databanks, as well as shop, make payments, develop social networks, and play games. It has spawned new companies such as Google, Amazon, and eBay, as well as Web sites such as FaceBook, Twitter, and YouTube. The Iranian government was accused of election fraud in their presidential election of 2009, and the world was kept informed of the protests and violence by messages from ordinary residents in Iran using Twitter. The 2011 uprising of Arab people against their governments in Tunisia, Egypt, and Libya were all instantaneously reported by Al Jazeera and other international media, and aroused international indignation and condemnation from daily reports of graphic violence. Thomas Friedman wrote his book, *The World Is Flat*, to explain how the Internet made possible greater world integration and changed the nature of competition on a level playing ground. It is particularly striking that global outsourcing has given professional service providers in India the chance to participate in global economic growth without becoming immigrants in the United States or Europe, by tutoring mathematics, compiling income tax returns, and analyzing medical X-ray pictures.

The information revolution has also created a new inequality: the elite early users become better informed and better at handling complex data, which makes them even more wealthy and powerful in comparison with the late users and non-users. For the much older technology of television, where the passive users did not need much knowledge and experience, there was a near universal adoption in all nations except the lowest income nations. For the newer technology of personal computers and the Internet, which need more knowledge and experience for active participation, the rate of penetration is sharply differentiated according to income levels. Thus, the poorer people are falling even further behind in the knowledge and skill required to compete on the highest level for productivity, prosperity, security, and influence in the world.

REFERENCES

Agar, J. "Turing and the Universal Machine", Icon Books, Cambridge, 2001.

Beker, H. and F. Piper. "Cipher Systems: The Protection of Communications", Wiley, New York, 1982.

Box G. E. P., J. S. Hunter, and W. G. Hunter. "Statistics for Experimenters", Wiley-Interscience, New York, 2005.

Bruen, A. A. and M. A. Forcinito. "Cryptography, Information Theory and Error-Correction", Wiley-Interscience, New York, 2005.

Crocker, B. "Betty Crocker's Cookbook", Golden Press, New York, 1972.

Davis, M. "The Universal Computer: The Road from Leibniz to Turing", W. W. Norton, New York, 2000.

DeVore, J. L. "Probability and Statistics for Engineers and the Sciences", Wadsworth Publishing, Belmont, CA, 1995.

Doyle, A. C. "The Return of Sherlock Holmes: The Adventure of the Dancing Men", Oxford University Press, London, 1993.

Edwards, E. "Information Transmission", Chapman and Hall, London, 1964.

Eisenstein, E. L. "The Printing Revolution in Early Modern Europe", Cambridge University Press, Cambridge, 1983.

Friedman, T. L. "The World is Flat: A Brief History of the Twenty-First Century", Farrar, Straus and Giroux, New York, 2006.

Goldstine, H. "The Computer from Pascal to von Neumann", Princeton University Press, Princeton, NJ, 1972.

Guyton, A. C. "Basic Human Physiology", W. B. Saunders Company, Philadelphia, PA, 1971.

Heims, S. J. "John von Neumann and Norbert Wiener: From Mathematics to the Technologies of Life and Death", MIT Press, Cambridge, MA, 1980.

Hodges, A. "Alan Turing, the Enigma", Simon and Schuster, New York, 1983.

Hyvarinen, L. P. "Information Theory for Systems Engineers", Springer-Verlag, Berlin, 1970.

Ifrah, G. "The Universal History of Computing", Wiley, New York, 2001.

Jain, A. K. "Fundamentals of Digital Image Processing", Prentice Hall, Englewood Cliffs, NJ, 1989.

Konheim, A. G. "Cryptography: A Primer", Wiley-Interscience, New York, 1981.

Leavitt, D. "The Man Who Knew too Much: Alan Turing and the Invention of the Computer", W. W. Norton, New York, 2006.

Macrae, N. "John von Neumann", Pantheon Books, New York, 1992.

Raisbeck, G. "Information Theory: An Introduction for Scientists and Engineers", MIT Press, Cambridge, MA, 1964.

Robinson, A. "The Story of Writing", Thames and Hudson, London, 1995.

Shen, K."Meng Xi Bi Tan: Dream Pool Essays (in Chinese)". Gutenberg Project. Available on http://www.gutenberg.org.

Singh, S. "The Code Book", Anchor Books, New York, 1999.

Stinson, D. R. "Cryptography: Theory and Practice", CRC Press, Boca Raton, FL, 1956.

Teuscher, C. editor, "Alan Turing: Life and Legacy of a Great Thinker", Springer, Berlin, 2004.

Tsien, T. H. "Joseph Needham, Science and Civilisation in China, Vol. 5, part I: Paper and Printing", Cambridge University Press, Cambridge, 1984.

Wang, W. S. Y. "The Emergency of Language: Development and Revolution", W. H. Freeman, New York, 1991.

Willard, H. H., L. L. Merritt, and J. A. Dean. "Instrumental Methods of Analysis", D. Van Nostrand Company, New York, 1974.

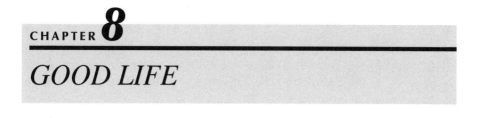

CHAPTER 8

GOOD LIFE

After the chores are done and the basic necessities of life have been adequately satisfied, we can turn to less urgent and discretionary pursuits that are fun to do, make us feel good, and elevate our minds. The good life can take an earthly direction captured by the words "Wine, Women and Song" as in a Johann Strauss waltz, as well as an intellectual and spiritual direction toward refinement, wisdom, and enlightenment. We want to party with people and have fun, to compete and excel in games of cards and dice, and to partake in athletic contests. We want to enjoy beautiful and refined things and we want to impress and dazzle the world with our exclusive possessions. We also want knowledge, beauty, and truth; and we search for answers about the meaning of life, about supreme rulers of the universe, and life after death. These are pursuits that rank highest in Maslow's hierarchy of needs. Some of these needs do not require specialized inventions, but others require the support of specialized inventions.

Economists use money and the level of consumption as the measure of the importance of our needs and desires. The measure of money is convenient but not always an accurate reflection of needs; for instance, air for breathing is even more important than food and drinks for survival, but we do not have to pay for it. The good life involves discretionary needs that are not essential, and people buy them only after their stomachs are full, their health is good, and they have extra income. The economic measures for these items are called income elasticity of demand (IED) and price elasticity of demand (PED). An average family with a budget of $40,000 per year may customarily buy $400 of whiskey and beer. When the family income increases by 10% to $44,000, the family may splurge and increase whiskey purchase by 20% to $480. The IED of whiskey is 2.0, which is the ratio of a 20% increase in expenditure divided by the 10% increase in income. For luxury goods, such as aged whiskey and diamonds, the elasticity should be much larger than one; but for necessary goods such as salt, which is indispensable to health and well-being but for which there is no urge to increase consumption, the elasticity may be close to zero. On the other hand, when there is a loss of income, such as from unemployment and economic depression, we cannot buy less bread but we can buy less whiskey. The PED measures the case when the family income has not changed, but the price of whiskey goes down by 10%, so the family may buy 20% more. When the prices drop, we are more likely to dine out at an expensive restaurant, attend theatrical entertainment, travel for pleasure, buy new cars and electronic gadgets, and have dental work done.

Great Inventions that Changed the World, First Edition. James Wei.
© 2012 John Wiley & Sons, Inc. Published 2012 by John Wiley & Sons, Inc.

People everywhere have the same basic necessities of food, clothing, and shelter, but there is a great variety in the discretionary activities called leisure and the good life. In the United States, less than 5% of adults attend classical music performances, go to horse races, play chess, and make scale models. Some discretionary activities are not costly, such as sewing, which is engaged by many families regardless of their income; but more expensive activities such as attending an opera or classical music performances are very sensitive to income levels. How many of your peers drink beer at least once a year, and how many have a Steinway grand piano at home? It can be compared with the famous declaration of Tolstoy who said at the beginning of *Anna Karenina*: "Happy families are all alike; every unhappy family is unhappy in its own way."

8.1   PARTY AND PLAY

Leisure activities involve doing things that are fun and not related to making a living or doing essential domestic chores. They may be solitary, such as reading books, doing crossword puzzles, taking photographs, playing solo musical instruments, drawing pictures, and making ship models. They may also be social activities of getting together with family and friends or even total strangers; they may involve drinking and dining together, dancing and singing, going to the beach or a concert, marching in a parade, or watching fireworks. Then there are competitive activities structured as a game or a contest, where different sides strive to become the winner.

Johan Huizinga coined the term *homo ludens* or "Man the Player" to emphasize the role of play in human culture. He observed that all animals, especially the young, have elements of play in their everyday lives. For a limited space and time, we pretend that an important event is taking place in a fantasy world that obeys its own peculiar rules, and we play roles till we return to the ordinary "real world." Perhaps play has a purpose in training the young for the serious work that life will demand later on, and for working together as a team. Children play with dolls in preparation for taking care of their own children when they become adults. Theater performances are play-acting about real life, and the audience may go through the catharsis of emotional cleansing and purification. Other types of plays may emphasize an urge to dominate and to boost personal ego. Both sports and chess have objectives that are shared with military training: the former in the physical exercise of running and throwing, and the latter in the intellectual strategy of moving troops around for attack and defense. The Duke of Wellington was quoted as saying that "The Battle of Waterloo was won on the playing-fields of Eton." Many leisure activities require dedicated inventions and technologies, such as alcoholic beverages for partying and building structures like the Coliseum for the Roman gladiator games.

8.1.1   Beer, Wine, and Whiskey

Alcohol is universally associated with having a good time, relaxing, going to parties, meeting people, singing and dancing, and fellowship. Despite legal restrictions, what is Friday night without beer, wine, and whiskey? With wine, we loosen our inhibitions, do things that we would otherwise not dare to do, and think thoughts that we otherwise would not have. An average American family spends $309 per year on alcohol, which is about 1% of total family expenditure. Alcohol is also associated with inspiration for poets, toasting at joyous celebrations and weddings, and launching new ships.

Scenes of wine making can be found in many Egyptian tombs as early as 2000 BCE. Wine is associated with many solemn and sacred religious ceremonies. Dionysus was the god of wine in Greek mythology; he came from Lydia in present-day Turkey and fled to Thrace in northern Greece. Wine spread all over the Greek world, from Fergana and Parthia in the east to the Pillars of Hercules in the west. The Greeks think that a balanced life needs something Apollonian, which is cool, rational, and intellectual, as well as something Dionysic, which is passionate, emotional, and dynamic—and associated with wine. It is also an integral part of Jewish laws and traditions and part of their sacrificial service. In Christianity, wine is used in the rite of the Eucharist, in which Jesus shared bread and wine with disciples to "Do this in remembrance of me," as the bread and wine symbolize or change into the real body and blood of Christ. In Colonial America, Benjamin Franklin was a strong advocate of wine as good for health. Thomas Jefferson was the most knowledgeable connoisseur of wine in his time, and he was the wine advisor to Presidents Washington, Adams, Madison, and Monroe. In the ancient world, there was another reason for the consumption of beer and wine, as they were likely to be much safer to drink than water, which may be swarming with bacteria and amoebae that may be disease causing. The alcohol content of beer and wine, even at a relatively low level, would discourage the formation and proliferation of microbes. On the other hand, the use of wine is prohibited by some religious groups.

Alcohol has many competitors as stimulants that are psychoactive, meaning that they act on the central nervous system to reduce pain, to stimulate, to tranquilize, and to cause hallucinations hallucinate. Some stimulants are highly dangerous, addictive, and illegal, and it is estimated that 5% of the world's adult population are users. The international production and trafficking of narcotics is highly lucrative, as the street price of heroin can be three times higher than that of gold in dollars per gram, which attracts drug lords and crime syndicates.

- Caffeine is the mildest form of stimulant and is found in coffee and tea, and the closely related substance theobromine is found in chocolate.

- Nicotine in tobacco was brought back to Europe by visitors to North America, which is used by smoking or chewing, and is addictive and associated with lung cancer.

- Cannabis is also known as hashish or marijuana, which can be a stimulant and cause a feeling of "high" or be a depressant with the resultant feeling of being "stoned." It was celebrated in Hindu Vedas as the "food of the gods." The Assassins was a religious sect from Syria to Persia, described by Marco Polo as being inspired by the use of cannabis.

- Cocaine is found in the coca leaves of South America and had been chewed by the indigenous people for a thousand years to give them "great contentment." It can also be smoked, sniffed, or injected. It was the vice indulged in by the fictional detective Sherlock Holmes, and is addictive.

- Morphine is found in the opium poppy, was used by Sumerians as early as 4000 BCE to relieve pain, and can be modified to give the even more dangerous heroin as well as the pain-relieving codeine. In sixteenth century Anatolia, opium was eaten to give ecstasy to dervishes, to give courage to soldiers, and to give bliss to dignitaries. The British fought two wars with China in 1840 and in 1858 to force China to buy opium grown in British India. Morphine is highly addictive.

- Famous hallucinogens include ergot, cactus, mushroom, henbane, mandrake, and deadly nightshade. They were used in religious and shamanic ceremonies in Egypt, Mycenae, Greece, India, Maya, Inca, and Aztec.

- Synthetic drugs such as Speed (amphetamine) and Ecstasy (MDMA) were first created in the research laboratories, and have become important in the streets and parties since the 1960s.

Returning to alcohol, it has been naturally produced by fermentation long before history began. The starting material should be rich in sugar, and in ancient times, perhaps honey and nectar were the most available natural sugar-rich substances and the source for early alcohol drinks. Even without human intervention, the fermentation process would take place when wild yeast floating in air settled on sugar and split it into alcohol and carbon dioxide gas. When wine making became a planned event, specific strains of approved yeast would be used to ensure quality, so the brewer would save the remains of a previous batch of brewing to use as the starter in a new batch of brewing. Wine yeast is similar to the yeast involved in the baking of bread to make it rise and in making vinegar or acetic acid. The alcohol content of wine does not go above 15%.

Wine is more often produced from fruits that are rich in sugar, such as grapes, dates, and apples. The grapes are harvested and thrown into wooden vats and the juice is squeezed by foot treading accompanied by music. After decanting the grape juice, the solid residues are placed in a press to squeeze out more juice of an inferior quality. This process is similar to pressing olives to produce oil, where the first cold press is called extra virgin and the second cold press is called virgin, to be followed by hot press with steam to produce lower quality oil. The amphora is a clay vessel used to carry wine or olive oil, with two handles and a point at the bottom to rest in sand or soft ground. Many amphorae were found in a cargo ship of 200 BCE that had been shipwrecked off Marseilles, which were used to ship wine and olive oil all over the Mediterranean Sea.

Beer, on the other hand, starts from grains that contain starch instead of sugar and requires a longer two-step fermentation procedure: the malting process turns starch into a sugary liquid called wort and then the fermentation process begins to convert the wort to alcohol. A brewer's yeast is involved, and the alcohol content of beer generally does not go above 10%. A flavoring material such as hops is often used. It is a distinctly everyday drink, and not associated with religion and ceremony. If the husks are not removed from the brew, they will float to the top of the beer, and the drinkers have to resort to using tubes or straws to drink from the bottom, which is a scene shown in many Mesopotamian and Syrian seals. In the Epic of

Gilgamesh, the wild man Enkidu was given beer so that "his heart grew light, his face glowed, and he sang out with joy."

In China, the sacred and formal *jiu* is brewed from rice instead of fruit juice; however, it is usually translated into English as "wine" instead of beer or ale. It was produced as early as 5000 BCE, and archaeologists have found many elaborate vessels of pottery and bronze for fermenting, filtering, and drinking wine. Jiu was the drink presented to the gods and ancestors at ritual offerings, and is used in toasting at banquets and weddings. It is also the drink that inspired thousands of poems, written by emperors, scholars, and common people. The Chinese wine is warmed up before being served, which is just the opposite of the practice in the West, where beer or white wine are served chilled and red wine is served at room temperature. The elegant Shang dynasty bronze cup called the "jue" has a ring for the right index finger and a channel to pass the wine into the mouth. It also has three legs, perhaps for placing over a fire to warm up the wine.

Champagne has launched thousands of ships, toasted countless weddings, and inaugurated innumerable New Year parties. Grapes harvested in the fall in Northern Europe were crushed and fermented in wooden tubs to release alcohol and carbon dioxide, but fermentation stopped when the weather turned cold. If the unfinished product was bottled until it became warmer in the next spring, the yeast would revive and begin to ferment again to release more alcohol and gas. The gas bubbles could not escape and would dissolve in the wine, increasing the pressure.

The fizz in the sparkling wine was not favored for a long time, but became popular later when fashions changed. A major role was played by a 29-year-old Benedictine monk by the name of Dom Pierre Pérignon, who was appointed the treasurer of the Hautvillers Abbey near Reims in 1668. The process of secondary fermentation could be increased by taking still wine from casks and adding sugar and yeast before putting the mixture into bottles, which would create even more fizz and alcohol. Dom Pérignon made many improvements in the making of champagne, including changing the stopper from a rag or wooden pegs to a cork, which kept more bubbles in solution. With the cork, the pressure of bubbles could go up to several atmospheres and cause the bottles to explode, so Pérignon also had to introduce stronger glass bottles. Each type of grape has its own distinct flavor and bouquet, and he mixed many of them to create a superior blend.

There was one problem with secondary fermentation to make fizz: the dead yeast cells would form sediment and make the wine cloudy, preventing the delightful sight of bubbles rising in crystal clear wine. Various methods were used to remove the sediment, such as decanting the wine into another bottle and leaving the sediment behind, but this meant losing much of the treasured gas and wasting some wine. The modern method of *remuage* is attributed to Barbe-Nicole Clicquot Ponsardin, better known as Veuve Clicquot or Widow Clicquot, the first businesswoman who created and presided over an international commercial empire, starting in 1815. Her picture is shown as Figure 8.1. She was widowed at the age of 27 and inherited a wine house, but had no training in business or wine making. She became an entrepreneur, defied conventions, and took great risks to start an international marketing effort for champagne to reach as far as Russia. She also did experiments

FIGURE 8.1 The inventor Veuve Cliquot.

in improving the process by turning the bottles upside down, so that the sediment would fall into the neck and the bottles would be given a quarter turn every day for a few months. A quick flick of the cork would allow the residue to shoot out; this process was followed by a rapid recorking to prevent the wine and pressure from being lost. But the modern method involves dipping the neck of the bottle into a cold bath to freeze the wine with the sediment into a plug of ice, which can be cut with a knife and removed with even less loss of wine and pressure. Her firm, Veuve Clicquot Ponsardin, is operating even today as one of the most prestigious and successful champagne makers.

A faster way to approach euphoria and intoxication is to consume a drink with a much higher alcohol content. Ogden Nash said that to break ice at a party, *Candy is dandy, but liquor is quicker*. A higher alcohol content can be attained by the process of separation, which is fundamental to many chemical and related processes and which creates the desired purified products as well as various amounts of by-products. One looks for differences in one or more properties between the desired product and the by-products and separates the mixture into two streams— one called the raffinate that is richer in the desired product and the other that is poorer. There are many methods of separation, which are described as follows:

Mechanical separation involves screening to separate fine from coarse particles, filtration serves to separate dust particles from gas or liquid, sedimentation involves settling the heavier gold dust from lighter sand grains, decanting is to separate lighter oil from heavier water, and crystallization is a way to obtain salt from seawater.

SCIENCE AND TECHNOLOGY: WINE

In the simpler case when the raw material is fruit juices with soluble sugar in the form of glucose or fructose, alcohol is produced by yeast working on sugar to produce carbon dioxide and alcohol.

$$C_6H_{12}O_6 \rightarrow 2C_2H_5OH + 2CO_2$$

The discovery of this chemical reaction was made by Louis Pasteur in 1860. Yeast is a single-cell plant that is 3–4 μm in diameter. The best wine yeast is *Saccharomyces cerevisiae*, which works best at a temperature of 25–30°C. To ensure that the right yeast is at work instead of a wild yeast from the air, most modern breweries eliminate the wild microbes from the raw material by pasteurization and then inoculate the material with a starter culture from previous batches of successful brew. The alcohol content seldom goes above 15% in wine.

When the raw material is grain with insoluble starch, the production of alcohol requires an additional preparation stage to convert starch to sugar. Starch is an insoluble polymer of sugar and can consist of hundreds to millions of units of sugar, $C_6H_{12}O_6$, joined together. It is broken down by a process called hydrolysis with the enzyme *amylases* into a two-unit sugar called the maltose:

$$(C_{12}H_{20}O_{10})_n + nH_2O \rightarrow nC_{12}H_{22}O_{11}$$

The malting enzyme is also in human saliva, so that the chewing and spitting of starchy food such as bread or potato is a favored method for home brew in South America. But in the Orient where the raw material is rice, the appropriate *amylase* is derived from a mold *Aspergillus oryzae*. When complete, the beer has an alcohol content of 5–7% alcohol.

A distilled spirit or liquor has at least 30% alcohol, such as brandy, gin, rum, tequila, and vodka. Whiskey is legally defined as distilled spirit from grains, and aged in oak barrels. The usual grains involved are barley, rye, wheat, and corn. Whiskey dates from the 1400s in Ireland and Scotland. The process of distillation starts by placing the beer or wine with about 10% alcohol in a container held over a flame. The vapor that rises over the beer has a concentration of about 40% alcohol, which moves into a condenser cooled by air or water. To obtain an even stronger drink, one can redistill the 40% alcohol liquid in another container placed over a flame, so that the vapor that emerges has an even higher alcohol content of 60%. The next stage in the production of high-quality whiskey is aging in an oak barrel.

When water is selectively removed from wine, the concentration of alcohol increases in the remainder. In China in the year 700, the making of "ice wine" was reported by freezing wine on ice; as the crystals of ice that form in wine do not contain any alcohol, the liquid that remains in wine will have a higher concentration of alcohol. A more convenient method is to heat the fermented wine to produce a

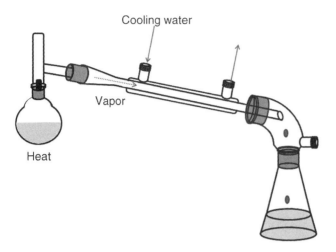

Cooling water

Vapor

Heat

FIGURE 8.2 Distillation still.

vapor, which is higher in alcohol content than the wine, and then condense the vapor by cooling. The earliest mention of *xiao jiu*, or burnt wine, took place around 800 in Tang Dynasty poetry and at increasing frequency thereafter. The production of this drink involves mixing strong wine with the fermentation residues and putting the mixture into a steamer. Upon steaming the mesh with alcohol, the vapor is made to rise and condense against a cooling vessel and the condensation droplets are collected in a catch bowl, which is shown in Figure 8.2. The product is as clear as water and its taste is extremely strong. Similar stills were used for distilling perfumes and spices.

The Arab alchemist Jabir ibn Heyyan invented the alembic still in 800 to produce acids. The Iranian scientist Avicenna (980–1037) extracted oil from flowers by distillation. In 1100, the medical school at Salerno was a center of chemistry experimentation and is reputed to be the place where distilled alcohol was discovered in the West. The name given for a product containing enough alcohol to burn was *aqua ardens*, or *aqua vitae*, the water of life. These distillation methods spread to Western Europe in 1370, and the word "alcohol" is an Arabic word with a similar origin as "alchemy" and "alkali." The simplest European still consisted of a kettle with a closed lid, which was placed over a flame to heat the liquid to boiling.

In the United States, the manufacturing of distilled spirits is heavily taxed and is under the supervision of the government. Gin is distilled with juniper berries to give it a flavor, brandy is distilled from wine, bourbon comes from corn, and rye whiskey comes from rye. They are then aged, and the best spirits are aged in charred new white oak barrels in bonded warehouses at 18–29°C for a number of years. Some of the water in the whiskey is absorbed into the oak or leaked out of the barrel, and the government allows a small shrinkage of no more than 8% for the first year. Illegal whiskey brewed at home without the attention of "revenooers" or government tax collectors is called "moonshine," and is a much loved tradition of the backwoods of Appalachia in America. When a backwoodsman from the hills enters a hardware store to buy a length of copper coil, everyone knows what it would be used for.

Beer, wine, and whiskey are responsible for many joyous moments in our lives, as well as much mischief. Around the year 800, the great Chinese poet Li Bai said,

> *Do you not see, the water of Yellow River comes from heaven,*
> *Running to the sea and never to return,*
> *Do you not see sorrow in front of bright mirrors in high halls,*
> *Hair like black silk in the morning, and like snow in the evening.*
> *Success should be celebrated, let no wine vessel be empty under the moon.*
> *Do not fret master, about shortage of cash, as we need more wine.*
> *Five-colored horse, thousand-gold-coin fur coat,*
> *Go exchange them for wine, and we shall quench the sorrow of ages.*

Edward Fitzgerald translated the Persian poems of Omar Khayyam from 1120, and called it the *Rubaiyat*. He said,

> *A book of verses underneath the bough,*
> *A jug of wine, a loaf of bread - and thou,*
> *Beside me, singing in the Wilderness -,*
> *Oh, Wilderness were paradise enow!*

Alexander Hamilton argued in favor of taxing drinkers to help finance the new federal government. He said that a levy on spirits would benefit the morals and the health of society by curbing the national extravagance of consuming spirits. The resulting alcohol tax was a burden on the settlers of the western frontier, where whiskey was a profitable product. The resulting turmoil was called the Whiskey Rebellion in the Appalachian region, which was finally quelled by George Washington in 1794 when he marched an army into western Pennsylvania. The widely evaded tax had collected little revenue, and Congress eventually repealed it under Thomas Jefferson.

America consumes a staggering amount of alcoholic beverages. In the year 1995, our population was 262 million and the alcoholic beverages produced were as follows:

Beer	7200 million gallons (27,250 million L)
Still wine	412 million gallons (1560 million L)
Distilled spirit	104 million gallons (393 million L)
Whiskey	69 million gallons (261 million L)

The world consumption of pure alcohol in annual liters per capita varies a great deal among nations, from 13.6 in Ireland to 0 in Saudi Arabia.

There is also a dark side of drinking alcohol, especially in alcohol's adverse health effects and its association with accidents. Michelangelo depicted the drunken state of Noah, on the ceiling of the Sistine Chapel in the Vatican, to illustrate the evils of alcohol. The euphoric, lighthearted, jovial, inspirational sensations associated with drinking beer and wine are due to the absorption of a moderate quantity of alcohol into the blood; heavier use leads to depression and stupor. The blood alcohol concentration can determine the mood of the user:

Blood alcohol%	Mood and physiology
0.06	Relaxed, joy, well-being, and loss of inhibition
0.08	US legal limit, approximately two drinks
0.10	Impaired coordination and judgment
0.20	Boisterous, angry, and depressed
0.30	Stupor and blackout

At the present time, all states in America have rules that minors cannot drink, and the legal drinking age throughout the United States is 21. In 1997, 623,000 patients were treated with alcohol problems. In 1995, 56,000 drivers were involved in fatal automobile crashes, and 19.3% of them had a blood alcohol concentration of 0.10% or higher. The total deaths from alcohol-induced causes were 20,000 in 1955, of which the male-to-female ratio was 3:1. The World Health Organization estimates that 140 million people throughout the world suffer from alcohol dependence.

The drinking of alcohol was entirely banned in America during the period from 1920 to 1933, known as Prohibition or the "Noble Experiment." It was mandated in the Eighteenth Amendment to the US Constitution, which passed over the veto of President Woodrow Wilson. Prohibition became increasing unpopular during the Great Depression, and "bootlegging" or the running of illegal rum from places such as Canada became widespread. In 1933, the Twenty-First Amendment was passed to repeal the Eighteenth Amendment. Many novels and movies have discussed the effects of Prohibition on society, and perhaps the most famous is F. Scott Fitzgerald's novel, *The Great Gatsby*, who made his fortune in rum running. The "Demon Rum" was denounced by organizations such as the Women's Christian Temperance Union, and was depicted in popular movies such as "Guys and Dolls," "Days of Wine and Roses," and "The Lost Weekend."

8.1.2   Games

Competition between two animals of the same species for limited resources such as food, water, mating rights, and territory often leads to conflict. Rams butt heads together to determine the winners and losers. Young goats often imitate their elders and engage in head butting perhaps as an exercise in preparation for things to come. Charles Darwin proposed the theory of sexual selection to explain some evolutionary traits as "the struggle between the individuals of one sex, generally the males, for the possession of the other sex." Many of our athletic contests have their origins in military action, toned down to be less deadly. In the *Iliad*, Achilles held the funeral games to honor Patrocleus, which consisted of a chariot race, a foot race, throwing of iron lumps and spears, archery, boxing, wrestling, and dueling with spears and shields. The spoils of a real war can be the possession of kingdoms and slaves; the trophies for the games were scaled down to include cauldron tripods, stallions, cattle, and lovely women.

Modern games take many forms, and the most popular are athletic games and tabletop games—the former stresses strength and speed, and the latter stresses agility and thinking. A game requires not only a set of equipment but also a set of rules. In a foot race, no one starts till a signal is given. When two goats lock horns to determine who is stronger, they are not allowed to charge till the opponent is ready; once defeat is admitted, the loser is not allowed to sneak in another attack when the winner is not prepared. Expulsion and disgrace are in store for a game player who participates in unsportsmanlike conduct.

A rolling object appeals to a kitten or puppy, as well as to a baby. Ball games such as soccer, football, basketball, tennis, cricket, and ping-pong are certainly among the most important of athletic games. Chinese emperors played "cuqiu" or kick ball, from the period of the Warring Nations in the third century BCE. The Olmecs and Aztecs in Mexico played an elaborate ball game with great courts, equipped with stone rings that were high above the ground, where two teams would try to bounce solid rubber balls through a ring. This may have been both a religious ceremony of great solemnity and a ball game for fun. The Persians played a game of polo (the King of Games) on horseback and the Mongols continued the game a thousand years later, using the headless carcasses of sheep instead of balls. Most ball games involve kicking, throwing, or hitting a ball with a stick, resulting in a ball in flight that should be long and accurate. After the invention of a ball game and its rules, many inventions can follow to make the ball's trajectory longer and more accurate.

Egyptian footballs were made of soft leather or fine linen and stuffed with reeds or straw. By the third century BCE, the Chinese had a ball game where the ball was made of an inflated bladder, usually from a pig, and covered with hide. In front of the emperor's palace, the opposing teams would try to kick the ball through holes in silk nets. The original Scottish golf balls were made of wood, which were replaced in the seventeenth century with hand sewn leather pouches stuffed with chicken or goose feathers.

When a ball flies rapidly through the air, it meets significant resistance so that a light ball cannot fly very far. With the exception of ping-pong that has a short trajectory, most of the small balls for organized sports are solid, such as those in golf and baseball, but the larger balls can be filled with air, such as basketball and football. An air-inflated bladder has very good elasticity to store the energy of impact for later release to become the energy of flight; wood and stuffed feathers, however, do not have the best elasticity and give very inferior bounce. A far superior solid stuffing emerged with the invention of vulcanized rubber. We have noted, in Chapter 2, about the origin of rubber balls in the Americas, as observed by the followers of Christopher Columbus, and the transformation of raw and unsatisfactory rubber by Goodyear, through vulcanization with sulfur under heat into the all-weather vulcanized rubber that remains elastic in hot or cold weather. In 1898, Coburn Haskell of Cleveland, Ohio, made a solid golf ball wound with layers of rubber thread and covered with a thin outer shell. It had a superb bounce characteristic and easily stored the energy of impact from the golf club and released the energy in speed. Today, synthetic polyurethane is often the favored material for filling the solid cores.

Several methods have been invented to overcome the aerodynamic resistance on a ball when it flies through air. A thrilling attack in an American football game involves the forward pass hand thrown over a long distance. This feat would not be possible with a spherical ball that would encounter an aerodynamic drag that would cause it to tumble and lose its directional precision. The modern American football is oval in shape and can be thrown to spin about its axis to stabilize the direction, much as a gyroscope in submarines and airplanes. A quarterback can throw a "Hail Mary" forward pass for 100 yards or 92 m, with enough accuracy to reach a running receiver.

It was accidentally discovered that when a golf ball had nicks and scrapes, the ball would give a truer flight than would a perfect sphere. Golf balls began to be deliberately nicked by hammering. In 1905, William Temple registered a patent for a dimple design for golf balls. The symmetrical dimples on golf balls cause the airflow behind the ball to be more turbulent, thus reducing the drag. This allows the golf balls to be lofted higher and fly longer distances. A golfer can easily reach 230 m with a dimpled golf ball, but could not reach more than 92 m with a smooth ball.

In addition to sports balls, other kinds of sports equipment have benefited from technological advances. The handles of tennis racquets and the shafts of golf clubs are required to be light but stiff, and were originally made of wood. Howard Head, a former engineer in the aerospace industry, introduced in 1976 the Prince brand of oversized tennis racquet with an aluminum frame, which made possible a combination of lightweight with a large sweet spot to increase power. The aluminum frame was later replaced with modern materials, especially carbon fibers with plastic resins to bind them together. The stiffness of the frame can be adjusted by the ratio of fibers to resins: the stiffer rod is stronger and more accurate, but gives a stronger shock on impact, while the softer rod is gentler on the hand. These modern materials also have very long lives compared to wood. Athletic records, such as in the Olympics, are constantly being broken, partly due to better health and training methods of the athletes and partly due to better equipment.

Tabletop games of skill and chance are also central to our civilization. Some of the oldest board games found in the ancient tombs of Egypt and Mesopotamia include Senet, the Royal Game of Ur, and Backgammon. Some games are won purely by chance, depending only on the outcome of the roll of a dice or the turn of a card; there are also games that depend only on superior skill and strategy such as chess; in between there are far more games that depend on all of these factors, such as bridge. The game of Senet dates from 3100 BCE and represents the journey of the dead, played on a grid of 3 rows of 10 squares where the movement of the pieces depends on throwing two-sided sticks to determine the number of squares to move at each turn. The oldest game of chance may be the dice that were found in Egyptian tombs around 2000 BCE and may have originated using the knuckle-bones of animals, thus the popular expression "roll them bones." The random outcome of rolling dice is used not only for gambling, which often depends more on luck than on skill, but also for divination and fortune telling. Loading the dice to make them roll in your favor was an ancient illegal practice, which was more or less suppressed with the modern transparent plastic dice. There are many magnificent ancient Greek vases called amphoras, with scenes of Achilles and Ajax playing dice together.

On the other extreme, chess is a game of skill, where chance plays no role. Chess originated in India during the Gupta Empire in the sixth centuryBCE and was known as Chaturanga, which means the four divisions of the military: infantry (pawn), cavalry (knight), elephants (bishop), and chariots (rook). It was taken up by the Persians, where Shah is the word for king that mutated later to chess. Evidence of the Persian influence is found in the declaration "checkmate," which comes from the "Shahmat," the Persian word for death of the king. The modern European chess was codified around 1475 and became part of the noble culture in the Renaissance to teach war strategy, which was described by Baldassare Castiglione in *The Book of the Courtiers*. The board is an 8×8 arrangement of 64 squares, and each side starts the game with 16 pieces. Benjamin Franklin endorsed chess, writing, "It teaches several very valuable qualities of the mind, useful in the course of human life. For life is a kind of Chess, in which we have often points to gain, and competitors or adversaries to contend with, and in which there is a vast variety of good and ill events. By playing chess, we may learn foresight, to consider the consequences from an action; circumspection, to survey the whole chess-board or scene of action; and caution, not to make our moves too hastily." A chess-playing machine, known as the Turk, was built in 1770 as a clever automaton, but it was later exposed as an elaborate hoax. True chess-playing machines were built later based on computers. The Deep Blue was an IBM computer that could evaluate 200 million positions per second and anticipate the consequences of 6–8 moves, and defeated the reigning world champion Gary Kasparov in 1997.

The Chinese board game of Wei-qi, which is also known as Go, has been played since 2300 BCE; it is played on a board of 19×19 gridlines with an unlimited number of pieces that are all identical pawns. It is also a military game about capturing territory instead of pieces, which begins with an empty board. The pieces are called stones and two players alternate in placing one stone at a time on an empty intersection, where the stones cannot be moved on the board after being played. When your stones completely surround a group of your opponent's stones, you capture and remove them from the board. At the end of the game, the player with more territory, defined by stones, is the winner.

Playing cards often take up an intermediate position, as chance is involved in shuffling and dealing the cards, but skill can be involved in the strategy of the game after the cards have been dealt. Cards were played in the Tang dynasty during the period when paper had started being used as a writing surface and for paper money. There were four "suits" in playing cards in China around 1120, and similar playing cards spread to Europe in 1200. Back in modern China, these suits have survived in the game of mahjongg that does not use paper cards but is played with solid pieces called tiles. Tarot cards were used widely for fortune telling as well as for games, and the modern version has 78 cards divided into the Minor Arcana with 56 cards organized into 4 suites of 14 cards each and the Major Arcana with 22 cards bearing names such as the Magician, the Pope, the Wheel, the Devil, and Death. The traditional names of the four suits are swords (clubs), batons (spades), coins (diamonds), and cups (hearts). Modern playing cards retained only the Minor Arcana, and simplified it to four suits of 13 cards each. Poker and contract bridge are among the best known of the card games.

Throughout history, many new games have been invented and have enjoyed popularity. Today Monopoly is one of the most famous games, which was developed in 1934 during the Great Depression by Charles Darrow, and which is claimed to be the most played commercial board game in the world. Computer and video games have become more important in recent years, including the mysterious game of Myst and the family game of Sims.

8.2 LUXURY

Entertainment and recreation involve discretionary activities that are not necessary for survival; luxury and vanity involve discretionary consumption that goes beyond serving a function of displaying elegance. The design of a house must be functional to keep out wind and rain, but can also be for showing the status of the owner. Thorstein Veblen said that besides private enjoyment, we also want other people to be impressed and envious of what we have, thus we display our wealth in "Conspicuous Consumption." In this view, a mansion with a butler and a limousine with a chauffeur are not only for the private enjoyment of the owner but also for impressing others and gaining status. A butler with a white starched collar and black tailcoat obviously cannot do productive work like planting potatoes or fixing cars, so the owner who can afford such an impractical servant must be wealthy indeed.

In the *Epic of Gilgamesh*, the earliest story in the world, the standard of living was sufficiently high for the people to indulge in discretionary consumption. They had robes dyed purple from the famous mollusks of Lebanon, many jewels and trinkets, and crowns for the exalted. The wealthy people bathed in water with tamarisk and soapwort and enjoyed music, dancing, and festivals. They were full of civic pride on the beauty and magnificence of their city of Uruk, of its great walls, temples, palaces, and public squares; there was no equal in the world. Gilgamesh decided to take a long journey to the cedar forest of Lebanon to kill the demon Humbaba, not because Humbaba was a menace to Uruk but because this feat would increase Gilgamesh's fame in the world.

When we care about how we present ourselves, we look at ourselves in a mirror and smell the perfumes that we wear. A long time ago, the only mirror from nature was smooth and still water; later we invented polished stones or metals, like obsidian and copper. Even today, the symbol for woman ♀ is the mirror of Venus, who was from Cyprus that is the home of copper. As long as the mirrored surface was made of copper or colored glass, the reflected image would have an imposed color and not entirely satisfactory. The Venetians solved the problem of colored glass by adding decolorizing material and obtaining clear and transparent glass. In 1291, Venice moved its glass manufacturing to the guarded island of Murano and maintained a strict monopoly over the secrets of Venetian glassmaking.

The invention of the silver-glass mirror is credited to the German chemist Justus von Liebig, in 1835. His process involved the deposition of a thin layer of metallic silver onto glass through the chemical reduction of silver nitrate. Mirrors today are more often produced by the deposition of aluminum on glass. How do we improve upon our image that we see in the mirror? We can use soap and shampoo to clean ourselves, brush and comb our hair or use a wig, and apply cosmetics.

There was a religious and magical origin to the use of cosmetics and perfume in the ancient Near East. The Bible says, "Ointment and perfume rejoice the heart." The election of a king was confirmed by anointing with holy oil, and the embalming of a dead body was anointed with unguents. The burning of incense to make smoke was a means for purification and to fumigate insects and worms. The three Magi Kings visited the infant Jesus, not with rattles and teddy bears but with luxury gifts.

> After coming into the house, they saw the Child with Mary His mother; and they fell to the ground and worshiped Him. Then, opening their treasures, they presented to Him gifts of gold, frankincense, and myrrh. (Matthew 2:11)

A modern perfume consists of three ingredients:

(1) Vehicle, which is the solvent of alcohol and water.

(2) Odor material, which are volatile molecules that give the scent.

(3) Fixative, which is a heavier substance than the odor material, used to retard evaporation, and originating from animals (civet, musk, and ambergris) or resins (hard gum, soft myrrh, and balsam).

The sources of odor material are berries, seeds, bark, wood, rhizome, resin, flowers, peel, and animals; they are extracts from plants and animals by distillation and solvent extraction. A modern perfume is designed as an ensemble of odors, some of which are released immediately upon application while others do not make their appearance till hours later. They are divided as follows: top note, which is highly volatile and perceived immediately on application, with small light molecules that evaporate quickly and create the initial impression (citrus and ginger); middle note, which forms the "heart" or main body of the perfume and is more mellow and rounded (lavender and rose); and bottom note, which gives depth and solidity to a perfume, with large heavy molecules that evaporate slowly, and is "deep" and not perceived until 30 min after application. This ensemble of odors can be compared with the musical harmony of instruments from violins and flutes to bassoons and tubas.

The use of incense may have originated in Egypt, where the gums and resins of aromatic trees were imported from the Arabian and Somali coasts to be used in religious ceremonies. In ancient times, incense was used in Babylon, Greece, Rome, India, and China for religious offerings, and a major part of the commerce along the Silk Road involved incense. To release its aroma, a stick of incense is combined with a combustible material such as charcoal or wood powder, so that it can burn slowly and release the smoke over a period of time. Incense is also used to obscure other less desirable odors, such as in a funeral to conceal the scent of decay. Incense was so highly valued that the Nabatean tribe rose from the Arabian deserts to control the water supply at the crossing of the north–south Incense Route from South Arabia to Damascus and Gaza and the overland east–west route from the Persian Gulf to Aqaba and the Red Sea. Their control of these routes was so successful that around 100 BCE, they built as their capital the rose red city of Petra, now designated as a UNESCO World Heritage Site. There has always been a great market for innovations in luxury during prosperous times.

8.2.1 Color

The color displays of springtime are associated with the revival of the creative forces of nature and of courtship and the propagation of the species. The colors of flowers are designed to lure the bees, butterflies, and hummingbirds to share the nectar and to carry the pollens to fertilize other flowers. The colorful courtships of birds include peacock feathers, the bright plumage of the parrots, and the scarlet chests of the frigate birds. It would be a drab world indeed if humans did not have colors to brighten their lives, which also has the practical utility of promoting courtship and propagation.

The ancient Egyptians and the Mesopotamians enjoyed brightly colored clothes of wool and linen, as shown in their tomb pictures and other archaeological artifacts. Jacob gave his favorite son Joseph a coat of many colors, which made him the envy and the enemy of his brothers, according to the Bible and the rock musical, *Joseph and the Amazing Technicolor Dreamcoat*. If a coat of wool or linen is only intended to keep a person warm, any color would do just as well. So, the purpose of colors is not to keep warm, but to make a statement such as, "I have something that you do not have, I am more stylish than you, and I am more beloved." The desire to be better than others, to rank higher in some poll and ranking, or to score a psychological victory can often be served by an invention or a technology.

There is also a psychological element to our reaction to colors, based on associations that may be specific to individual civilizations. For instance, the Western world is accustomed to the following color associations: white—purity, cleanliness, winter, and cold; black—mystery and death; red—heat, passion, summer, and stop; pink—feminine, women, and girls; blue—sea, sky, noble, men, boys, and sadness; green—nature, vegetation, spring, and go; yellow—sunlight and earth; and purple—aristocracy and royalty.

The color of our food and drinks can make them appear better than they are. In the year 127, an illegal use of dye was reported by Pliny, the Roman scholar, involving vegetable extracts that had been added to young red wines to give them the appearance of mature claret, which was much higher in price. Some modern synthetic dyes make food more appealing, but contain substances that may be carcinogenic and are regulated by the Food and Drug Administration.

Pigments are insoluble colors that do not penetrate the surfaces to be covered, but adhere to the fabric or canvas surface with gum, wax, or drying oil. They are usually obtained from minerals such as the yellow and red ochre that are iron oxides. *Dyes* are water-soluble colors that can penetrate a fabric and are often of vegetable or animal origin. Most of them are not permanent and can wash away, unless treated after dyeing with a mordant such as alum to turn them insoluble. Plants are the source of many dyes, such as the blue woad that British warriors used to dye their bodies for a terrifying appearance in battle and indigo, which is another form of blue dye. The red madder comes from a root, which was mentioned by Pliny the Elder and also in the Merovingian chronicles. The color brown comes from walnuts, and yellow from safflower. Animal dyes, such as sepia from cuttlefish, were used less frequently. The finest red is crimson from the cochineal insect.

Color	Inorganic pigment	Organic dye
Red	Red ochre, iron oxide, vermilion	Cochineal, madder, crimson, alizarin
Yellow	Yellow ochre, cadmium yellow	Saffron, turmeric
Green	Malachite, copper carbonate, viridian	Janus green, Lincoln green
Blue	Turquoise, lapis lazuli, ultramarine	Woad, indigo
Purple	Han purple	Murex, gentian violet
Brown	Raw umber, raw sienna	Juglone
Black	Charcoal	
White	Chalk, gypsum, lead white, titanium white	

The most celebrated animal dye in the ancient world, Tyrian purple, was from the land that the Greeks called Phoenicia—or purple people. Tyrian purple is obtained from a sack in the head of the murex shell, which is colorless while the murex snail is alive but oxidizes in air to turn to a brilliant purple color. Returning to the *Epic of Gilgamesh*, the hero and his companion walked 15 days from Sumer and reached the cedar forest in search for glory in the land of the purple people. This dye was more precious than gold, so only a Roman emperor could afford a purple cloak, and a Roman senator was allowed only a purple stripe on his cloak. The volcanic rock porphyry has a purple grain and was used in Byzantine imperial monuments such as the Hagia Sophia. It was used in the official birthing room of Byzantine empresses, and children that were delivered in this room were called *porphyrogeniture* or "born to the purple." The art of dyeing with purple was a Byzantine secret, and was lost after the fall of Constantinople in 1453.

In China, the color purple was also associated with royalty and was used to color the famous terra-cotta soldiers at the tomb of the First Emperor in Xi'an. The official name of the Forbidden City in Beijing is Zi-jin-cheng or the Purple Forbidden City. The first novel written in Japan was the *Tales of Genji* by Lady Murasaki, whose name means purple. The Chinese color of Han Purple has more blue than red, which differs from the Tyrian purple that has more red than blue. According to the Ming Dynasty history of technology *Tian-gong Kai-wu*, the color purple is achieved by using sap pan wood as a red base, and then dyeing again with green vitriol or ferrous sulfate.

The modern chemical and pharmaceutical industry began with the accidental discovery by an 18-year-old schoolboy in London. William Perkin was a schoolboy at the Royal College of Chemistry in London, studying as a laboratory assistant under the guidance of August von Hofmann. In 1856, he undertook the synthesis of quinine for the treatment of malaria, which had been a leading cause of death in many continents, particularly in tropical countries. At that time, quinine was derived solely from the bark of the cinchona tree, which grows on plantations in Southeast Asia. Over the Easter vacation in 1856, Perkin worked at his home laboratory to oxidize aniline from coal tar with potassium dichromate, and he produced instead a black precipitate.

Treating it with ethyl alcohol, he dipped a piece of silk in it and found a wonderful purple color. This was the beginning of the modern dye industry, as well as the modern chemical industry, and his discovery was named aniline purple or mauve. This dye has nearly the same hue as Tyrian purple, but is a great deal less saturated, thus a lighter shade of color. At that time, all the dyes had natural origins, from plants, animals, and minerals. Perkin had accidentally found a substitute to the imperial purple. In 1856, Perkin obtained a patent for manufacturing the dye; the very next year, with the aid of his father and his brother Thomas, he developed a manufacturing process, procured raw material, and worked with textile manufacturers to produce the first synthetic dye in history.

The conservative Scottish textile manufacturers were slow to adapt to this new dye, so the leadership of market innovation was captured by the French textile manufacturers by default. Perkin's breakthrough came when Empress Eugenie of France wore mauve in public; it impressed Queen Victoria of England enough to wear mauve at her daughter's wedding, so it started a wave of enthusiasm for synthetic dyes. The mauve decade took place in the 1860s, when the dye became the rage in Europe. Perkin became very successful in commerce, and at the age of 36, he sold his business and returned to chemistry research. He was knighted in 1906 at the 50th anniversary of his discovery, and he died in 1907. Figure 8.3 shows Perkin when he was a very prosperous man. The Perkin Medal is named after him,

FIGURE 8.3   William Perkin. Reprinted with permission © National Portrait Gallery, London.

and is given each year for the most successful invention in applied chemistry. The award is celebrated every year with a formal dinner, except that the attendees wear bowties that are not in the usual black but in mauve to honor his discovery. Apparently, it has been very difficult to reproduce that delicate shade of color. At a Perkin Medal dinner, the CEO of DuPont, Dr Edward Jefferson, reported that his best chemists were not able to reproduce the precise shades of mauve of previous years.

Mauve had a short commercial life, as its early success led to a host of followers who did further chemical experiments that produced many superior dyes, particularly in Germany where many university and company researchers toiled to make modifications on existing dyes and to discover new families of dyes. It is ironic that Britain was the leader in the discovery and commercialization of synthetic dyes, but the marketing leadership was soon captured by France, and finally, the leadership in research and new dye technology was captured by Germany.

The most important attributes of synthetic dyes are their tremendous variety of shades to suit every taste and their low cost that makes them affordable to every person. Colorful clothing evolved gradually from a mark of class distinction for the high and mighty patricians to becoming the common way to express individuality and personal taste even for the lowly plebeians. Everyone gained when the world went from drab to rainbow, which also opened the doors to other synthetic materials. Since some synthetic dyes had colors that were much more dazzling than natural dyes, it logically followed that perhaps other man-made materials might be synthesized that would have properties that were very different or perhaps much better than those of natural materials. This logic led to the modern industry of synthetic chemicals, including plastics, rubber, fibers, and drugs. The global consumption of textiles has been estimated at 50 million tons per year, and the use of dye is estimated to be 750,000 tons.

SCIENCE AND TECHNOLOGY: COLOR

Visible light is a vibration of electromagnetic waves, with wavelengths between 4.0 (violet) and 7.5 (red) nm or nanometers. Isaac Newton discovered the power of the prism in splitting a beam of sunlight into a rainbow of colors, separated by their wavelengths. This separation is due to the fact that the shorter wavelength beams of blue and violet are bent more by the prism than the longer beams of red and orange. Can any particular color be represented as the sum of a few fundamental colors? Johann Wolfgang von Goethe, the great poet, developed one of the earliest theories of color, which is based on only three primary colors. The modern color theory was created by Young and Helmholtz. The color of any beam of light can be considered as the sum or addition of various amounts of primary colors. There is a physiological basis to the theory of three primary colors. The human eye has 120 million rods that are sensitive to black and white in low light and 6 million cones that function only in bright lights and are divided into three sets. The first set is sensitive to the longer waves that correspond to yellow light at 570 nm, the second set is sensitive to the medium waves that correspond to green light at 540 nm, and the third set is sensitive to the shorter waves that correspond to blue light at 440 nm. White is created by adding a great deal of all three lights, and black is created by removing all three lights.

There are two modern methods of creating colors. The *additive* method works by adding various proportions of lights of red–green–blue; the more you add, the brighter is the resulting beam. In a television or a computer screen with a cathode ray tube or light-emitting diodes, there are three phosphors that give red–green–blue lights. Red light plus green light gives yellow light, and the further addition of blue light gives white light. Black can be created by taking all three colored lights away. The *subtractive* method works by using absorbing dyes such as in an ink-jet printer, with usually three sets of inks that contain dyes that are yellow–magenta–cyan; the more you add, the darker the result. White can be created by adding no dye, and black can be created by adding larger amounts of all three dyes, but often a black ink is used as well to ensure that the result has no tint.

An even more important consequence of the Perkin invention is in the subsequent use of dyes for scientific and medical purposes. Under the microscope, different slices of tissue in biological specimens tend to have very similar flesh color and transparency, so that they are difficult to recognize and to distinguish. Histologists started using dyes to stain samples and found that some dyes adhere to bacteria much more strongly than to mammalian tissue; this selective dyeing is very helpful for identification, and leads to clear photographs of the microbes and organelles in the human body. In 1882, Walther Flemming found that the cell nucleus observed in microscopes has colored thread-like bundles of cells, which are now called the chromosomes. The mapping of the human genome would be impossible without appropriate dyes. Robert Koch used the dye methylene blue to discover the tuberculosis bacillus, for which he won the Nobel Prize in 1905. Coal tar derivatives were crucial in the work of Paul Ehrlich in his development of chemotherapy. It occurred to him that if some dyes adhere more strongly to bacteria than to mammalian cells, one could make the dye more lethal to deliver death to the bacteria; this is based on the principle of selective poisoning—to harm the bacteria more than the mammalian cells. Modern medicine has developed advanced dyes for tumors, which can then be blasted with a high-precision laser beam.

People associate colors with flavors and the color of food can influence the perceived flavor. Food color is used to make a product more appealing, and even oranges and salmon are sometimes dyed. The size of the global market for food coloring is around $1.2 billion per year, and falls under the regulation of the Food and Drug Administration in the United States. There are a number of permitted natural food dyes such as caramel and saffron, and there are seven permitted artificial colors, including the dye Red No. 40 responsible for the vibrant color of maraschino cherries. However, there is a growing concern about the consumption of synthetic dyes and increasing efforts are being made to replace them with natural dyes in the coloring of food and drinks.

8.2.2 Monuments

At the end of the Epic of Gilgamesh, the hero knew that he would not achieve immortality and he returned to his home. He said these noble lines:

> This is the wall of Uruk, which no city on earth can equal. See how its ramparts gleam like copper in the sun. Climb the strong staircase, more ancient than the mind can imagine, approach the Eanna Temple, sacred to Ishtar; a temple that no

king has equaled in size or beauty; walk on the wall of Uruk, follow its course around the city, inspect its mighty foundations, examine its brickwork, how masterfully it is built, observe the land it encloses; the palm trees, the gardens, the orchards, the glorious palaces and temples, the shops and marketplaces, the houses, the public squares.

These lines represent the civic pride about the city of Uruk, with monuments that no other city can compare with. These monuments do have some utilitarian function to perform, but they are also designed to give pride to the inhabitants; they are so difficult and expensive to build that they send the message that their rulers are powerful and successful. The city magistrates who build and maintain these public monuments bask in the reflected glory. They are also designed to intimidate the foreigners and barbarians beyond the gates, who should be awed and humbled and thereby would not dare to raid or challenge the leadership.

Some of the oldest public monuments are the ancient megaliths that have been found in many parts of the world, which may have been used for religious, administrative, or funereal purposes. The Gobekli Tepe in southeastern Turkey was erected around 11,000 BCE, with monolithic pillars and walls to form circular and oval structures with diameters of up to 30 m. The monoliths were decorated with carved reliefs of animals and abstract pictograms that might represent sacred symbols. Stonehenge is the most famous European prehistoric monument, and is composed of concentric circular sets of standing stones that may be as old as 3000 BCE. Some of the stones are 7 m tall and are constructed as trilithons with two vertical columns supporting a horizontal lintel. The circles of stones have diameters as large as 120 m. The structure has an orientation toward the northeast, and the Heel Stone was apparently placed to mark the position of sunset at the summer solstice on June 21. Constructing these monuments must have involved the immensely difficult technology of quarrying the megalithic stones and transporting them to the site, as well as standing them upright and lifting and placing the lintels on top of them. Without the benefit of steel tools, dynamite, lifting cranes, and engines, even modern observers are awed by these monuments to say nothing of the ancient visitors.

The Seven Wonders of the World was the name of a list of tourist attractions, first compiled by Antipater of Sidon in the second century BCE, of monuments in the eastern Mediterranean. The oldest of these Wonders were the Great Pyramids of Giza dating from 2550 BCE, which were covered with Greek graffiti at the time of Alexander and are the only surviving wonder today. Chronologically, the five Wonders in the middle of the list range from the Hanging Gardens of Babylon (600 BCE) to the Colossus of Rhodes (292–280 BCE). The youngest was the Light-house of Alexandria (280 BCE), which had a significant practical function of guiding ships at sea and was not designed purely for vanity and glory. These monuments all involved the technology of quarrying, transporting, and erecting large blocks of stone. The overall effect was calculated to impress upon the observer that the rulers who decreed this monument must be great leaders with tremendous resources, who should be obeyed and followed.

There are many other famous monuments that are located outside the eastern Mediterranean, such as in Persia, India, China, and the Americas. A list of the greatest man-made tourist attractions today would include the Great Wall of China, the Taj Mahal in India, the Cathedral of Notre Dame of Paris, Saint Peter's Basilica within Rome, Buckingham Palace in London, and the Golden Gate Bridge of San Francisco.

Towers and domes are the most spectacular structures that serve to inspire awe more than they function as architecture. In the biblical story, the Tower of Babel was a symbol of human pride that invited the punishment of god and was built in the tradition of the ziggurats of Sumer and Akkad. The Minaret of Samarra of 848 has a spiral staircase to its summit, which is one of the best preserved of that tradition. The functional purpose of a tower could be to survey troop movements, send signals that would be visible over a large distance, serve as a jail, and serve as a fortress to deter attackers by showering with stones or boiling oil from the top.

The traditional towers of India were called stupa; in China, they were called pagodas, such as the Great Goose Pagoda in Xi'an. In Europe, the tradition of a tall tower was revived during the Renaissance by the construction of bell towers such as the St. Mark's Campanile in Venice and Giotto's Campanile in Florence. The modern descendants are the Eiffel Tower of Paris and the Empire State Building in New York with a height of 380 m. There are a number of buildings vying for the short-lived distinction of being the tallest building in the world: the Petronas Tower of Malaysia at 452 m tall, the CN Tower of Toronto at 554 m, and the Burj Khalifa of Dubai at 830 m. Why do people commission tall skyscrapers with more than a 100 stories, when buildings of 50 stories would be just as functional and much more economical to build and to maintain? Is it from vanity and pride, to earn bragging rights?

The dome has a function in housing a large assembly indoors under a roof, and is often more impressive on the inside than on the outside. Samuel Taylor Coleridge wrote:

> *In Xanaadu did Kubla Khan*
> *A stately pleasure-dome decree:*
>
> *. . .*
>
> *That with music loud and long, I would build that dome in air,*
> *That sunny pleasure-dome! those caves of ice!*

An arch is designed to span a doorway, and a row of arches can be designed to form a barrel vault to span a rectangular space; or the arch can be rotated to form a dome and span a circular space. The simplest man-made domes were small dwellings of branches and straw, such as wigwams or igloos. The Treasury of Atreus in Mycenae was a tomb with a false dome or corbel, covered with a mound of earth. The corbel is made of a number of overlapping horizontal stones. The greatest ancient dome was the Pantheon, commissioned in 126 by the Emperor Hadrian and designed by the architect Apollodorus of Damascus. The Pantheon's interior encloses a spherical space that is 43 m in diameter: the dome itself is a hemisphere with a diameter of 43 m resting on a circular support that is 22 m high. It is made of unreinforced concrete resting on a circular wall that is 6.5 m thick at the base and there is an opening of 8 m in diameter at the top of the roof that is open to the sky. The biggest technical problem the builders had to solve was the tendency of a heavy dome to spread and to slump down. They solved the problem by having eight massive buttresses pushing against the dome. The size of this gigantic diameter was not challenged for the next 12 centuries, but some smaller domes were built in the meantime. When the Emperor Justinian built the Hagia Sophia in 537, which is only 32 m in diameter, he is said to have murmured: "Solomon, I have surpassed you." The Muslim Caliphs built the Dome of the Rock in Jerusalem in 685, which is a wood dome covered with gold leaves.

The Italian Renaissance brought a new dynamism and ambition to surpass the ancients and the Pantheon. The city council of Florence commissioned a dome with a large diameter at the Santa Maria del Fiore, which was to be raised on a drum so that it would be more visible from far away. They scorned the idea of buttresses and wanted a dome on the top of a tower, but they had no idea whether it was feasible. Filipo Brunelleschi (1377–1446) had a literary and mathematical education, and had enrolled in the silk-maker guild. He also took a trip to Rome with the sculptor Donatello to study antiquities. Brunelleschi was a man of many talents and accomplishments, and in 1421 he also received one of the first patents ever granted to anyone by the Council of Florence for his design of a hoisting machine that gave him a monopoly.

Brunelleschi won the competition to design the newly commissioned dome, which was to originate from 52 m above the ground and span 45 m in diameter. He could not use low and massive buttresses as in the case of the Pantheon, because his dome needed to rest on an octagonal cylinder to make it higher and more visible from a distance. His ingenious solution was to place eight massive piers forming an octagon to resist the downward thrust, which was then buttressed by side chapels and the nave. The octagonal dome made of brick and mortar rests on top of the drum and is ovoid in shape so that it is taller than it is wide, creating more downward pressure than outward spreading. Brunelleschi put four sets of heavy stone and iron chain around the dome, which act as a hoop around a barrel to hold it together and keep it from spreading. It is crowned with an octagonal lantern and reaches a height of 114 m. The Florence cathedral with the Brunelleschi Dome is shown in Figure 8.4.

The elegance of raising a dome high in the air without visible buttresses was considered miraculous and sparked many followers—Saint Peter's in Rome of 1626, the Taj Mahal in India in 1632, Saint Paul's Cathedral in London in 1677, and the US Capitol that was made of cast iron in 1855. But the diameter of the Florence dome was not challenged till the rise of modern steel beams that can resist flexing much better than stone and concrete can, and can thus be used to span longer distances. The first geodesic dome was built by Walther Bauersfeld in Jean, Germany in 1926, with subsequent developments by Buckminster Fuller. The geodesic dome is a spherical or partially spherical shell based on a network of great circles on the surface that intersect to form triangles and hexagons. It can be made of lightweight material, and can have a shape ranging from a hemisphere to a sphere. The current record for the largest permanent dome is the Louisiana Superdome in New Orleans with a diameter of 207 m.

FIGURE 8.4   Florence Cathedral with Brunelleschi's Dome.

8.3 ARTS

Our highest aspirations include the pursuit for beauty, truth, and enlightenment. We are concerned with inventions to increase material wealth and comfort, but we also need inventions to support the arts and sciences to uplift the spirit and to enlarge the mind. What is a liberal education for a civilized person? The ancient Greeks had nine Muses to inspire the creation of arts: three muses for epic, lyric, and choral poetry; two muses for tragedy and comedy in the theater; two for music and dance in performance; and two for history and astronomy. Note that there were no muses for painting, sculpture, philosophy, or mathematics. In Imperial China, a cultivated scholar had to master the four arts of calligraphy, painting, music, and chess. In medieval Europe, the school curriculum consisted of the seven liberal arts: the Trivium of grammar, rhetoric, and logic and the Quadrivium of geometry, arithmetic, music, and astronomy.

Many of these activities are served by technologies common to other pursuits. To create and preserve literature and poetry, we need to write them down on clay tablets or paper, print them for wide distribution, and organize them to be accessible in libraries and electronic databases. Other creative arts and sciences require specialized inventions to make them successful. Astronomy and the observation of celestial motion benefited from many centuries of observatory instrumentation and methods: the Almagest of Ptolemy in the second century discussed many of the critical instruments for observing the sky, and more than 1,400 years later, Galileo's achievements from observing the sky depended on his telescope. In the visual arts of painting and sculpture and also in the performing arts of music and theater, it would be impossible to make progress without the support of a large set of specialized inventions and technologies.

8.3.1 Drawing and Painting

Painting, sculpture, and architecture are considered the most important of the visual arts. Prehistoric art is known to us through cave paintings and small figurines that date from 50,000 years ago. They may have been created as magical aides for hunting, as spiritual and religious offerings, or simply as beautiful things for enjoyment in the long and cold winter nights.

Prehistoric drawings were made by rubbing a dry stick of color on a surface, and thus creating lines. The images created consist of only the colors of the sticks and the surface, which could be nonuniform. One of the oldest painted caves is the Chauvet Cave in Southern France that dates from 32,000 years ago. This was followed by the Altamira in Spain and the Lascaux in France that are half as old. These cave images consist of lines of black and brown mineral pigments rubbed onto the cave walls. They show a profusion of wild animals, antelope, sheep, camels, human figures, and handprints. The images of the human hand were thought to be made by placing their hands against the wall as a stencil, then spitting red ochre which they held in their mouths. Many other famous prehistoric wall paintings are found in Knossos, Crete; the caves of Dunhuang, China; and Ajanta and Ellora, India and also from Canada to Patagonia. Drawings often form an initial sketch and serve as a basis for more elaborate and finished versions of art.

There is a long evolution of many methods of drawing and painting, with different material properties. A drawing can be made simply by using a stick of charcoal or graphite to create black lines against a background of the original surface of paper. How would you make lines with different shades from a pale gray to a dark gray and to black? The invention of pencils involved mixing clay and graphite in different proportions to make sticks from the very black 6B to the ordinary HB, and then to the very pale 6H. Today, these sticks are enclosed in wood cylinders to make the traditional pencils, and the addition of an eraser at the other end makes corrections easier. Colored pencils are made by mixing pigments with clay in different proportions and are then encased in wood cylinders. The pigments can also be mixed with many other binders such as gum Arabic to make pastels, gypsum to make chalks, and wax to make crayons. When you have a box with 12 colored pencils, how would you create a color that is midway between red and blue, and how would you cover an area with a uniform color, or an area which is darker on the left and becomes gradually lighter on the right? If you decide that a given line is a mistake, can you cover it by putting another line or area on top? These challenges are difficult to meet by drawing with solid dry sticks, but they can be handled with the development of liquid paints, which are made by grinding and mixing pigments in fluid media such as water and oil. This makes it possible for the painter to mix colors on a palette or on the ground, and spread them in a thin line or over a wide area. Painting an image on a flat surface requires the following materials:

- *Pigment:* This gives the color in the painting and may be inorganic such as iron oxide or can be organic or synthetic.
- *Media:* Water or oil to disperse the pigments, often enforced with binders to improve adherence to the substrate; the drying oils have components that solidify to form thick layers.
- *Instrument:* Hand tool to apply the color that can be a pencil, crayon, charcoal, pastel, brush, or paint knife.
- *Substrate:* Surface or support base such as papyrus, parchment, silk, paper, wall, wood panel, and canvas; they are sometimes covered with a porous base such as gesso to improve the penetration and adhering of pigments.

Fresco is a water-based paint applied to fresh plaster on walls or ceilings. The pigment is mixed with water and painted on a thin wet layer of fresh lime mortar or plaster. The pigment in water sinks into the porous wet plaster, which absorbs the pigment. The pigment and plaster dry after about 10–12 h and react with the air to set permanently. The painter can control the intensity and saturation of a stroke by mixing more or less water to a unit of pigment. It is also possible to produce different shades of color by mixing two or more pigments together before application. If a mistake is made that needs to be covered up, the surface can be roughened and a second plaster layer is applied and repainted after drying. One of the oldest frescoes is found on the island of Crete in Greece, which contains a famous scene of boys and girls jumping over the backs of large bulls. The Ajanta Caves in India hold frescoes on the ceilings and walls that date from 200 BCE to 600. Many of the famous wall paintings in Pompeii as well as the magnificent ceiling of the Sistine Chapel done by Michelangelo are in fresco.

The modern watercolor is a very popular method to produce small portable art, usually on paper or silk. The water-soluble dyes or insoluble pigments are dispersed in water and mixed with a binder such as gum Arabic, which is an edible glue that comes from acacia trees in Africa and Arabia. The result is a picture with transparent washes, which allows much of the original surface to show underneath. If a mistake is made, there is no way to cover up the mistake by another layer, and a fresh painting must be made. This lack of covering power can be remedied by the closely related technology of *gouache*, which also uses pigments in water with gum, but is reinforced with an opaque powder such as chalk or titanium dioxide. When a stroke is made with this medium, the surface and the lines to be corrected are underneath and are not visible due to the opaque powder. Watercolor is a subtle and poetic form of painting, which is not employed to make dramatic statements. Durer,

Homer, and Cezanne were painters who have left us with enduring art in this medium. Chinese paintings evolved from Chinese calligraphy and are executed with a brush applying watercolor to paper or silk. In this type of "ink and wash" painting, the outlines and stems of a leaf would be lines, and the web of the leaf would be washes.

Tempera was an important painting method on dry plaster, found in Egyptian sarcophagi and on mummy portraits. It involves grinding dry colored pigments with a water-soluble binder medium, usually egg yolk, glue, and oil, to be applied to the wall. Tempera dries rapidly after being applied in thin semiopaque or transparent layers—which make possible the display of a higher concentration of pigments in a brilliant range of colors, far beyond what is attainable by surface adhesion. The base is inflexible Italian gesso, which is gypsum mixed with rabbit-skin glue and then spread as absorbent primer coat on a wood panel. Tempera is particularly appropriate for panels to hang on the wall, and remained popular from Byzantine times to the Italian Renaissance. There are many great works by masters such as Botticelli and Raphael.

Oil-based paintings give artists additional freedom, as by adding a resin or drying oil, the paint sets into a thick solid layer after exposure to air, which makes it possible to hold a high concentration of pigment. *Oil paintings* are done with pigments bound with a medium of drying oil, especially linseed oil, which reacts in air to turn into a solid holding the pigment and adhering to the surface. It was first used in western Afghanistan from the fifth to the ninth centuries and became popular in Europe during the Middle Ages. Vasari credits Jan van Eyck from The Netherlands with the invention of painting with oil media on wood panels. It is traditional to cover every square inch of the surface with paint, in contrast with watercolor where most of the surface is not covered. Oil painting eventually became the principal method of painting in Europe, starting from The Netherlands and then spreading to the Italy during the Renaissance.

The early oil paintings involved grinding pigments in oil and then painting on the surface with paintbrushes, palette knives, and rags. The surface was often a canvas made of linen or cotton fabric, nailed onto a wooden frame called a stretcher. The canvas had to be sized to isolate the fibers of the canvas from the acidic paint. The canvas would be coated with a layer of animal glue (size) from rabbit skin and then primed with lead white or chalk. Oil paintings have a number of significant defects: the flammability of the oil and turpentine thinner, the irritating and foul-smelling fumes, and the long drying time of 2–14 days that also delays efforts to cover a mistake with a subsequent layer.

Acrylic paint is actually a water-based paint, but shares many of the characteristics of oil paint. In addition to pigment and water, acrylic paint contains a water-soluble acrylic polymer resin that replaces the drying oil. There is no need of gesso to bind to the canvas and thinning is done by adding water instead of turpentine for easy spreading. After painting acrylic on the surface, the water evaporates and leaves the resin and pigments to form thick layers on the canvas. Thus, acrylic paint has the same brilliance and three-dimensional structure of oil paint, but it does not have the flammability and smell. Since it dries in about 10 min, it is also possible to put a correction layer on top of mistakes in a very short time.

Before attending to the creative work of observing a model and making a composition, a traditional painter had to grind and mix the pigments into water or oil, as well as add resins and gums. Pig bladders and glass syringes were sometimes used to store extra paints for the next day, as it might be difficult to reproduce a very good mixture. The paint tube was invented in 1841, so paints could be produced in bulk and sold in tin tubes with caps. An artist could also buy empty tubes and fill them with customized paints produced previously. In a modern paint tube, the top of the tube has a screw cap, the filling is done through the bottom, and then the bottom is folded and crimped. Painters can now make quick choices among numerous tubes of color that are reproducible, which was particularly useful for Vincent Van Gogh, who loved to do outdoor painting. All these modern inventions in painting technology give modern painters an incomparable toolbox far beyond what was available to the painters of the Lascaux cave. Pierre-August Renoir once said, "Without tubes of paint, there would have been no Impressionism."

8.3.2 Piano

When you walk into a living room with a grand piano that looks sleek and elegant next to the picture window, you would be likely to associate the owner with good social standing and culture. The owner must be able to afford the large cost and space that a piano demands, and some family member might have undertaken expensive and time-consuming piano lessons for many years. This family can gather around the piano and sing together, give solo performances in the living room or perform with other instruments. The modern piano is the result of a very long history of innovations and inventions.

Music and dance are among the oldest expressions of joys and sorrows by a few performers for a larger audience or by the entire community. Primitive people had simple music to express their joys in celebrations and their sorrows in mourning, which may begin with little more than rhythmic clapping or drumming to accompany dances and processions. Melody and harmony were added later to make music more complex and moving. Music today has three elements: rhythm or beat, such as the dactyl in a waltz (-..-..-..), melody or tune such as the *Twinkle, Twinkle, Little Star* (do do so so la la so), and harmony of several notes together at the same time such as the major triad (do mi so).

In classical mythology, the inventors and users of musical instruments were the gods, including Hermes, who invented the lyre when he found a dried-out tortoise on the banks of the Nile; Apollo, who played the lyre that was the divine instrument; and Pan, who played the panpipe that was a common instrument. David played music on the harp to chase the demon from the soul of Saul, and Joshua's trumpet brought down the walls of Jericho.

The oldest known musical instrument is a bone-flute created 35,000 years ago in southwest Germany, which was constructed from a vulture leg bone with four holes. From the excavations in the first cities in Sumer and Akkad were found flutes, trumpets, and drums. Harps and lyres were depicted in paintings on walls and objects. Egyptian tomb paintings showed many musical instruments. The number of strings in a lyre or kithara varied from 3 to 12. Plato suggested that music should be one of the basic influences in Greek education, as music shapes the soul and character.

Confucius said, "It is by the odes that the mind is aroused. It is by the Rules of Propriety that the character is established. It is from music that the finish is received." Good music guarantees a well-ordered community, whereas bad music brings the state into danger. The oldest book of Chinese poetry is the *Shi Jing* that dates from 1100 BCE, and the first poem is "Guan Ju:"

> *Guan, guan, sang the great birds*
> > *On an island in the river*
> > *Modest, retiring young lady*
> > *Sought by the accomplished gentleman*
> > *Modest, retiring young lady*
> > *We befriend with qin and she*
> > *Modest, retiring young lady*
> > *We rejoice with zhong and gu*

The "qin" is a fretted (like the guitar with bars across the fingerboard) long instrument with 7 strings; the "she" is unfretted (without bars, like the violin where the pitch can change continuously) long zither with around 25 open strings in pentatonic tuning; the "zhong" is a bell, and the "gu" is a drum.

The main types of musical instruments can be divided by the vibrating element:

- *Strings:* Put into vibration by plucking (harp), bowing (violin), or hammering (piano).
- *Air columns:* Put into vibration by blowing, woodwind (flute, oboe), brass (trumpet), and organ.
- *Solid objects:* Put into vibration by percussion or striking (bell, drum, cymbal, and xylophone).

Music is produced by vibrating instruments creating sound waves in air, of alternating compressions and expansions from the instruments to the ears of the listeners. We recognize and distinguish a musical note mainly by three qualities: pitch, loudness, and timbre.

MUSICAL NOTES

A music pitch is the frequency of vibration in the air, such as the note of A (or la) that is 440 Hz or 440 vibrations/s. When the pitch is doubled or halved, we recognize it as the same note but higher or lower by "an octave," such as 110, 220, 440, 880, and 1760 Hz. The piano keyboard has 88 keys and is divided into 7 octaves: the first octave is the lowest and the fourth is at the center. The male voice ranges from the second to the fourth octaves, and the female voice ranges from the third to the fifth octaves. The ear is normally most sensitive to sounds in the fifth to seventh octaves, which is part of the explanation of the dominant importance of the violin, the flute, and the soprano voice in music melody. Galileo recognized that the pitch of a vibrating string v depends on the length L, the tension T, and the density ρ of the string, according to the formula:

$v = (\pi/L)\sqrt{(T/\rho)}$. So to make the pitch go up, you can increase tension on the string, decrease the weight of the string, or make the string shorter. A violin has four strings, and they are tuned to G3, D4, A4, and E5—where the numbers refer to the relevant octaves. Since the four strings of violin are of the same length, they are designed and tuned by making the high E string lighter and tenser. On playing a violin, you can make a string shorter by pressing and sliding your finger on the string, so the frequency can change continuously from E to F and G, and so on. On a guitar, there are metal frets to help your finger to find the right length for E, but you cannot achieve an arbitrary length between E and F. On a piano, the length of a string is fixed.

The loudness is related to the amplitude of the sound wave, which causes the eardrums to vibrate, and is measured in decibels (dB). We are normally exposed to a range of loudness from 0 to 120 dB, where 0 dB is barely perceptible and 120 dB is during a jet takeoff, near the threshold of pain.

The timbre or tone quality involves the complexity of waves. When a string vibrates, it produces not only the lowest fundamental note but also a number of "higher harmonics" whose frequencies are multiples of the fundamental frequency by 1, 2, 3, ... Hermann von Helmholtz discovered that the tone quality of a complex tone depends on the complexities of the harmonics that it contains. A flute sounds more "pure" as it is made of mainly a single wave, but an oboe sounds more reedy and colorful as it has a much richer set of overtones.

Music requires melody, which is a sequence of notes to make a tune, such as do-do-so-so-la-la-so for Twinkle, Twinkle, Little Star. Music also requires the harmony of several notes at the same time as in a duet or a quartet. When two notes sound at the same time, do they sound harmonious or discordant? Pythagoras, the great Greek mathematician, declared that two notes sound harmonious together when their frequencies are in the ratio of small whole numbers such as 2/1 that is recognized as the same note and monotonous, and 3/2 that is do-so, regarded as perfect consonance. We can make the palette richer by adding other intervals such as 4/3, 5/3, and 5/4, regarded as imperfect consonance. Things get worse and more dissonant when we consider even higher ratios such as 13/7. So, if we have only a few notes, the music is more harmonious but monotonous; and if we have more notes, the music is richer but more dissonant. How many notes should we have in an octave to make music both rich and harmonious? After centuries of evolution, the dominant scales in modern Western music are the Pentatonic Scale with 5 notes (such as all the black keys on the piano), the Diatonic Scale with 7 notes (such as all the white keys on the piano, designated as C, D, E, F, G, A, and B), and the Chromatic Scale with 12 notes (such as both the white and black keys, designated by adding sharps ♯ and flats ♭).

A composer can write a piece of music in any key, such as D or G, but a performer can transpose it and play it in another key. How should you tune the piano so that the notes will be as harmonious as possible and still be able to transpose a piece of music by a scale change? The best harmony is achieved by using exclusively the Pythagoras ratio of 3/2 and repeating the ratio till we have all the notes; the problem with this method is that the higher notes are increasingly more dissonant, especially when you change scale. A compromise is the well-tempered scale, which makes the interval between two adjacent notes all the same. Advancing

12 times makes the pitch go up by a factor of 2 in an octave, so the size of each individual step must be $2^{1/12} = 1.0594631$. This is what we find on a piano today, which makes it very easy to change scale; the drawback of this method is that none of the scales are strictly harmonious. For instance, the interval from C to G becomes $2^{7/12} = 1.4983$ instead of $3/2 = 1.5000$. This error may not be too evident for most people who do not have perfect pitch, but it could bother those with perfect pitch, and would be a disappointment to Pythagoras who believed that the heavenly harmony of music must be perfect.

Most musical instruments can make only one note at a time, such as the flute or the trumpet, and can carry out a melody as a sequence of notes. They are not often used for solo performance, as they cannot simultaneously produce many notes to provide harmony or complex polyphonic rounds. Some instruments can manage to do two notes at the same time, such as plucking two strings on a harp or banging two sticks on a xylophone. A keyboard instrument can play up to 10 notes at the same time using 10 fingers, which can simulate the complexities and harmonies of an orchestra of many instruments or a choir of voices from soprano to bass. It is suitable as a solo instrument, as well as to provide accompaniment to other instruments or singers.

During the Renaissance (1400–1600), the most important keyboard instruments were the clavichord and the harpsichord for small ensembles and the organ for churches. Each of these instruments has unique ways to handle the problems of activation of the vibrating strings or pipes, the pitch of the sounds, and the loudness and duration of the notes. The pipe organ has many pipes of different lengths, such as 16, 8, and 4 ft, which determine the pitch. When the organist's finger pushes down on a key such as C, a pipe is activated by air blowing from the wind box at a slot, and the volume depends on the pressure in the wind box. The sound continues with no change in volume as long as the finger depresses the key, but the sound stops when the finger is removed and the wind no longer blows into the pipe. All the pipes that are fed from the same wind box will have the same pressure and thus similar volumes, so that an organ cannot easily vary the volume from soft to loud.

The clavichord has vibrating strings attached to a soundboard. The pitch of each string is controlled by the length of the string, the weight of the string per length, and the tuning that adjusts the tension on the string. When a key is pushed down, a jack at the far end is raised that throws a hammer or a pick against the string. The clavichord uses a triangular piece of metal that is called a "tangent," which strikes the string and stays in contact with the string as long as the key is held down. The volume of sound created is louder when the force is greater, but the clavichord is a quiet instrument that cannot be heard over a loud orchestra. It has a damper for each string that rises, allowing the string to vibrate without restraint when the key is down, but when the key is released the damper drops to rest upon and silence the string.

In the harpsichord, each string has a pick that is made of quill or leather, which plucks the string on its upward motion. The sound has good volume, but it is fixed and does not depend on how hard the finger pushes the key. This constant volume gives the harpsichord a disadvantage when the music requires an emotional range from a mighty roar to a soft whisper; so at the emotional climax of a piece, the player can only play faster instead of louder!

Both the harpsichord and the clavichord function well with small ensembles in aristocratic drawing rooms with a small audience, but function less well in large, modern concert halls. During the Romantic Period (1750–1900), the middle class began to rise in wealth and influence and thus began to attend the large public concerts that took place in concert halls instead of in small aristocratic salons. When a single violin is not loud enough, we can put 10 or 40 violins in the orchestra. Orchestras continued to increase in size, but the existing keyboard instruments could not produce enough volume to keep up with the trend of larger audiences and concert halls. Thus, the world needed a keyboard instrument that could play many notes at the same time and could play with enough dynamic range to compete with a large modern symphony orchestra. The piano was invented as a keyboard instrument capable of complex harmony like a harpsichord, while also having the ability to be both loud and soft like a violin.

Bartolommeo Cristofori was a harpsichord maker from Padua who traveled to Florence to work for Ferdinando Maria de' Medici, then Prince of Tuscany and descended from Lorenzo the Magnificent. The Prince had 40 harpsichords and spinets, and hired Cristofori to take care of his instruments. By 1700, Cristofori had produced the first piano, which he called the "Arpicemablo...di nuova inventione, che fa' il piano, e il forte"—a new invention that can be played soft or loud. Most of our information about the invention came from an article written in 1711 by Marquis Scipione Maffei, who was the publisher of a Venetian magazine, *Giornale de' Letterati d'Italia*. Maffei said,

> Everyone who enjoys music knows that one of the principal sources from which those skilled in this art derive the secret of especially delighting their listeners is the alternation of soft and loud. This may come either in a theme and its response, or it may be when the tone is artfully allowed to diminish little by little, and then at one stroke made to return to full vigor - an artifice which has often been used and with wonderful success, at the great concerts in Rome.

> Now of all this diversity and variation of tone, in which the bowed instruments excel among all others, the harpsichord is entirely deprived, and one might have considered it the vainest of fancies to propose constructing one in such a manner as to have this gift. Such a bold invention, nevertheless, has been no less cleverly thought out than executed in Florence, by Mr. Bartolommeo Cristofori, a Paduan, a harpsichord player in the employ of His Serene Highness the Prince of Tuscany. . . . . The bringing out of greater or less degree of sound depends upon the varying force with which the keys are pressed by the player; and by regulating this, may not only loud and soft be heard, but also gradations and diversity of sounds, as if it were a violoncello.

The first Cristofori piano had 49 keys, and had neither legs nor pedals. Cristofori declared the piano to be his invention as an intellectual property, but he did not take out a patent, despite the fact that patents had been granted in Florence three centuries earlier to Filipo Brunelleschi the architect. The Maffei article gave details on its construction in words and attached a drawing, so that any skilled harpsichord maker could study the document and produce their own pianos. So why did Cristofori authorize this publication so that a competitor could copy his invention without paying a fee or royalty? Perhaps the patent law of his day would cover only the Tuscan area, but had no force outside. Perhaps he was more motivated by receiving credit and fame than profit, as long as he was working for a generous Prince and had a guaranteed income. Cristofori could only produce one piano at a time, upon request or commission from individual patrons, followed by painstaking craftsmanship since he had no means to mass produce for large profit. However, he did achieve glory across Europe for this invention.

Instead of the quill jacks of a harpsichord, the Cristofori piano has hammers that strike the strings from below. Each hammer has a head covered with deer leather and hangs from a rack. When a key is hit by the finger, the hammer is thrown upward to strike the string and then falls back a short distance so as not to dampen the vibrations. At the same time, the damper on the string is raised, so that the vibration will not be hampered as long as the key is held down; but when the finger on the key is released, the damper comes down to muffle the sound. The strings in the piano must resist being knocked out of tune, which requires greater tension. Cristofori used the same string lengths as in a harpsichord, but used a heavier string weight to withstand the greater tension. The original soundboard was made of cypress wood, but later piano makers shifted to spruce as a better material. Cristofori used two strings per note, which were made of metal such as brass and steel. One of his pianos from 1722 is now in a museum in Rome, looking plain and without adornment, and very much like a harpsichord with no more than four octaves, and lacking pedals of any sort.

Numerous developments came later to make the piano more versatile and powerful. Modern pianos have three pedals to give the player more control over the sound from the piano. The pedals seem to have been invented in England and arrived in Germany around 1780 in time to delight Wolfgang Mozart. In a grand piano, the right pedal (damper) is designed to lift the dampers from all the strings so they can all vibrate freely. The left pedal (soft or *una corda*) shifts the entire carriage of hammers to the right by fraction of an inch, so that for notes that have two or three strings, the hammer will hit only one string and make less sound. This *una corda* mechanism was invented by Cristofori, but before pedals were added, it was operated by hand. The center pedal (*sostenuto*) selectively lifts only the dampers for the strings that belong to keys that are depressed at the time, so that other strings cannot vibrate in sympathy. When larger volumes of sound were required from the piano, the solution was to employ heavier strings under higher tension, so that the wood frame would no longer suffice. In 1820, William Allen and James Thom introduced the iron frame for the strings, which can stand much higher tension and has much greater dimensional stability. Henry Steinway in 1859 introduced cross-stringing, where the base string crosses over at an angle on top of the treble strings, allowing the piano to become shorter in length.

During the Baroque period (1600–1700), the best composers, such as Vivaldi and Bach, wrote for the harpsichord and the clavichord. By the Classical Period (1730–1820), the transition to the piano had taken place, which totally eclipsed the two predecessors of harpsichord and clavichord. The German piano maker Gottfried Silbermann in Dresden showed a piano to Johann Sebastian Bach in 1736, who criticized it as an unsuitable instrument. The Silbermann piano was improved later and presented to Frederick the Great, who was reputed to have bought 15 grand pianos. In 1747, when Bach visited Frederick at Potsdam, he was shown the improved pianos and received a theme on which to compose a fugue, resulting in his masterpiece, *Musical Offering*. It was now possible for a keyboard composer to mark the music *piano* and *forte*, with *crescendo* and *diminuendo* to indicate when the music should become gradually louder or softer. Later the piano became increasingly more powerful. In 1791, Haydn (1732–1809), who was accustomed to conducting his symphonies while playing the harpsichord, switched to conducting from the piano. Mozart (1756–1791) followed Haydn's lead and wrote many pieces of music for the piano.

Toward the end of the 1700s, the switch to piano was complete in Western Europe and North America. When 28-year old Thomas Jefferson was planning to get married in 1771, he sent his agent a list of things to buy in Europe. He wrote,

> I must alter one article in the invoice. I wrote therein for a Clavichord. I have since seen a Forte-piano and am charmed with it. Send me this instrument instead of the Clavichord; let the case be of fine mahogany, solid, not veneered, the compass from double G to F in alt, a plenty of spare strings; and the workmanship of the whole very handsome and worthy of the acceptance of a lady for whom I intend it.

Ludwig van Beethoven (1770–1827) had a much more emotional and stormy approach to music, which is very suited to the piano. His dynamic music was one of the main drivers for the replacement of harpsichords and clavichords by the piano. Beethoven wrote 32 piano sonatas for solo performance, 10 sonatas for piano and violin, and 5 sonatas for piano and cello. He also wrote five piano concertos in comparison to a single violin concerto. Beethoven's music would sound totally out of place on a harpsichord, and must be played on a piano because of its superb dynamic range. He was particularly fond of a Broadwood grand piano in 1818 that has six octaves.

This was also the period that music became much more important in the western world, and the status of musicians rose from that of a servant in the time of Haydn in the Esterhazy household and of Mozart at the house of the bishop of Salzburg to superstar at the time of Franz Liszt. When the virtuoso Franz Liszt (1811–1886) arrived on the scene, there was no more mention of the older harpsichords and clavichords. When Liszt was giving a concert in Paris, he knocked some strings out of tune and broke others, so the concert had to be stopped while the strings were replaced and tuned. Heinrich Heine wrote about Liszt in 1844 stating, "He is here, the Attila, the Scourge of God, for all Erard's pianos, which trembled at the news of his coming, which writhe, bleed and wail under his hands." The arrival of the jazz pianists gave another glamorous chapter to the history of the piano, with names such as Count Basie, Dave Brubeck, Scott Joplin, Nat King Cole, Duke Ellington, and George Gershwin.

The piano is the most widely used musical instrument today, be it as a solo instrument, as an accompaniment to singing, as a part of chamber music, or with a symphony orchestra in piano concertos. George Bernard Shaw said, "The pianoforte is the most important of all musical instruments; its invention was to music what the invention of printing was to poetry." The piano became a symbol of elegant social behavior and an obligatory part of social life. When the middle-class families in Europe and America became more prosperous, they spent more time on culture and embraced the piano as a symbol of sophistication and upward mobility. Many families put a piano in the living room and sent their children to piano lessons to demonstrate that they had arrived at a life of culture and social standing. Girls learned to play the piano because it was central to the criteria of marriage prospects. Music at home, performed by the adults or the children, was a major form of home entertainment, especially during weekends and holidays.

In the early days, only the wealthy and the nobility could afford to own a piano and the instrument workshops were small and low volume. During this aristocratic age, a piano would be custom-built after the customer had discussed any special requirements with the piano maker. The initial reception of the piano was slow because of its expense. Broader acceptance of the piano came only in the 1760s, when general economic prosperity and cheaper pianos combined to make it possible for more people to acquire one. Johannes Zumpe designed the piano for the first time as a popular instrument rather than a courtly ornament. It was priced at 16–18 guineas, about one-third the cost of a good harpsichord, and it was exceedingly reliable due to its simplicity of design. It was written that "There was scarcely a house in the kingdom where a keyed instrument had ever had admission, but was supplied with one of Zumpe's piano-fortes for which there was nearly as much call in France as in England. In short he could not make them fast enough to gratify the craving of the public."

In the bourgeois age, a piano maker would build many pianos of standard design and then offer his inventory for sale. In London, the harpsichord maker John Broadwood sold his first square piano in 1780, and his first grand piano in 1785. He had two categories of customers: (i) the regular customers, made up of nobility and gentry, and (ii) the chance trade, which were the people who walked into his showroom on Great Pulteney Street. The chance trade included newly middle-class parents and daughters eager to emulate a more refined style of life.

In England, the price of a fine grand piano in the early 1800s was 84 pounds. In New York in 1854, a Steinway square sold for $550, and in comparison the annual wage of a skilled worker was $625–$1000. The number of pianos sold increased from 25,000 in 1869 to 350,000 in 1910, through propaganda by the makers of high-end pianos and the aggressive selling methods of the makers of commercial pianos. Installment buying was introduced by Kimball, while mail order was introduced by Sears, Roebuck and Montgomery Ward. The makers advertised to markets that pianos were now "within the reach of farmer on his prairie, the miner in his cabin, the fisherman in his hut, the cultivated mechanic in his neat cottage in the thriving town," claiming that a piano would provide "an influence which refines his home, educates his children, and gladdens his daily life like a constant ray of sunshine on his hearth." Manufacturers, such as Steinway, would sell a few high-end pianos for prestige and advertising, but depended on large volumes of low-end pianos sold to the middle class for greater profits.

The dominant status of the piano in middle-class homes remained secure until broadcast and recording technology made possible the enjoyment of music in more passive forms with less investment of money and time. First came the player piano in 1900 that could run on punch cards just like the Jacquard loom. Then came the home phonograph, which became popular just before World War I; this was followed by the radio in the 1920s, which became a more important form of home recreation. During the Great Depression of the 1930s, piano sales dropped sharply and many manufacturers went out of business. In the 1980s, another blow to the industry was the widespread acceptance of the electronic keyboard. These keyboards were initially considered to be a poor substitute for the tonal quality of a good piano, but they were cheaper, more flexible, able to imitate various instruments, and more suited to the performance of popular music. As a consequence, the acoustic pianos sold today have migrated back to the higher end of the marketplace and tend to be of higher quality and cost than those of several decades ago. They are now more often found in the homes of wealthier and better-educated members of the middle class. Many parents feel today that piano lessons teach their children concentration and self-discipline, and open a door into the world of classical music. The piano continues to enjoy a superior social standing, as it requires hours of lessons and practice that denote dedication and discipline in comparison to the passive entertainment of radio and television. Magazine advertisements of Toll Brothers luxury homes often show a photograph of a living room with a grand piano, but the piano no longer has the monopoly of music at home.

REFERENCES

Bailey, J. and D. F. Ollis. "Biochemical Engineering Fundamentals", McGraw-Hill, New York, 1977.

Beer, T. "The Mauve Decade: American Life at the End of the Nineteenth Century", A. A. Knopf, New York, 1926.

Blom, E. "The Romance of the Piano", Da Capo Press, New York, 1969.

Gaines, J. R. editor, "The Lives of the Piano", Holt, Rinehart and Winston, New York, 1981.

Garfield, S. "Mauve: How One Man Invented a Color That Changed the World", Norton, New York, 2000.

Good, E. M. "Giraffes, Black Dragons, and Other Pianos: A Technological History from Cristofori to the Modern Concert Grand", Stanford University Press, Stanford, CA, 2001.

Huizinga, J. "Homo Ludens: A Study of the Play-Element in Culture" (translation of German edition 1944), Routledge, London, 1949.

Johnson, H. "The Story of Wine", Mitchell Beazley, London, 1989.

King, C. J. "Separation Processes", McGraw-Hill, New York, 1980.

Kladstrup, D. and P. Kladstrup. "Champagne: How the World's Most Glamorous Wine Triumphed over War and Hard Times", Harper Collins, New York, 2005.

Kottick, E. L. "A History of the Harpsichord", Indiana University Press, Bloomington, IN, 2003.

Liger-Belair, G. "Uncorked: The Science of Champagne", Princeton University Press, Princeton, NJ, 2004.

Loesser, A. "Men, Women and Pianos: A Social History", Dover Publications, New York, 1954.

Mazzeo, T. J. "The Widow Clicquot: The Story of a Champagne Empire and the Woman Who Ruled It", Harper Collins, New York, 2008.

McGrayne, S. B. "Color and W.H. Perkin", "Prometheans in the Lab", McGraw-Hill, New York, 2001.

Parakilas, J. editor, "Piano Roles: Three Hundred Years of Life with the Piano", Yale University Press, New Haven, CT, 1999.

Phillips, R. "A Short History of Wine", Harper Collins, New York, 2000.

Sachs, C. "The History of Musical Instruments", W. W. Norton & Company, New York, 1940.

Shreve, R. N. "Chemical Process Industries", McGraw-Hill, New York, 1967.

Smits, A. J. "A Physical Introduction to Fluid Mechanics", John Wiley & Sons, Inc., New York, 2000.

Stanier, R. Y., M. Doudoroff, and E. A. Adelberg. "The Microbial World", Prentice-Hall, Englewood Cliffs, NJ, 1963.

Wang, D. I. C., C. L. Cooney, A. L. Demain, P. Dunnhill, A. E. Humphrey, and M. D. Lilly. "Fermentation and Enzyme Technology", John Wiley & Sons, Inc., New York, 1979.

FUTURE CHALLENGES

Inventions since the first stone axe 2 million years ago have served us well, by giving us capabilities to accomplish astonishing feats, to work more productively; and to provide for basic necessities of food and shelter, health care and security, transportation and information, and support for the good life. In recent years, "We have replaced tape recorders with iPods, maps with GPS, pay phones with cell phones, two-dimensional X-rays with three-dimensional CT scans, paperbacks with electronic books, slide rules with computers, and much more." This statement was declared in the report, *"Rising Above the Gathering Storm."* The rate of progress in the last few decades has been particularly striking, and raises the expectations in many people that this momentum will be maintained if not accelerated. There are many urgent and long-term needs to be addressed, and there are many scientific advances presenting new opportunities. Instead of resting on their laurels, inventors are urged to work ever harder and more productively. However, inventions depend critically on social investments in education and research, and there are signs of distress that the levels of support will not be adequate to support expectations.

9.1 FUTURE NEEDS AND OPPORTUNITIES

Which human needs have the highest priority, and are most worthy of investigation? Which investigations at scientific frontiers are the most likely to yield inventions? Inventions are made by inventors who are experts in technology, for the benefit of society who specifies the needs. The leaders in government and business who control the funding for education and research need to examine this vast menu and rank the top priorities for inventions within their budgets.

9.1.1 Market-Pull Needs

What are the most urgent unsolved problems of today? A starting point is the Human Development Index of the United Nations Development Program that ranks the quality of life in each nation according to three criteria:

(1) Population health and longevity, measured by life expectancy at birth.

(2) Education and knowledge, measured by the adult literacy rate and school enrollment.

(3) Standard of living, measured by GDP per capita.

Improvements in these quality-of-life criteria need new inventions and technologies.

We all desire a long and healthy life, but that is threatened by infant mortality, malaria, HIV/AIDS and other infectious diseases, and drug-resistant microbes. In the developed nations, the big killers are stroke, cancer, heart attack, and lung diseases. The Bill and Melinda Gates Foundation issued the "Grand Challenges in Global Health" with seven goals to develop: improved vaccines, new vaccines, control of insect vectors, improved nutrition, reduced drug resistance, cures for infections, and ways to measure health status. Longer lives also lead to new problems, as the quality of life of the aged becomes more significant.

The most important basic resources to sustaining life are food, water, and energy. We cannot rest even when synthetic fertilizers and new plant species help to produce four times more food per acre of land than before. As long as the human population and income continue to expand to consume all resources available, we remain perpetually in a Malthusian race between expanding population and demands and increasing resources from new inventions. Fresh water has become increasingly scarce in many arid countries, and the problem has become more acute due to global warming. So far the main solution has been to divert water from wetter regions to drier regions. We know how to desalinate brackish water and seawater for human drinking, but the process is too costly for irrigation and washing. We have been rapidly depleting the fossil fuels of coal, oil, and natural gas, which cannot be replenished. We need to develop abundant and cheap energy supplies from renewable biomass, wind, solar, hydro, and nuclear sources. All these new energy supplies also generate problems that need solutions.

The wrath of nature produces numerous sudden geohazards to people, such as tsunamis, earthquakes, volcanic eruptions, hurricanes, and river and coastal flooding. The forces unleashed by these geohazards are probably too large to be controlled in the foreseeable future, and the best solution may be early prediction and detection, leading to early warning and evacuation to safer ground. The destruction caused by these calamities pales in comparison with the destruction from long-term changes in patterns of rainfall, which result in droughts and floods and turn breadbaskets into deserts, or raise the ocean level to inundate low-lying ground in places like Bangladesh and oceanic islands. Geoengineering is a set of ideas that would lower the level of global warming without the costly step of lowering carbon dioxide emission. Some of the ideas include putting mirrors into space orbits to deflect some sunlight, spraying sulfur in the air to create clouds, or fertilizing the ocean to encourage the growth of phytoplankton for photosynthesis and absorption of carbon dioxide. These proposals also have the potential to create major problems for the environment, and must be tried out first on a small scale and monitored carefully.

Perhaps the most haunting threats to humans are from human actions, such as terrorists obtaining nuclear weapons or devices and detonating them in major metropolitan areas. Decaying nuclear installations in Russia are a potential source of nuclear material for terrorists. This threat is increased when many unstable nations acquire nuclear capabilities, such as North Korea, Iran, and Pakistan. Besides guarding us from nuclear threats, another form of security is needed to protect us from hackers and attackers invading the files and communications of national security, financial institutions, and personal accounts.

We have altered the environment in many ways, which frequently causes harm. The earliest environmental concerns were about water and air pollution. Our environment is frequently clogged with pesticides, trace metals, chemical mutagens, and ionizing radiation. We also disrupted many natural ecological systems and damaged sustainability and biodiversity. The ozone hole would be relatively easy to cure by banning chlorofluorocarbons. Solving global warming by controlling the production of greenhouse gases will be enormously costly and disruptive unless we have bold new inventions, including the controversial geoengineering.

Sometimes today's solutions become tomorrow's problems, so we often have to fix the last solution by finding another solution, which may create yet another set of problems. This may be stated as a case of Hegel Dialectics that a highly successful thesis leads to the contradictions of an antithesis, and the tension between the two must be resolved by a synthesis. But the synthesis becomes a new thesis, which would in turn leads to a new antithesis and would require a new synthesis. We also have needs associated with powerful long-term trends that are changing the world, and we must find ways to adjust. Thus, there will never be an end to our reliance on new inventions.

The world population in 2011 is 6.8 billion people, and is growing at a rate of 1.4% per year; however for the next 10 years, the growth rate is predicted to slow down to 1.1% per year. In the last century, the nations with higher incomes have had a lower growth rate than nations with lower incomes.

Region	1990–2006	2006–2015
World	1.4	1.1
Low income	2.0	1.7
Lower middle	1.1	0.8
Upper middle	0.9	0.6
High income	0.7	0.4

The population growth rates are highest in Sub-Saharan Africa and Central America at the level of 2.5–3.5% per year. The lowest growth rates are found in Eastern European nations such as Estonia, Russia, and Ukraine, which range from 0 to −1.0% per year. A slow growth rate of population can be due to a low birth rate, or high emigration to other nations due to poverty or war.

Inventions in modern medicine have played a major role in the trends of lower birth rates and longer life expectancies, which in turn have led to a change in the age distribution of the population, toward fewer young people and more ageing seniors. In the fast growing nations of Algeria and Uganda, children between the ages of 0–15 make up 50% of the population, but only 2% of the population were 65 or older. However, in the most industrial nations of Western Europe and Japan, 14% are children, which is fewer than the seniors who make up 20% of the population. The growing share of seniors in high-income nations requires more attention to inventions of geriatric goods and services, which are very different from the juvenile goods and services for children.

Another consequence of the changing age distribution is the shortage of young workers, which is greatly needed for many types of industry and business. If a developed nation has a liberal policy, a shortage of youthful workers can be partially filled by immigrants from less developed nations, which sometimes leads to cultural conflicts. With this long-term shortage of young workers in our future, inventions such as intelligent robots could lessen this impact.

Ten thousand years ago, the Neolithic Revolution changed the main occupation on earth from hunter-gatherers to farmers, as farming is far more productive per hour of work. A similar change occurred in the Industrial Revolution of 1800, which released millions of underemployed farm workers into the cities for industrial and manufacturing jobs, which were more productive and thus paid better wages. In 1810, 84% of American labor was engaged in agriculture, 8% was employed in industry, and the remaining 8% were in services, such as business, government, the professions, and domestics (see Fig 9.1). From 1810 to 1929, the share of American labor in farming declined steadily to 21%, while industry rose to 30%. This process of industrialization in America reached its peak in 1965 when employment in farming dropped to 6% and industry rose to 33%. From that point in time, America has steadily deindustrialized so that by the late 1900s, farming had

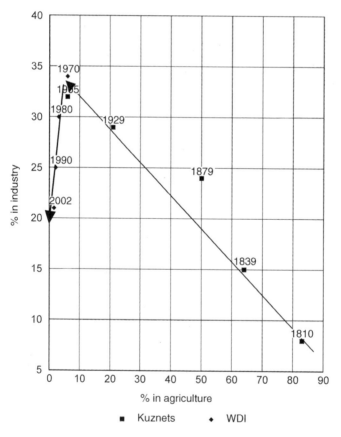

FIGURE 9.1 US labor structure, 1810–2002.

declined further to 2.5% and industry labor had declined to 19%. This deindustrialization was primarily caused by the phenomenon of outsourcing, which involves moving manufacturing plants overseas where labor wages are much lower, and large companies like Wal-Mart are principally sustained by manufactured goods from abroad.

When we look at a snapshot of the division of labor in the world in the year 2000, we find a striking resemblance to the historic path of the United States from 1810 to 2002 (see Fig 9.2). Specifically, the labor structures of Bhutan to Somali very much resemble that of the United States in 1810, and China today resembles the United States in 1860, while Singapore and Japan resemble the United States in 1965 at the peak of industrialization. Hong Kong today is even more deindustrialized than the United States, as land in Hong Kong is scarce for farming and factories, wages there are high, and the region is devoted mainly to shipping, finance, and

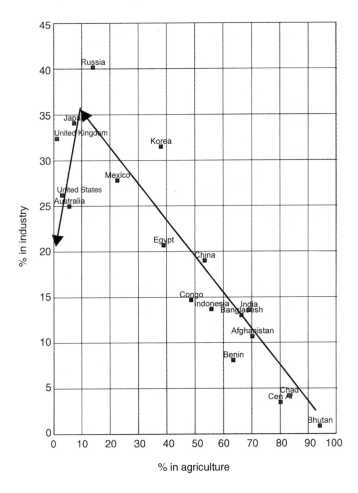

FIGURE 9.2 World labor structure in 2000.

trade. In a postindustrial world, inventions for the growing service sector are needed more than for the shrinking industry and agriculture sectors.

In the developing nations such as China, industry and service are many times more productive than agriculture.

Sector	National labor (%)	National output (%)	Output/labor
Agriculture	29	6	0.21
Industry	21	28	1.34
Services	50	67	1.33

It would be very profitable for the underemployed agricultural labor to move to the cities, but education and training are required to change agricultural labor into effective and productive industry and service workers. Such a migration would also increase the national GDP/capita.

9.1.2 Tech-Push Opportunities

The frontiers of science and technology present many leads and opportunities for future inventions based on recent discoveries that promise benefits to mankind. George Bednorz and Alex Muller shared the 1987 Nobel Physics Prize for high-temperature superconductivity of ceramic materials, which cause little or no loss of electric energy in conduction. It was widely predicted that superconductors would revolutionize many branches of technology, such as magnetic levitation trains, but its practical impact has been minimal up to this time. There used to be three forms of carbon: amorphous soot, graphite, and diamond. But Curl, Kroto, and Smalley shared the 1996 Nobel Chemistry Prize for their discovery of fullerenes, another form of carbon with marvelous and unprecedented properties of strength and flexibility. Plastics have been used for decades to insulate copper wires. But Heeger, MacDiarmid, and Shirakawa shared the 2000 Nobel Chemistry Prize for the discovery and development of conductive plastics. These two inventions represent the unfulfilled potential for future inventions.

Nanotechnology involves manipulating and manufacturing materials and devices that are smaller than 100 nanometers, or 0.1 millionth of a meter. The importance of small devices was explained by Richard Feynman who said, "There is plenty of room at the bottom." One of the most powerful nanotechnology techniques recently developed is called molecular self-assembly, in which molecules assemble themselves into a designed pattern. Nanotechnology has great potential in medicine for diagnostics, drug delivery, and tissue engineering—the production of tissues and organs by growing one cell at a time. Nanotechnology may also find applications in information and communications, as memory devices, displays, and optoelectric devices.

The computer and the Internet have revolutionized our search for information and our communication with each other. The meteoric rise of companies such as Microsoft, Apple, Amazon, and Google, showed how the modern world depends on the information and communications technology (ICT). They are producing a revolution in games, entertainment, virtual reality, health informatics, enhanced virtual reality, and advanced personalized learning. An OECD report said that ICT is spending 2.5 times more money on R&D than the automotive industry, and three times more than the pharmaceutical industry. Much more can be expected in the future.

Advances in biology and genetics made a revolutionary leap when Watson and Crick deciphered the molecular basis of genetics, as paired strands of DNA molecules. Genomics is the determination of the sequence of DNA molecules, from virus to the exciting Human Genome Project. The knowledge of the entire blueprints of plants and animals has the potential to make advances in agriculture, such as improved crop yield, reduced vulnerability to environmental stresses, and increased nutritional values. Knowledge of the genes of each individual can make possible "designer medicine" to target a person who has unusual genes and may be sensitive to things in a different way from the average person.

9.2 FUTURE SOURCES OF INVENTIONS

Where are future inventions most likely to come from? Who will sponsor future inventions, and would the support be adequate for so many needs? Would the current domination of inventions in a few rich nations be replaced by "polycentric innovation," where many players around the globe will make significant contributions?

9.2.1 Future Cradles of Inventions

The cradle of invention for the stone axe was in East Africa 2 million years ago. Later, the centers for inventions moved to Mesopotamia and Egypt, and then to China, Greece, and Rome. After 1500, Western Europe (United Kingdom, France, Germany) and later North America became the most important sources of modern inventions. There is an expectation that in the future, emerging nations such as China, India, Korea, Israel, and Singapore, will increase their share of contributions to future inventions.

Let us take a look at the GNI/capita and the GDP growth rates of the 20 largest economies of the world. Recently, there has been a clear tendency for poorer nations to experience faster economic growth in playing catch-up, at rates of 6–8%, in comparison to modest growth for richer nations, at rates of 2–3%, which makes the world more equal. The purchasing power and needs of the poorer nations will grow to become a larger share of the world economic pie. Some of them will have the capability to compete to become leaders in inventions as well.

The difference in growth rates between 2% and 7% may not seem like very much, but if you measure their effect in the number of years it would take to double national income, then the difference becomes enormous. At a growth rate of 2% per year, it takes 35 years for national income to double, and 115 years to grow by a factor of 10; however at a growth rate of 7% per year, it takes only 10 years to double and only 33 years to grow 10 times. The United States has a GDP of $13.4 trillion compared to $2.6 trillion for China, but if the United States continues to grow at 2% and China at 10%, the GDP of China will be larger than that of the United States by the year 2028. There is no assurance that these growth rates are sustainable, but these are the calculated consequences of compound interest over time.

Enrolment of young adults in colleges and universities and investments in research and development are strongly dependent on national income. In 2006, the World Bank gave the following ratios in 2006 for different national income groups.

Income group	College enrolment (%)	R&D (% of GDP)
High	67	2.38
Upper middle	40	0.72
Lower middle	23	1.03
Lower	9	0.57

The college enrolment ratio of the United States is 82%, but only 16% of the graduates have a degree in natural science and engineering. In the developing nations, the college enrolment is much lower, but a much higher fraction are in natural science and engineering, which is perceived to be more relevant to national modernization and prestigious careers. In 2007, the cost of R&D for the United States was $368 billion, which was 2.7% of the national GDP. This percentage is higher than the 2.4% average for high-income nations, but is less than the leaders: Israel 5.0%, Sweden 3.9%, Japan 3.2%, and Korea 3.0%. The US ratio is still higher than the R&D per GDP for the emerging powers: China 1.3%, Russia 1.1%, Brazil 0.9%, and India 0.6%.

We are interested in indicators that could be used to make credible predictions on the rate of future inventions. Some indicators are called lagging indicators, since they usually come after the successful inventions. Examples of lagging indicators include the number of Nobel Prize winners by the country where they work, the export of high-tech products, and the income derived from royalties and patents. Some are called leading indicators, since they may have taken place before the successful inventions, such as the number of researchers in R&D, the level of research funding as a percent of GDP, the level of higher education and training in research universities, and the number of patents filed by residents.

The statistics on patents compiled by the OECD shows that in 2006, there were 51,579 Triadic Patents filed in all three major world centers that consist of the European Patent Office, the Japan Patent Office, and the US Patent and Trademark Office. The largest number were filed from East Asia (Japan, Korea), followed by North America (United States, Canada), and Western Europe (Germany, France, United Kingdom). But it is more difficult to measure the impact of an invention than to gather patent statistics.

Largest region	Patents	Largest nation	Patents
East Asia	17.0	Japan	14.2
North America	16.7	United States	15.9
West Europe	13.8	Germany	6.2

There is a heavy concentration of inventions in ICT, then in electrical and electronic equipment, followed by biotechnology and nanotechnology. It is to be expected that the targeted inventions from the LDC would be influenced by local needs and conditions, which may be significantly different from the markets in the MDC.

Within the United States, most of the inventions were made in the major industrial states with high-tech industries.

State	Patents	State	Patents
California	19600	Michigan	3141
Texas	5733	Illinois	2894
New York	5007	New Jersey	2693
Massachusetts	3510	Minnesota	2554
Washington	3228	Pennsylvania	2500

The most famous incubator of new inventions is Silicon Valley south of San Francisco, which is home to Stanford University, Apple, Hewlett-Packard, Google, and Intel. High-tech research centers like Silicon Valley share the following characteristics: proximity to a major research university, talented pools of educated and experienced scientists and engineers, availability of venture capital and entrepreneurial people, major metropolitan area with good "quality of life"—climate, scenery, recreation, and extensive infrastructure for communication and transportation. The most important ingredient is people who are well educated in the sciences and have the creativity and persistence to succeed. Many of them are immigrants who thrive far away from their original homes. Many other nations understand the importance of inventions to their wealth and power, and are trying to repeat this successful formula by promoting their own local Silicon Valleys.

The United States is still the undisputed leader in the performance of basic and applied research, which is supported by many of the indicators. But the United States is gradually losing some degree of dominance in specific fields to other countries, such as Singapore, Hong Kong, Japan, and Germany. Perhaps catching up is inevitable for the latecomers of Germany and Japan, and even the latecomers of Singapore and China. There are even ambitious national programs to promote inventions in Israel, Korea, and Singapore.

The United States now has a negative trade balance for high-technology products. The international competition for talented researchers is another source of concern. In the United States, half of the current doctoral recipients in engineering and computer science are immigrants. The United States after World War II became a great magnet for international talent, and attracted roughly half of the Nobel laureates, innovators in Silicon Valley, and engineering professors in research universities. Some high-tech innovation centers can be compared to the prestigious musical institutions like the New York Philharmonic Orchestra or the Metropolitan Opera, where American-born artists are in the minority. China and India are beginning to lure many of their talented expatriate researchers in Europe and America back to their homelands. We should continue to lay down the welcome mat for talented persons from other countries. Many leading international high-tech companies are also investing heavily in new R&D centers in China and India, as the cost of research is much lower and the results are more readily applicable in local consumption and manufacturing.

9.2.2 Sponsorship of Inventions

Of the $398 billion that the United States spent on R&D in 2008, industry paid for $268 billion, the federal government $104 billion, universities $11 billion, and nonprofits $15 billion. But out of its $104 billion, the federal government makes many grants to industry and universities where the research is carried out. The US government in-house research budget in 2005 showed the following distribution

Department	Obligation (% of total)
Defense	56
Health	20
Energy	7
Aeronautics and Space	8
National Science Foundation	3
Other	6

Around the globe, the principal public sponsors of research are national governments, who have the responsibility to oversee all aspects of national prosperity and power. The research budgets of national governments are usually dominated by military concerns.

The budget of the National Science Foundation, that has general responsibility for basic research and the advancement of sciences, constitutes no more than 3% of the total obligations. The US government becomes even more important to scientific research in times of war, such as in the invention and deployment of the atomic bomb, radar, and penicillin. In the battlefield, superior inventions and technology have frequently played and will undoubtedly continue to play critical roles in victory and defeat.

The agendas of corporate R&D are *market-driven*, toward serving large groups of wealthy consumers that can lead to large sales and generous profits. This is often aimed at domestic consumption first, and then for export to other wealthy countries. This type of investment will be driven by the purchasing power or GNP, of wealthy countries, and by the wealthier consumers. Most market-driven inventions concentrate on the needs of high-income groups, and pay little attention to the needs of the low-income groups. If a disease affects less than 1 million patients in the world, it would not be profitable for a pharmaceutical company to set up research and production, as there may not be enough sales and profits to cover the costs involved in research and building new plants. It is said that, for maximum revenues, the ideal drug to invent should not be designed to cure an acute and deadly disease, which would no longer be needed after the patient recovers; but should be designed for long-term chronic diseases, such as high blood pressure and diabetes, so the patient must be dependent on the drug till death.

Governments and philanthropic foundations can sometimes step in when there is not enough market demand, to do good deeds and to compensate with funds. The UN Millennium Development Goals is a list of international objectives to eradicate extreme hunger and poverty, achieve universal education, promote equality, reduce child mortality, combat infectious diseases, and ensure environmental stability. Unfortunately, these laudable goals have no funding to support the research activities necessary to meet these objectives. The Bill and Melinda Gates Foundation has an ambitious agenda to provide funding for many similar health initiatives.

Manufacturing is less than one-fourth of the US economy, but it produced nearly all the patents issued. In 1997, there were 111,983 patents issued to residents of the United States. They were dominated by

Electronics and electrical equipment	27,640
Machinery, except electrical	23,547
Chemical	17,210
Instrument	15,726
Fabricated metal	5,952
Transportation equipment	4,498
Rubber, plastics	4,074

These industries do not have the largest market revenues in comparison with Wholesale and Retail Trade. But their profitability depends more on advances in technology. They have management with optimistic views on technological progress, the opportunity to build on the platform of new scientific advances and technology developments, and can hire research scientists and engineers with the required education and skill. In the future postindustrial society, where services are much more important to the economy than industry, the patents devoted to services are expected to increase, particularly in the high technology sectors of information technology and health care.

The financial support of science education and basic research by governments and companies has been declining steadily in the last few decades in the United States and Western Europe. A sizable segment of the public regard a constant stream of innovative inventions as an entitlement, instead of as a much needed investment of current resources to ensure future prosperity.

The influential study *"Rising Above the Gathering Storm: Energizing and Employing America for a Brighter Economic Future,"* was published in 2006 by the United States National Academies, to explain the necessity of inventions and the need to nurture them for future generations. It was followed by the sequel "Rising Above the Gathering Storm Revisited: Rapidly Approaching Category 5" in 2010. It was written by a committee of outstanding national leaders: presidents of research universities, Nobel laureates, and CEOs of Fortune 100 corporations, and was chaired by Norman Augustine. The reports stressed the importance of education as the creator of "Human Capital," and research expenditure as creating "Knowledge Capital," which are essential for innovation, the creation of quality jobs, economic prosperity, and competition in the world.

The Executive Summary begins with this paragraph:

> The United States takes deserved pride in the vitality of its economy, which forms the foundation of our high quality of life, our national security, and our hope that our children and grandchildren will inherit ever greater opportunities. That vitality is derived in large part from the productivity of well-trained people and the steady stream of scientific and technical innovations they produce. Without high-quality, knowledge-intensive jobs and the innovative enterprises that lead to discovery and new technology, our economy will suffer and our people will face a lower standard of living. Economic studies conducted even before the information-technology revolution have shown that as much as 85% of measured growth in US income per capita was due to technological change.

So why are science and technology critical to America's prosperity in the 21st century? This report declares in Chapter 2 that

> Since the Industrial Revolution, the growth of economies throughout the world has been driven largely by the pursuit of scientific understanding, the application of engineering solutions, and continual technological innovation. Today, much of everyday life in the United States and other industrialized nations, as evidenced in transportation, communication, agriculture, education, health, defense, and jobs, is the product of investments in research and in the education of scientists and engineers. One need only think about how different our daily lives would be without the technological innovations of the last century or so.

> The products of the scientific, engineering, and health communities are, in fact, easily visible—the work-saving conveniences in our homes; medical help summoned in emergencies; the vast infrastructure of electric power, communication, sanitation, transportation, and safe drinking water we take for granted. To many of us, that universe of products and services defines modern life, freeing most of us from the harsh manual labor, infectious diseases, and threats to life and property that our forebears routinely faced. Now, few families know the suffering caused by smallpox, tuberculosis (TB), polio, diphtheria, cholera, typhoid, or whooping cough. All those diseases have been greatly suppressed or eliminated by vaccines. We enjoy and rely on world travel, inexpensive and nutritious food, easy digital access to the arts and entertainment, laptop computers, graphite tennis rackets, hip replacements, and quartz watches.

The closing words of the "Gathering Storm" leave us with the following sobering thoughts:

> It is easy to be complacent about US competitiveness and pre-eminence in science and technology. We have led the world for decades, and we continue to do so in many fields. But the world is changing rapidly, and our advantages are no longer unique. Without a renewed effort to bolster the foundations of our competitiveness, it is possible that we could lose our privileged position over the coming decades. For the first time in generations, our children could face poorer prospects for jobs, healthcare, security, and overall standard of living than have their parents and grandparents. We owe our current prosperity, security, and good health to the investments of past generations. We are obliged to renew those commitments to ensure that the US people will continue to benefit from the remarkable opportunities being opened by the rapid development of the global economy.

We must pledge that we will not be blind to the sources of our past and future prosperity by adequately supporting science education and research and that we will do what is necessary to ensure the best future for our grandchildren and generations to come.

REFERENCES

Augustine, N. editor, "Rising Above the Gathering Storm: Energizing and Employing America for a Brighter Economic Future", National Academy Press, Washington DC, 2005. "Rising Above the Gathering Storm Revisited: Rapidly Approaching Category 5", 2010.

Ausubel, J. H. "Five worthy ways to spend large amounts of money for research on environment and resources". *The Bridge* 29(3), 4–16, 1999.

Bent, R. editor, "Energy: Science, Policy, and the Pursuit of Sustainability", Island Press, Covelo, CA, 2002.

Carson, R. "Silent Spring", Fawcett World Library, New York, 1962.

Ehrlich, P. R., A. H. Ehrlich, and J. P. Holdren. "Ecoscience: Population, Resources, Environment", WH Freeman, San Francisco, 1977.

Gladwell, M. "In the air: who says big ideas are rare?" The New Yorker, May 12, 2008.

Global Challenges for Humanity, UN Millennium Development Project, United Nations University. Available at http://www.acunu.org/millennium/challeng.html.

Inter-governmental Panel for Climate Change, "Fourth Assessment Report", "Climate Change 2007: Summary for Policymakers", 2007. Available at http://www.ipcc.ch/pdf/assessment-report/ar4/syr/ar4.

National Academy of Engineering, "Grand Challenges for Engineering". 2009. Available at http://www.engineeringchallenges.org.

Omenn, G. S. "Grand challenges and great opportunities in science, technology, and public policy", presidential address. *Science* 314, 1696–1704, 2006.

Zedillo, E. editor, "Global Warming: Looking Beyond Kyoto", Brookings Institution Press, Washington DC, 2008.

INDEX

Great Inventions that Changed the World, First Edition. James Wei.
© 2012 John Wiley & Sons, Inc. Published 2012 by John Wiley & Sons, Inc.